Dan Bahadur Pal, Amit Kumar Rathoure, Nishant Kumar Singh (Eds.)

**Green Hydrogen and Fuel Cells**

# Green Tribology and Tribochemistry

Edited by
Yashvir Singh and Nishant Kumar Singh

## Volume 3

# Green Hydrogen and Fuel Cells

Integrating Sustainability, Efficiency, and Green Tribology

Edited by
Dan Bahadur Pal, Amit Kumar Rathoure
and Nishant Kumar Singh

DE GRUYTER

**Editors**
**Dr. Dan Bahadur Pal**
Department of Chemical Engineering
Harcourt Butler Technical University
Kanpur 208002, Uttar Pradesh, India
danbahadur.hbtu@gmail.com

**Dr. Amit Kumar Rathoure**
Department of Chemical Engineering
Harcourt Butler Technical University
Kanpur 208002, Uttar Pradesh, India
amitrathour04@gmail.com

**Dr. Nishant Kumar Singh**
Department of Mechanical Engineering
Harcourt Butler Technical University
Kanpur 208002, Uttar Pradesh, India
nishant.singh78@gmail.com

ISSN 2940-0376
ISBN 978-3-11-215916-3
ISBN 978-3-11-218107-2 (PDF)
ISBN 978-3-11-219036-4 (EPUB)
DOI https://doi.org/10.1515/9783112181072

Library of Congress Control Number: 2026933840

Bibliographic information published by the Deutsche Nationalbibliothek
The Deutsche Nationalbibliothek lists this publication in the Deutsche Nationalbibliografie; detailed bibliographic data are available on the internet at http://dnb.dnb.de.



De Gruyter and Walter de Gruyter GmbH are part of De Gruyter Brill.
www.degruyterbrill.com

Questions about General Product Safety Regulation:
productsafety@degruyterbrill.com

Cover illustration: Scharfsinn86/iStock/Getty Images Plus
Printing: CPI books GmbH, Leck

FSC www.fsc.org MIX Papier | Fördert gute Waldnutzung FSC® C083411

# Preface

The world has genuine issues such as climate change and excessive use of fossil fuels. Cleaner and more sustainable energy sources are what we require to address these challenges. One of the most hopeful solutions is hydrogen energy, particularly green hydrogen, produced from renewable sources such as the sun and wind. This book presents an unmistakable and comprehensive picture of green hydrogen, how it is produced, stored, and utilized, and how it can contribute to the design of tomorrow's energy systems. Chapter 1 discusses why green hydrogen matters in terms of pollution reduction and the promotion of clean energy. It illustrates how it can assist industries, transport, and power systems in leaving behind fossil fuels. Chapter 2 discusses how hydrogen can be produced through water being split with electricity in machines known as electrolyzers. It contrasts various forms of electrolyzers and their advantages, disadvantages, as well as how they can be used with renewable energy. Chapter 3 discusses hybrid systems that integrate various renewable sources, such as sun, wind, and biomass, to produce hydrogen. These ensure a consistent source of energy even when the weather is not so cooperative. Chapter 4 considers how hydrogen can be stored as a liquid, either by extremely cold temperatures or by special organic liquids. It details how these processes operate and are safe and convenient. Chapter 5 discusses fuel cells, which convert hydrogen into electricity. It addresses new and more affordable materials that can be used to replace costly metals, thus making fuel cells cheaper and more effective. Chapter 6 looks into small, portable fuel cells applied in gadgets such as electronics, medical devices, and military applications. It describes how they are manufactured and how performance can be enhanced. Chapter 7 addresses hydrogen vehicles (fuel cell electric vehicles) and compares them with electric vehicles. It also discusses challenges in the construction of hydrogen infrastructure and the necessity of solid support to encourage these vehicles. Chapter 8 addresses the environmental impact of green hydrogen. Although it is cleaner than fossil fuels, it also addresses the resources required and how we can make it more sustainable. Chapter 9 describes the technical issues hydrogen can create, such as material damage. It illustrates how new materials and coatings can reduce the lifespan of equipment and improve its performance. Chapter 10 provides an overview of the development of green hydrogen worldwide. It discusses existing methods, their limitations, and how they can be optimized in the future. Chapter 11 discusses smart energy systems that integrate hydrogen with renewable energy sources. It emphasizes the importance of planning and collaboration to make these systems efficient and cost-effective. Chapter 12 concludes by comparing various forms of hydrogen (blue, gray, and green) and examines storage, transport, cost, and public acceptance. It indicates how policies and emerging technologies can ensure hydrogen plays a large role in a cleaner energy future.

This book is for students, researchers, engineers, policymakers, and anyone interested in clean energy. Regardless of whether you are just starting to learn or already

 | https://doi.org/10.1515/9783112181072-202

work in the field, it provides clear and useful information about how hydrogen can assist in creating a better energy future. As the world increasingly prioritizes climate change and clean energy, knowledge about hydrogen is extremely important. We hope this book inspires new research, ideas, and collaboration to make hydrogen a significant component of a greener, cleaner world.

Acknowledgments

We sincerely thank all contributors and researchers whose expertise enriched this book. We are grateful to the Harcourt Butler Technical University, Kanpur, India, for technical facilities. We are grateful to the publishers for their support and to the academic community for providing the foundation of knowledge in this field.

Special thanks to our families, friends, and colleagues for their encouragement, and to our readers, whose curiosity inspires such efforts. We hope this book serves as a valuable resource for further exploration in nanobioenergy.

With sincere gratitude,
*The Editors*

 | https://doi.org/10.1515/9783112181072-203

# Contents

# About the Editors

**Dr. Dan Bahadur Pal** is currently working as an associate professor in the Department of Chemical Engineering at Harcourt Butler Technical University, Kanpur 208002, Uttar Pradesh, India. He received his M.Tech. in 2011 and Ph.D. in 2017 in the field of chemical engineering from Indian Institute of Technology (BHU) Varanasi, Uttar Pradesh, India. Before that, he has completed his B.Tech. in chemical engineering from UPTU, Lucknow, in 2006. He completed his doctorate degree in the field of nanotechnology and catalysis. These nanofibers have very promising potential to provide benefits to nanotechnologies, energy, environment, catalysts, sensors, etc. He published 99 articles in the international refereed journals, 6 books, and 30 book chapters. From all publications, he had over 3,940 citations with an h-index of 28 and an i-10-index of 55. His research interests include nanotechnology, catalysis, energy, and environment and waste management with a special focus in developing process and materials by using waste as raw materials. He also prefers to work on biowaste processing and value addition.

**Dr. Amit Kumar Rathoure** is currently serving as an associate professor in the Department of Chemical Engineering at Harcourt Butler Technical University (HBTU), Kanpur, Uttar Pradesh, India. He earned his M.Tech. in 2012 and his Ph.D. in 2017 from the Department of Chemical Engineering, Indian Institute of Technology (BHU) Varanasi, Uttar Pradesh, India.

His research focuses on advancing high-demand sectors, particularly fuel cell science and technology, green energy, and energy storage technologies. His work emphasizes innovative solutions for sustainable energy systems, reflecting his dedication to addressing critical challenges in modern energy landscapes.

**Dr. Nishant Kumar Singh** is an associate professor in the Department of Mechanical Engineering at Harcourt Butler Technical University, Kanpur, Uttar Pradesh, India. He earned his Ph.D. in modern machining processes from the Indian Institute of Technology (ISM), Dhanbad, India, and his B.Tech. from Delhi Technical University (formerly Delhi College of Engineering), Delhi, India. With 19 years of experience in academia and research, he has established himself as a prominent figure in his field. He has published approximately 120 research papers in high-impact SCI, Scopus, and Web of Science indexed journals. He serves as a reviewer for several reputable journals and is also a member of the editorial team for others. He has mentored numerous M.Tech. and Ph.D. students and is the author of 10 books published by international houses such as CRC Press, Nova Publishers, River Publishers, and De Gruyter. Furthermore, he is a book series editor for both CRC Press and De Gruyter. He holds over 20 patents and has been recognized by Stanford University as being among the top 2% of scientists worldwide.

# List of Contributors

**Vishal Kumar Sandhwar**
Department of Chemical Engineering
Parul Instiitute of Technology
Parul University
Limda, Waghodia, Vadodara, Gujarat, 391760
India
Chapter 1

**Alok Tiwari**
Department of Chemical Engineering
Parul Instiitute of Technology
Parul University
Limda, Waghodia, Vadodara, Gujarat, 391760
India
alok.tiwari30232@paruluniversity.ac.in
Chapter 1, 12

**Shivendu Saxena**
Department of Chemical Engineering
Parul Institiute of Technology
Parul University
Limda, Waghodia, Vadodara, Gujarat, 391760
India
Chapter 1, 12

**Diksha Saxena**
Department of Chemical Engineering
Parul Institiute of Technology
Limda, Waghodia, Vadodara, Gujarat, 391760
India
Chapter 1, 12

**Ashutosh Kumar**
Department of Chemical Engineering
Energy Cluster
University of Petroleum and Energy Studies
Dehradun 248007
Uttarakhand
India.
ashutosh.kumar@ddn.upes.ac.in
Chapter 1

**Pankaj Kumar**
Department of Mechanical Engineering
BIT Sindri
Dhanbad, Sindri, Jharkhand-828123
India
Chapter 1

**Manoj Kumar**
Department of Mechanical Engineering
Govt. Engg. College Palamu
Jharkhand 822118
India
Chapter 1

**Gajendra Kumar Gaurav**
Ministry of Environment
Forest and Climate Change (MoEFCC)
Indira Paryavaran Bhawan Jorbagh Road
New Delhi – 110 003
India
Chapter 1, 12

**Neeraj Kumar Yadav**
Department of Chemical Engineering
Government Polytechnic Mohammadpur Bahraich
Board of Technical Education
Uttar Pradesh, 271801
India
Chapter 2, 5

**Abhay Pratap Singh**
Department of Chemical Engineering
Government Polytechnic Mohammadpur
Bahraich
Board of Technical Education
Uttar Pradesh, 271801
India
Chapter 2, 5

**Deoashish Panjiara**
Department of Chemical and Petrochemical
Engineering
Ganpat University
Mehsana, 384012, Gujarat
India
Chapter 2, 5

 | https://doi.org/10.1515/9783112181072-206

**Abhay Kumar Choudhary**
Department of Chemical Science and Technology
National Institute of Technology Patna
Bihar 800005
India.
abhay.ch@nitp.ac.in
Chapter 2, 5

**Mohammed Irfan**
Department of Petro-Chemical Engineering
Rajasthan Technical University
Kota, Rajasthan
India
Chapter 3

**Rahul**
Department of Paint Technology
Government Polytechnic Bindki
Fatehpur, UP
India
Chapter 3

**Shiv Charan Prajapati**
Department of Paint Technology
Government Polytechnic Bindki
Fatehpur, UP
India
Chapter 3

**Ritesh Patidar**
Department of Petroleum Engineering
Rajasthan Technical University
Kota, Rajasthan
India.
riteshpatidar@gmail.com
Chapter 3

**Ambneesh Mishra**
Faculty of Biology and Environmental Protection
University of Lodz
Lodz
Poland
Chapter 4, 11

**Samanshi Rohilla**
Department of Environmental Studies
University of Delhi
Delhi
India
Chapter 4

**Shraddha Awasthi**
shraddhaawasthi02@gmail.com
Chapter 4, 11

**Deoashish Panjiara**
Department of Chemical and Petrochemical Engineering
Ganpat University
Mehsana, 384012, Gujarat
India
Chapter 4, 11

**Avanish kumar**
Department of chemical Engineering
Harcourt Butler Technical University Kanpur
Uttar Pradesh
India
danbahadur.chem@gmail.com
Chapter 6

**Amit Kumar Rathoure**
Department of chemical Engineering
Harcourt Butler Technical University Kanpur
Uttar Pradesh
India
danbahadur.chem@gmail.com
Chapter 6, 7, 11

**Dan Bahadur Pal**
Department of chemical Engineering
Harcourt Butler Technical University
Kanpur-208002, Uttar Pradesh
India
danbahadur.chem@gmail.com
Chapter 4, 6, 11

**Mradul Kumar**
Department of Chemical Engineering
Harcourt Butler Technical University
Kanpur-208002, Uttar Pradesh
India
amitrathour04@gmail.com
Chapter 7

**Amit Kumar Rathoure**
Department of Chemical Engineering
Harcourt Butler Technical University
Kanpur-208002, Uttar Pradesh
India
amitrathour04@gmail.com
Chapter 6, 7, 11

**Chetan Chauhan**
Department of Chemistry
Sardar Patel University
Mandi, H.P
chetan@spumandi.ac.in
Chapter 8

**Virendar Kumar**
Department of Mechanical Engineering
MMUT
Gorakhpur (India)
Chapter 9

**Sanjeev Kumar**
Department of Electrical Engineering
SKITM
Jaipur (India)
Chapter 9, 10

**Devendar Singh**
Department of Mechanical Engineering
G L Bajaj Group of Institutions
Mathura (india)
Chapter 9

**Nishant K. Singh**
Department of Mechanical Engineering
Harcourt Butler Technical University
Kanpur (India)
nishant.singh99975@gmail.com
Chapter 9, 10

**Vijay Singh**
Department of Mechanical Engineering
K. K. University
Bihar Sharif (India)
Chapter 10

**Shraddha Awasthi**
Department of Environmental Studies
MLNCE
University of Delhi-110021
shraddha.iitbhu@gmail.com
Chapter 4, 11

**Gajendra Gaurav**
Department of Chemical Engineering
Parul Institute of Technology
Parul University
Vadodara
Chapter 1, 12

**Gourang Damle**
Department of Chemical Engineering
Parul Institute of Technology
Parul University
Vadodara
Chapter 12

Vishal Kumar Sandhwar, Alok Tiwari, Shivendu Saxena, Diksha Saxena,
Ashutosh Kumar, Pankaj Kumar, Manoj Kumar,
Gajendra Kumar Gaurav

# 1 Introduction to Green Hydrogen and Its Role in the Energy Transition

**Abstract:** This chapter deals with the accelerating effects of global climate change, principally caused by the burning of fossil fuels and the release of greenhouse gas emissions, which necessitate an immediate and fundamental change in the world's energy systems. The world presently derives approximately 80% of its energy from fossil fuels, which significantly contributes to the degradation of the environment and to public health issues. An interdisciplinary investigation of green hydrogen as a vital facilitator of a sustainable, low-carbon future is discussed. It explores hydrogen's promise as a clean, scalable, and versatile energy carrier with the capacity to decarbonize hard-to-abate sectors of industry, transport, and power generation. The existing and future production technologies cover solar- and wind-powered electrolysis and new biohydrogen approaches, along with considering life cycle emissions, technoeconomic challenges, gaps in infrastructure, and safety. A systems-level view connects these technologies to changing policy architectures and global market forces, as well as national initiatives. The discussion highlights the need for collaborative global action, strong regulatory environments, public-private funding, and incentive financing to support accelerated innovation and deployment. Green hydrogen facilitates both decarbonization and energy resilience by storing excess renewable energy and adding grid flexibility. Being a technological advancement, it is the cornerstone of equitable and sustainable energy transition. By facilitating deep decarbonization in hard-to-abate sectors and augmenting renewable energy integration, it bridges both environment and socio-economic imperatives. To fulfill a gap in advancement in energy security, drives industrial change, and encourages even access to clean energy across the globe. Green hydrogen is therefore not just a clean fuel but a force to drive a more inclusive, resilient, and climate-friendly energy future.

**Keywords:** Green hydrogen, energy transition, decarbonization, electrolysis, policy frameworks, sustainable energy systems

## 1.1 Introduction

The impacts of climate change are increasing at an appalling rate; the global landscape makes it necessary to have a swift transformation in energy systems. The drastic increase in greenhouse gas (GHG) emissions, mainly from fossil fuels, has the potential to damage significantly to environmental systems, as well as human health. As per the studies have

 | https://doi.org/10.1515/9783112181072-001

been done, it has been found that 80% (approx.) of the global energy supply is from fossil fuels. These degrade the environment significantly through water and air pollution and climate change [1, 2]. It has been emphasized by scientists and policymakers to bring transformative changes in sustainable energy systems. According to the Intergovernmental Panel on Climate Change (IPCC), it is practically impossible to meet the 1.5 °C global warming target without an immediate transition away from fossil fuels [3].

This chapter is an interdisciplinary and comprehensive effort showing the effects of green hydrogen in building a sustainable energy future. This study explores the capabilities of hydrogen as a clean, scalable, and adaptable energy carrier, being emphasized as a potential factor responsible for global decarbonization strategies. This chapter is an effort exploring the current and developing technologies for production of green hydrogen, including the electrolysis, which gets powered by solar and wind energy. Besides these, novel biohydrogen methods have been also included. Critical assessment on technoeconomic challenges along with life cycle emissions as well as preparedness of infrastructure has been provided. Besides presenting the technical and scientific overview, a novel systems-level perspective has been introduced in the chapter by combining technological development with changing policy environments, global regulatory trends, and marketplace mechanisms like carbon pricing and hydrogen certification schemes. Also discussed is the interaction between regional approaches (e.g., EU's Hydrogen Roadmap, India's National Green Hydrogen Mission) and global energy equity, providing readers with a unique comparative perspective.

A thorough examination of the role of hydrogen can play in the energy transition requires a pragmatic evaluation of existing global patterns of energy usage. The long-standing supremacy of fossil fuels within the energy market is now being questioned in view of their greenhouse emissions and increased operating expenses, fuelling the motivation to seek alternative energy options [4]. The ongoing use of traditional fossil fuel energy sources has resulted in two major environmental effects, which include biodiversity loss and the intensification of extreme weather patterns [4]. The research results demonstrate that a worldwide shift to new energy systems becomes essential right away. The adoption of renewable energy systems leads to two major benefits, which enhance future energy stability while fighting climate change and achieving sustainable development goals. The shift from carbon-based energy systems to clean alternatives provides two essential advantages by maintaining ecological stability and protecting socioeconomic systems from climate-related dangers [5].

Green hydrogen functions as a crucial component for sustainable energy transition because it utilizes renewable power to execute water electrolysis [6]. The production process of green hydrogen through electrolysis differs from traditional steam methane reforming (SMR) since it creates an emission-free energy carrier. Green hydrogen stands out because it provides environmental benefits while serving multiple sectors including transportation and heavy industry and long-duration energy storage applications [5, 6]. Green hydrogen demonstrates flexibility because it can substitute fossil fuels in difficult decarbonization situations and maintain system stability when renewable energy generation fluctuates. Green hydrogen serves two purposes by substituting carbon-emitting fuels and by improving power system stability and resilience that rely on renewable energy sources. Green hydrogen demonstrates its greatest importance through its ability to address both decreasing fossil fuel reserves and increasing global energy demands. Green hydrogen has established itself as a primary energy solution because the world faces simultaneous challenges of obtaining energy resources while reducing GHG emissions. Its production from renewable sources offers a clean way of minimizing carbon emissions in industries that are generally hard to electrify, such as heavy industry, aviation, and long-distance haulage. By enabling deep, intersectoral emission cuts, green hydrogen supports countries in fulfilling their Paris Agreement commitments and drives the transition to a low-carbon world economy [7].

This chapter starts by presenting hydrogen as a clean and versatile energy carrier, setting the stage for a comprehensive examination of green hydrogen production technologies. The following sections delve into the diverse functions of green hydrogen within critical sectors, including transportation, industry, and energy storage. Additionally, they explore the latest market trends (Figure 1.1) and technological innovations that are driving its increased adoption. The concluding section offers a thorough assessment of the policy frameworks and governance structures necessary for facilitating a worldwide shift toward clean hydrogen. Green hydrogen stands as a pivotal element in the global energy transformation, necessitating the immediate advancement of technological innovation alongside the establishment of supportive policy frameworks to unlock its full potential. This chapter seeks to offer in-depth insights into the developing green hydrogen landscape by merging scientific principles, technological pathways, and regulatory perspectives, highlighting its essential role in promoting sustainable energy systems.

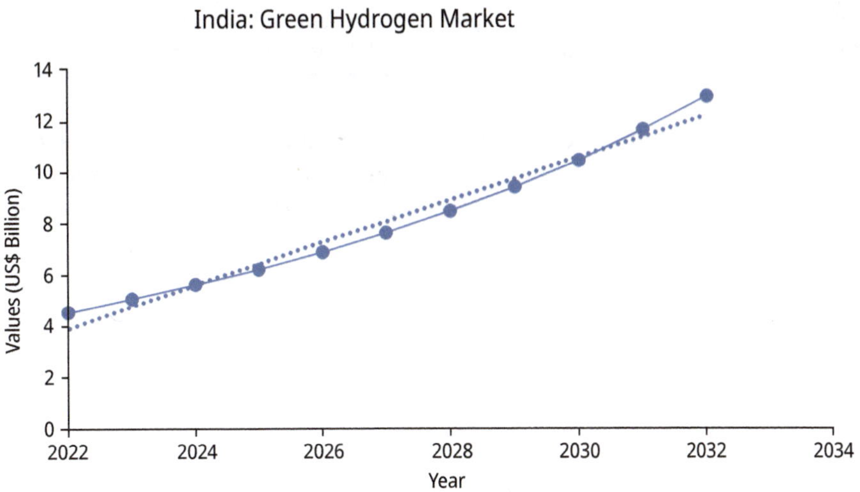


**Figure 1.1:** Chart shows the projected growth of India's green hydrogen market in between 2022 and 2033.

## 1.2 Green Hydrogen: Definition, Properties, and Production Technologies

The energy transition toward decarbonization and emission reduction benefits from green hydrogen production through renewable energy water electrolysis systems. Green hydrogen functions as an energy carrier due to its versatile properties such as colorless, odorless, and high energy density which is comparable to that of gasoline. Therefore, it is suitable for a wide array of applications across transportation systems, industrial processes, and energy storage frameworks [8]. Green hydrogen serves a dual role within energy system: as a clean energy vector and as a reservoir to buffer energy supply and demand fluctuations. The overview of green hydrogen is illustrated in Table 1.1.

**Table 1.1:** Overview of green hydrogen.

| Aspect | Description | Remarks | References |
|---|---|---|---|
| Definition | Hydrogen produced via electrolysis powered by renewable energy sources (solar, wind, and hydro) | Distinct from gray or blue hydrogen due to zero carbon emissions | [9] |
| Production process | Water is split into hydrogen and oxygen using electricity | Electrolyzers are key to this process | [10] |
| Energy source | Solar, wind, and hydroelectric power | The carbon footprint depends on the electricity source | [11] |
| Current share | <0.1% of total hydrogen production (as of 2021) | Green hydrogen is still in early development stages | [9] |
| Primary use cases | Industry (ammonia and steel), transportation (fuel cells), and energy storage | Seen as a decarbonization tool for hard-to-abate sectors | [12] |

Green hydrogen refers to hydrogen generated through the process of water electrolysis, utilizing renewable energy sources like solar, wind, and hydro, which leads to zero carbon emissions [9]. The method utilizes electrolyzers to decompose water into hydrogen and oxygen through the application of electricity [10] with the carbon footprint influenced by the electricity source employed [11]. Although it holds significant promise, green hydrogen presently represents less than 0.1% of worldwide hydrogen production [9], underscoring its nascent phase of development. Significant applications encompass industrial processes such as ammonia and steel production, fuel cell transportation, and renewable energy storage, establishing it as an essential solution for decarbonizing challenging sectors [12]. The advantages and disadvantages of green hydrogen based on environmental, economic, geopolitical, and technical aspects are demonstrated in Table 1.2.

**Table 1.2:** Advantages and disadvantages of green hydrogen.

| Category | Advantages | Disadvantages |
|---|---|---|
| Environmental | – Zero $CO_2$ emissions during use<br>– Enables decarbonization of heavy industries | – High water consumption<br>– Electrolyzer production emits $CO_2$ unless green-powered |
| Technical | – High energy density per mass<br>– Can be stored and transported | – Energy loss in production and conversion<br>– Storage and transport infrastructure challenges |
| Economic | – Potential job creation in new energy markets<br>– Decreasing cost of renewables helps affordability | – Currently expensive (~$4–6/kg vs. $1–2/kg for gray hydrogen) |
| Geopolitical | – Reduces reliance on fossil fuels<br>– Enhances energy security | – Could create new dependencies on electrolyzer tech and rare minerals |

Source: [9, 12, 13].

### 1.2.1 Electrolysis-Based Production

The fundamental approach to generating green hydrogen involves electrolysis, a technique that divides water ($H_2O$) into hydrogen ($H_2$) and oxygen ($O_2$) through the application of electricity [14]. The economic viability of green hydrogen depends on the cost, scalability, and operational effectiveness of this technology. The three leading electrolyzer technologies that dominate the market today are alkaline electrolysis (AEC), proton-exchange membrane (PEM) electrolysis, and solid oxide electrolysis cells (SOECs).

#### 1.2.1.1 Alkaline Electrolysis (AEC)

The AEC represents the current most commercially viable method for water electrolysis. The electrochemical reaction occurs through the use of potassium hydroxide as a liquid alkaline electrolyte. AEC operates at 60–80 °C temperatures while achieving 60–70% efficiency making it suitable for large-scale operations [15]. The process faces three main drawbacks, which include reduced current density and slow power response and electrolyte corrosion issues. Figure 1.2 shows the schematic of an alkaline water electrolyzer.

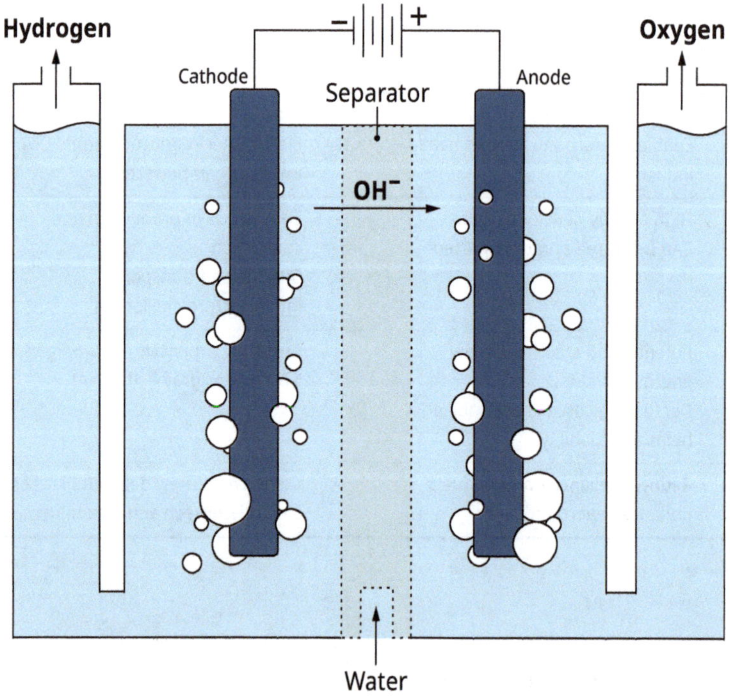


**Figure 1.2:** Schematic of an alkaline water electrolyzer.

#### 1.2.1.2 Proton-Exchange Membrane (PEM) Electrolysis

PEM electrolyzers employ a solid polymer electrolyte, allowing for greater current densities (>1 A/cm$^2$) and faster dynamic response, making them ideally suited for coupling with intermittent renewable sources of energy. They operate at temperatures between 50 and 80 °C and have efficiency between 70% and 90%, producing ultra-pure hydrogen in the form of more than 99.99% purity [16]. While these benefits are notable, the requirement for costly noble metal catalysts, such as platinum and iridium, increases capital expenditures and poses obstacles for widespread implementation [17]. Figure 1.3 shows the schematic of a proton-exchange membrane water electrolyzer.

#### 1.2.1.3 Solid Oxide Electrolysis Cells (SOECs)

SOECs operate effectively at high temperatures ranging from 800 to 1,000 °C, which allows for thermodynamically advantageous reaction conditions and the possibility of reaching efficiencies greater than 80% [14]. These systems utilize solid ceramic

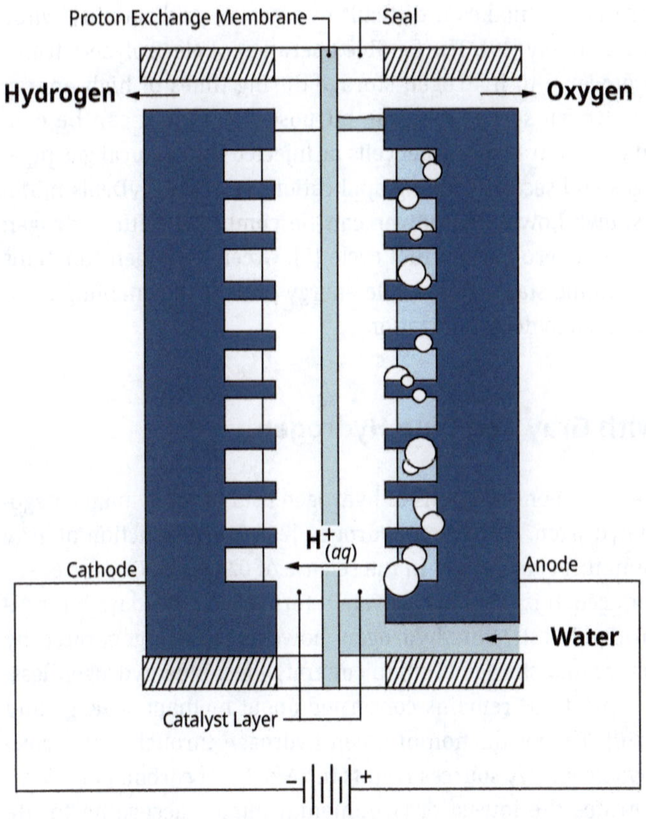


**Figure 1.3:** Schematic of a proton-exchange membrane water electrolyzer.

electrolytes and have the capability to integrate with thermal sources, such as biomass gasification. The commercialization of SOECs faces major technical challenges because material degradation and thermal cycling continue to be significant obstacles despite their potential advantages in energy efficiency and system integration [18].

The selection of electrolysis technology determines the commercial viability and scalability and environmental sustainability of green hydrogen manufacturing. Research efforts focus on developing new materials while reducing system costs and extending operational life.

### 1.2.2 Renewable Energy Integration

The deployment of green hydrogen at scale demands the combination of electrolysis systems with renewable power sources including solar, wind, and hydropower. The variable nature of these resources makes it difficult to preserve grid stability while delivering continuous power supply [19]. The flexible operation of electrolyzers transforms excess renewable power into hydrogen storage during times of high energy production. The stored hydrogen serves multiple purposes because it can be converted into electricity and used in transport fuel cells or injected into natural gas pipelines to enable power-to-gas and sector coupling applications [15]. The HyDeals initiative in South Australia shows how solar power can be combined with hydrogen production through a localized zero carbon fuel cycle [8]. Green hydrogen functions as a key element for maintaining stable renewable energy grids while enabling flexible and expandable energy system decarbonization.

### 1.2.3 Comparison with Gray and Blue Hydrogen

The production methods and carbon emissions of hydrogen lead to three main categories: gray hydrogen, blue hydrogen, and green hydrogen [20]. The production of gray hydrogen through SMR of natural gas results in the release of 0.96 kg of $CO_2$ for every kilogram of hydrogen produced. It is the most prevalent form used nowadays but contributes to emissions worldwide [21]. Blue hydrogen, however, is also generated by SMR but with carbon capture and storage (CCS) to cut emissions. Blue hydrogen lessens the environmental footprint but remains concerned about methane leakage and the effectiveness of CCS [20]. The production of green hydrogen through water electrolysis powered by renewable energy sources results in zero direct carbon emissions. Green hydrogen demonstrates the lowest environmental impact according to life cycle analyses when it is used with negative emission processes that include biogenic $CO_2$ capture [21].

The production cost of green hydrogen remains higher than both gray and blue hydrogen. The cost gap between green hydrogen and other hydrogen types is rapidly decreasing because of technology expansion and decreasing renewable energy costs and supportive policies [15]. The growing evidence from life cycle environmental impact assessments supports green hydrogen's position as it becomes a key component in national and global decarbonization strategies [20].

### 1.2.4 Global Energy Transition: Drivers and Challenges

The international energy transition represents a fundamental shift in energy production and distribution systems and consumption patterns which fight climate change while reducing fossil fuel dependence and building sustainable low-carbon energy systems. The deteriorating climate situation has made the transition more urgent because it produces more severe weather events and increasing ocean levels and disappearing species. The IPCC stresses the urgent need to decrease greenhouse emissions because it is vital to achieve the 1.5 °C global temperature target [22].

The worldwide energy transition depends on clean technologies including green hydrogen because they decrease emissions while building multiple energy sources and enhancing both domestic and international energy security [23]. This section evaluates the three fundamental factors which direct the worldwide energy transition by analyzing (1) climate goals and policy regimes, (2) decarbonization in hard-to-abate sectors, and (3) energy security issues. Table 1.3 summarizes the role of green hydrogen in the global energy transition.

**Table 1.3:** Role of green hydrogen in the global energy transition.

| Sector | Role of green hydrogen | Remarks | References |
|---|---|---|---|
| Power | Energy storage for renewables; backup power | Helps stabilize grids with high renewable penetration | [11] |
| Industry | Decarbonizing steel, cement, and ammonia production | Critical for hard-to-abate emissions | [10] |
| Transport | Fuel for heavy-duty trucks, ships, and aviation | Fuel cells offer longer range vs. batteries | [12] |
| Buildings | Used in fuel cells for heating and backup | Less common due to heat pump alternatives | [9] |

## 1.2.5 Climate Goals and Policy Frameworks

The United Nations Sustainable Development Goals through SDG 7: Affordable and Clean Energy and the Paris Agreement provide the strategic framework to achieve a low-carbon economy. The frameworks need countries to adopt cleaner technologies while changing carbon-intensive systems to reach their net zero targets [24]. The achievement of net zero GHG emissions by mid-century demands many countries to adopt green hydrogen as their scalable zero emission energy carriers. The worldwide transition to energy needs strong policy tools to move forward. The implementation of carbon pricing through carbon taxes and cap-and-trade systems functions as a fundamental policy instrument because it enables industries to create low-carbon substitutes by including environmental costs of GHG emissions [25]. The best practices for reviewing projection and execution of green hydrogen initiatives are presented in Table 1.4. The adoption of clean technologies requires green subsidies together with investment tax credits to reduce emerging solution costs and establish market equilibrium. In addition, renewable portfolio standards and green hydrogen requirements help by creating stable and reliable market conditions, promoting long-term investment in clean energy infrastructure and facilitating the large-scale integration of clean technologies. These tools are particularly suited to green hydrogen rollout. Because it is an adaptable energy carrier, green hydrogen is complementary to the intermittent nature of renewables by providing long-term energy storage and facilitating decarbonization in otherwise difficult-to-electrify industries [26, 27]. Developing policies increasingly endorse holistic green hydrogen approaches by connecting industrial decarbonization, grid balance, and climate adaptation goals.

### 1.2.6 Decarbonization of Hard-to-Abate Sectors

Perhaps the greatest challenge to the global energy transition is the decarbonization of difficult-to-abate industries like steel, cement, chemicals, aviation, and shipping. These sectors consume a lot of high-temperature heat and produce large quantities of process-related $CO_2$ that cannot be removed by electrification only [30]. Green hydrogen presents itself as a technically credible and scalable alternative by playing multiple roles across these industries. The fuel can replace coal or natural gas in cement and steel kilns for combustion while serving as a clean chemical feedstock for ammonia synthesis and providing green energy for hydrogen-powered aviation and seafaring vessels.

The possibility finds support through real-life case studies. The HYBRIT initiative in Sweden has proven successful in replacing coal with green hydrogen for steel production which enables near-zero carbon steel manufacturing [30]. The world's shipping industry shows significant progress in decarbonization through hydrogen-powered shipping transportation models according to Onyeneke et al. (2023). Liquid hydrogen serves as a green propulsion source for short-haul flights in current aerospace pilot projects according to Peng et al. [41]. The above examples emphasize that green hydrogen is not just a technological breakthrough but a foundational solution for making climate-neutral routes possible in some of the most emission-heavy sectors.

### 1.2.7 Energy Security and Geopolitical Realignments

Transitioning to the required energy form should be structured to tackle climate change suitably while, at the same time, improving energy security. Green hydrogen, particularly when locally produced from renewable energy sources, is vital for achieving this balance. Diversion away from imported fossil fuels shields national economies from the ups and downs of international energy markets and promotes a diversified and resilient energy mix [31, 32]. Along with its technical advantages, green hydrogen introduces new geopolitical challenges. Those nations with abundant renewable energy resources, such as the sun-drenched regions of the Middle East and North Africa or the windy heavens of the Nordic and coastal states, are positioning

**Table 1.4:** Best practices for reviewing and executing green hydrogen initiatives.

| Area | Best practices | Advantages | Disadvantages | Key facts | References | Remarks |
|---|---|---|---|---|---|---|
| **Market projection** | Integrated energy models are used for forecasting hydrogen demand | Data-driven decisions and scenario-based planning | Model assumptions can vary widely | Global demand projected to reach 500–600 Mt by 2050 | [9, 27] | Models should include sector coupling and renewable variability |
| **Policy planning** | Develop national hydrogen roadmaps with clear targets and funding | Encourages investment and aligns stakeholders | Risk of misalignment across sectors | Over 30 countries have published hydrogen strategies | [12] | Policy frameworks are key enablers |
| **Execution strategy** | Phase-wise deployment: pilots → demonstration → commercialization | Mitigates risk and enables learning | Time-consuming and high upfront cost | EU's Hydrogen Valley projects follow this path | [28] | Phased roll-out is globally accepted |
| **Public-private partnerships** | Joint ventures with risk sharing and clear governance | Access to capital and shared expertise | Risk of conflicting interests | Shell, Linde, ITM Power in joint $H_2$ projects | [29] | Ensure transparency and accountability |
| **Stakeholder engagement** | Early and continuous dialogue with communities and industries | Reduces opposition and improves design | Requires time and social effort | Public opposition stalled projects in Australia | [9] | Inclusion builds legitimacy and acceptance |

themselves as major future exporters of green hydrogen. This shift might essentially redefine existing global energy coalitions, away from the traditional oil and gas reliance and toward emerging hydrogen-based trade partnerships [31].

Against this shifting backdrop, strategic investments in hydrogen infrastructure such as dedicated pipelines, export terminals, large-capacity electrolyzers, and secure storage facilities are becoming increasingly integral components of national energy policy. The investments work to achieve decarbonization targets while boosting industrial competitiveness and building enduring economic stability [23]. The development of sustainable hydrogen supply chains needs international collaboration to establish them. The development of hydrogen supply chains requires three main focus areas which include safety standards maintenance and auditable green hydrogen certification systems and essential raw materials for electrolyzer production such as platinum, iridium, and nickel. The collaborative efforts create a secure path to the hydrogen economy that preserves both equity and geopolitical stability.

The worldwide energy transition demands solutions which tackle technological aspects together with economic and environmental and geopolitical transformations. Green hydrogen functions as a crucial element in this transition because it delivers a clean flexible and scalable energy system which supports international climate targets and enables industrial scale decarbonization and boosts energy independence. Green hydrogen offers promising opportunities to create a sustainable energy future despite ongoing challenges related to cost and infrastructure and policy consistency.

## 1.3 Role of Green Hydrogen in Decarbonization

The reduction of carbon dioxide ($CO_2$) emissions from human activities particularly in energy systems, industrial processes, and transport is important for addressing climate change and attaining global sustainability goals. These two objectives aim to decrease the levels of GHGs while facilitating sustainable development pathways which are in line with the Paris Agreement that seeks to keep the global temperature below 2 °C compared to pre-industrial levels.

Green hydrogen produced by electrolysis using renewable electricity represents an emerging technology that shows promise as a primary energy source which enables major decarbonization in sectors that are hard to decarbonize. Green hydrogen production and utilization produces zero emissions because it differs from conventional hydrogen production methods that use fossil fuels. The versatility of this method enables it to be used in many industrial operations, as well as electricity production and transportation systems provide a comprehensive solution to eliminate fossil fuel reliance and create a circular low-carbon economy.

### 1.3.1 Industrial Applications: Steel, Ammonia, and Refineries

The industrial sector emits the largest amount of $CO_2$ due to its consumption of high-temperature heat and fossil-based feedstocks. Green hydrogen provides a pathway to minimize emissions in essential sectors such as ammonia manufacturing and steel production and oil refining. Green hydrogen conversion in the ammonia industry enables significant emission reductions because the industry has traditionally used natural gas through SMR. The Haber-Bosch process becomes fossil fuel independent through the use of hydrogen produced by electrolysis which enables the achievement of sustainability goals worldwide [33, 34].

The steel industry which emits about 7% of total $CO_2$ emissions will experience a complete transformation through the implementation of direct reduced iron technologies that use green hydrogen. Hydrogen serves as a replacement for carbon-intensive coke as the reducing agent in this process. The Sweden-based SSAB HYBRIT project along with others demonstrates that it is possible to produce fossil-free steel by 2026 through the use of green hydrogen in low-emission metallurgical processes [35]. Refineries with their extensive hydrogen requirements for hydrocracking and desulfurization operations can become carbon neutral by switching to green hydrogen instead of gray hydrogen. The transition to green hydrogen allows refineries to both decrease their production emissions and establish a path toward achieving net zero status.

### 1.3.2 Power Sector and Grid Balancing

Renewable power integration into grids presents both intermittency and seasonal difficulties. Green hydrogen functions as a vital energy storage solution because it transforms excess renewable power into hydrogen through power-to-gas technologies. The stored hydrogen can be transformed back into electricity to meet peak demand periods which maintains grid stability and increases energy resilience [36].

Power systems benefit from hydrogen through its application in hydrogen fuel cells together with combustion turbines and combined heat and power systems. These power systems offer adaptable solutions that boost energy performance while minimizing carbon footprints during electricity and heat production [37].

The modification of existing natural gas infrastructure makes it possible for hydrogen turbines to function with minimal changes that facilitate the transition to cleaner power systems [38]. Green hydrogen functions as a strategic tool for maintaining long-term energy reliability in addition to its role in stabilizing power systems powered by renewables. Figure 1.4 shows the generic scheme of a power-to-power hydrogen value chain configuration.

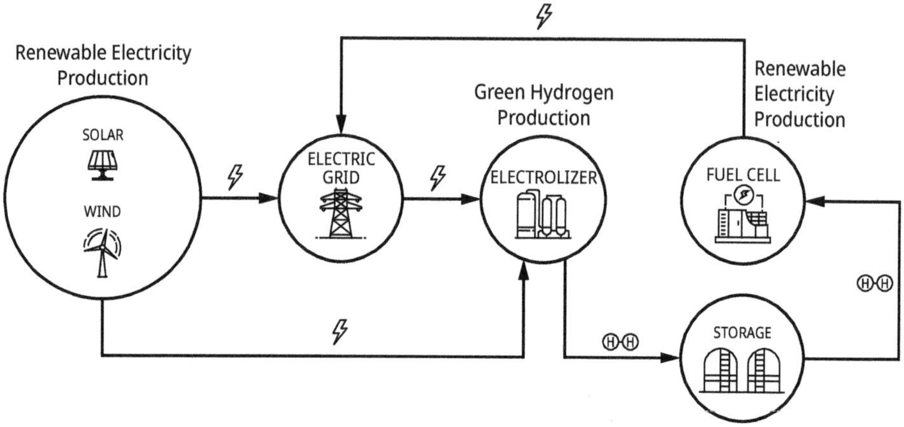


**Figure 1.4:** Generic scheme of a power-to-power hydrogen value chain configuration.

## 1.3.3 Transportation and Mobility

The transport industry, particularly its heavy-duty, maritime, and air transport segments, has a significant decarbonization challenge, driven mainly by the requirements of energy density and long operation ranges. Green hydrogen offers an attractive proposition to fuel cell electric vehicles, ships that use hydrogen as a fuel, and air transport solutions based on hydrogen. Fuel cell electric vehicles transform hydrogen into electricity internally, resulting in water as the sole by-product. In contrast to battery electric vehicles, hydrogen vehicles offer quicker refueling times and extended driving ranges, which are particularly beneficial for long-haul trucks and commercial fleets [39].

In the field of aviation, experimental initiatives are exploring liquid hydrogen propulsion systems aimed at minimizing $CO_2$ emissions while maintaining flight range [40, 41]. While hydrogen mobility holds significant promise, it encounters challenges related to infrastructure and efficiency. These encompass the necessity for comprehensive networks of hydrogen refueling stations and the energy losses linked to hydrogen production, compression, and conversion. Nonetheless, collaborative initiatives between government and industry are tackling these obstacles. Countries like Japan and Germany are making significant investments in hydrogen refueling infrastructure, showcasing the potential for sustainable scaling of hydrogen-powered transport systems [42].

The implementation of green hydrogen throughout industry power and transport sectors demonstrates its vital position in worldwide carbon emission reduction efforts. Figure 1.5 shows the generic scheme of a power-to-mobility hydrogen value chain configuration. The solution enables a complete transition away from fossil fuels which strengthens energy system stability while advancing climate targets and driving industrial progress. Green hydrogen functions as an integrated solution which unites clean energy production with storage and utilization for nations working to achieve net-zero emissions. The growing need for environmental protection alongside energy security requirements makes green hydrogen an essential element for building a sustainable energy system of the future. Green hydrogen can create a carbon-neutral economy through coordinated policy frameworks and technological innovation and infrastructure investment which will transform global energy systems.

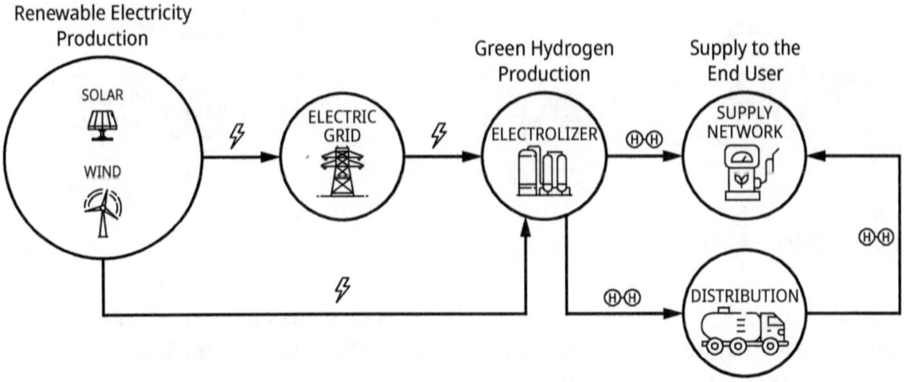


**Figure 1.5:** Generic scheme of a power-to-mobility hydrogen value chain configuration.

## 1.4 Technological and Economic Barriers to Green Hydrogen Deployment

The large-scale deployment of green hydrogen shows great potential for worldwide decarbonization yet faces multiple technological and economic obstacles that slow its advancement. The main obstacles to green hydrogen development include expensive production costs together with limited infrastructure and restricted storage capabilities and complicated regulatory distribution logistics. These issues significantly impact the availability, cost-effectiveness, and applicability of hydrogen in energy systems. This section further discusses these interrelated issues. The storage distribution flowchart is presented in Figure 1.6.

### 1.4.1 Cost and Infrastructure Limitations

The green hydrogen industry is now faced with one major economic concern: the high costs associated with electrolysis-based hydrogen production. As water electrolysis with renewable electricity offers a carbon-neutral option, the high capital in-

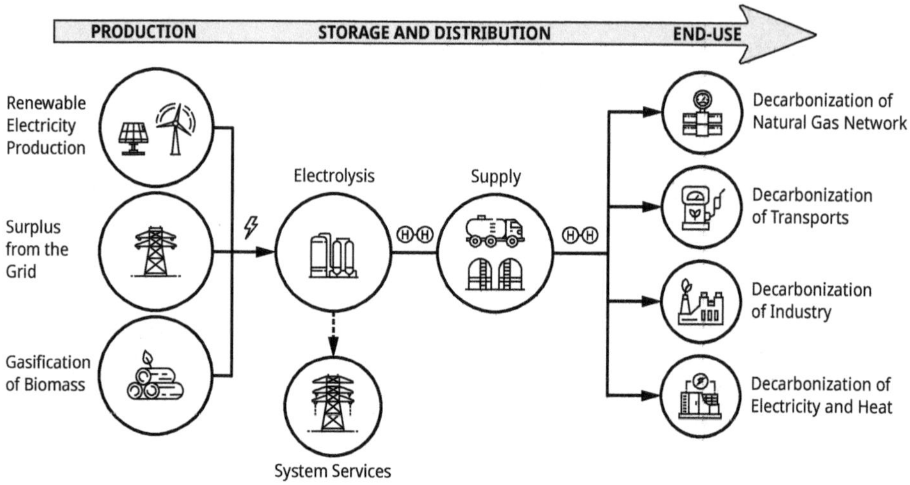


**Figure 1.6:** Storage distribution flowchart.

vestment (CAPEX) required in electrolyzers, as well as the attendant operational expenditure (OPEX), remains high. Current estimates indicate that the levelized cost of hydrogen (LCOH) generated through renewable-powered electrolysis varies between \$2.5 and \$6.0 per kilogram, influenced by factors such as electricity prices, electrolyzer efficiency, and utilization rates [43]. Scaling up the manufacturing and deployment of electrolyzers represents a crucial avenue for achieving cost reduction. Nonetheless, the scale-up process encounters obstacles such as supply chain bottlenecks, restricted production capacities, and insufficient standardization, which impede swift commercialization [44]. Implementing strategic policies, fostering public-private partnerships, and innovating in electrolyzer design are crucial for cost reduction and facilitating broader adoption.

A significant constraint pertains to the availability of renewable energy sources. The variability and intermittency of solar and wind resources present a significant challenge for maintaining consistent production of green hydrogen. The ideal scenario involves placing hydrogen plants in proximity to plentiful renewable resources; however, this is not always practical. Where renewable supply is low, reliance on grid electricity that might still contain fossil-based energy capable of generating greater production costs and higher carbon intensity [45]. Additionally, the lack of large-scale infrastructure for hydrogen transport, storage, and end-use greatly constrains market penetration. As of 2023, there are approximately 40 hydrogen refueling stations in the United States, with most being heavily concentrated in California [46]. This restricted network limits both supply and demand. Furthermore, the adaptation of current natural gas infrastructure for hydrogen transportation presents significant technical challenges, as the majority of pipelines are incompatible with hydrogen due to concerns like metal embrittlement and the potential for leakage [47]. Estimates indicate that the global hydrogen economy will necessitate substantial infrastructure investment, amounting to hundreds of billions of dollars. This investment will encompass pipelines, refueling stations, storage terminals, and port facilities to ensure viability and scalability [48]. Therefore, minimizing production expenses while concurrently developing supportive infrastructure is essential.

### 1.4.2 Storage and Distribution Challenges

The low volumetric energy density of hydrogen poses considerable technical challenges for its storage and distribution. For practical applications, hydrogen needs to be compressed, liquefied, or chemically stored, with each approach presenting various trade-offs related to energy input, cost, and safety considerations.

The storage of compressed hydrogen gas is prevalent because of its straightforward operation; nonetheless, it necessitates high-pressure containment (up to 700 bar) and sophisticated safety systems. Conversely, storing liquid hydrogen provides greater energy density; however, it requires maintaining extremely low cryogenic temperatures (−253 °C), which results in boil-off losses and heightened energy consumption [49]. Alternative storage methods like metal hydrides or underground salt caverns present long-term storage solutions; however, they are often either too expensive or limited by geographic factors.

The distribution presents a significant challenge. Pipeline transport is recognized as the most economical approach for handling large volumes; however, it necessitates the establishment of hydrogen-specific networks or costly modifications to current natural gas pipelines. The propensity of hydrogen to induce material fatigue and embrittlement in metals heightens the likelihood of leaks, thereby requiring the use of specialized pipeline materials and the implementation of stringent safety protocols [50]. The transportation of hydrogen through ship and truck, particularly over long distances, faces economic as well as technical issues resulting from the need for high-pressure storage cylinders or cryogenic tankers. The techniques utilized result in high costs per unit of hydrogen transported and exhibit reduced efficiency in contrast to pipeline transport for mass distribution [50].

The hydrogen supply chain faces multiple environmental and safety problems according to recent research together with logistical challenges. The energy efficiency of green hydrogen decreases substantially because of energy losses that occur during compression liquefaction and distribution processes. The increasing awareness of hydrogen air leaks has emerged as a secondary GHG threat because it enhances methane and ozone concentrations [51].

The complete realization of deep decarbonization depends on green hydrogen yet multiple technological and economic challenges need resolution. The establishment of a sustainable hydrogen economy requires the adoption of efficient production methods and robust infrastructure development alongside advanced storage solutions and secure distribution systems. The solution of these challenges requires sustained investment together with materials and systems engineering advancements and unified government-industry-international institutional collaboration. Cybulsky et al. (2024) argue that the future of the hydrogen economy depends on cost reduction, infrastructure preparation, and coordinated regulation to ensure long-term scalability and sustainability.

## 1.5 Policy Support and Global Initiatives

Green hydrogen is becoming increasingly recognized as a core component for sustainable energy transitions and long-term decarbonization plans. The successful implementation of its global potential depends on strong government policies together with cooperative international partnerships and effective financial approaches. The successful operation of green hydrogen technologies depends on unified policy structures and stable long-term funding and multiple stakeholder cooperation for large-scale implementation. The future hydrogen economy is being developed through national plans and public-private collaborations and global partnerships which are presented in Table 1.4.

### 1.5.1 National Roadmaps and Government Strategies

Multiple nations across the world have recognized green hydrogen as their essential component for reaching net zero objectives so they established extensive national programs to speed up its development. These roadmaps contain essential targets for electrolyzer capacity development alongside hydrogen infrastructure construction and market development and public-private partnership establishment.

The European Union Hydrogen Strategy establishes vital targets to reach annual renewable hydrogen production of 10 million tons by 2030. The strategy requires funding for developing cross-border infrastructure and storage systems and industrial decarbonization routes [52]. Germany's National Hydrogen Strategy outlines a 5 GW electrolysis capacity target for 2030, which receives funding from research and development projects and pilot cluster projects [53]. Japan stands as a leading nation in hydrogen adoption because it has incorporated hydrogen into its long-term energy policy framework. Under its Basic Hydrogen Strategy Japan dedicated substantial funding to develop hydrogen supply chains and construct liquefied hydrogen shipping terminals and advanced fuel cell vehicles [54].

The developing economies of India and Australia are creating detailed strategic plans. The National Hydrogen Mission of India works to build green hydrogen hubs while encouraging value chain expansion through clean hydrogen production incentives [55]. Australia plans to establish an export-oriented hydrogen strategy that aims to send 10 million tons of hydrogen annually until 2030 [56]. These strategies receive backing from policy instruments which include capital subsidies together with feed-in tariffs and tax incentives and public procurement requirements to reduce investment risks and encourage private sector involvement. The development of "Hydrogen Valleys" represents a notable example which combines public-private investments to create integrated regional systems that optimize hydrogen production and distribution and end-use applications [57]. In California, regulatory structures, pilot projects, and infrastructure investments have created a high-profile hydrogen transportation system that can be a model for state-level activity [58].

### 1.5.2 International Collaborations and Frameworks

Given the global integration of energy markets and the cross-border transfer of technology, it is essential to promote international collaborations to advance the production of green hydrogen and build end-to-end global supply chains. Co-operative structures allow for convergence of standards, reduce investment risk, and enhance knowledge transfer between developed and developing economies. Organizations such as the International Partnership for Hydrogen and Fuel Cells in the Economy (IPHE) encourage the sharing of knowledge and policy harmonization between member countries. Similarly, the Clean Hydrogen Mission, which is under Mission Innovation, encourages collective research and development activities along with investment to accelerate the development of the global market [57].

A focal point in global alignment is standardization and certification of hydrogen, particularly with regards to international trade. Tools like the Guarantees of Origin (GO), CertifHy, and similar traceability systems are pivotal in providing transparency, as well as confirming the carbon intensity of hydrogen along the value chain. The described processes play an essential role in developing stakeholder trust while promoting international hydrogen trade and preventing greenwashing [57]. The continuous worldwide momentum is demonstrated through concrete bilateral and multilateral ties. The Australia-Japan partnership serves as an example because Japan focuses on developing hydrogen liquefaction and shipping technology while Australia provides green hydrogen produced by solar-powered electrolysis [56]. The European Union implements Hydrogen Diplomacy through collaborative work with North African and Middle Eastern nations to build production and export facilities [55].

The alliances enable both technology sharing and joint funding of infrastructure development and position hydrogen as a key element in international geopolitical discussions. Countries now incorporate hydrogen into their foreign policy and trade strategies because they need energy security alongside export possibilities and climate resilience [56]. The commercialization process gains speed because the Hydrogen Council and the Global Hydrogen Ports Coalition work to merge industrial stakeholder agendas with government targets. The green hydrogen economy depends on a unified strategy that merges national plans with international partnerships for its development.

The development of the green hydrogen economy requires government incentives together with public funding and regulatory support to reduce costs and attract private sector participation. At the same time, collaborative efforts and the establishment of standardization frameworks are essential for fostering a strong global hydrogen market. Where climate policy, economic competitiveness, and energy diplomacy intersect, green hydrogen ascends to more than a technological fix, becoming a key strategic resource in the decarbonized future of the world.

## 1.6 Future Outlook and Recommendations

The worldwide adoption of green hydrogen demonstrates growing agreement about the immediate requirement to decarbonize energy systems because of the worsening climate crisis. The production of green hydrogen through electrolysis using renewable energy has become essential for building a low-carbon economy and providing industries with a flexible renewable substitute for fossil fuels. This strategy adds to renewable energy systems by providing a method for long-term energy storage and grid stabilization in times of intermittent supply. Realizing the full potential of green hydrogen as a transformational technology relies on future technological breakthroughs, strong policy environments, and international cooperation across multiple sectors.

### 1.6.1 Technological Advancements and Innovation Trajectories

The widespread use of green hydrogen is closely connected with continued advances in basic technologies such as electrolysis, hydrogen storage, and fuel cells. Advances in electrolysis are especially important since they directly influence the production cost of hydrogen. Present research on efficient electrocatalysts and next-generation membrane materials are expected to significantly reduce the energy requirements for water splitting. SOECs are attracting great interest due to their outstanding energy efficiency and reduced operating cost potential, making them prime candidates for industrial processes requiring steady performance. Concurrently, the development of hydrogen storage technologies is imperative in order to address safety, energy density, and transportability challenges. Substances such as metal hydrides and liquid organic hydrogen carriers and underground hydrogen storage systems are now being researched for their potential to store hydrogen safely in condensed forms for long durations of time. These developments have the potential to solve a number of the primary logistical issues involved with hydrogen distribution.

The commercialization of green hydrogen depends heavily on fuel cell technology which drives its adoption in transportation and stationary power generation sectors. PEMFCs lead the charge in this initiative because researchers focus on improving their durability and thermal management systems and reducing their costs. The future success of hydrogen utilization in practical applications depends on the progress made in this area. The commercialization of green hydrogen technology requires ongoing research and development funding from government agencies and industrial partners to achieve cost parity with fossil fuels and enable large-scale adoption.

### 1.6.2 Policy Evolution and Market Frameworks

The successful implementation of green hydrogen demands both technical progress and supportive market conditions and strong policy backing. Governments worldwide need to establish innovative policy structures which ensure market stability and attract investments through incentives and carbon pricing and public procurement agreements. Financial instruments including subsidies and green bonds and grants serve as essential tools to address the risks that come with deploying hydrogen infrastructure and technology during its initial development phase.

Public institutions need to create industrial strategies that support development and innovation and build public-private partnerships to establish a thriving hydrogen economy. The establishment of hydrogen clusters or "hydrogen valleys" serves as a strategy to create integrated production-distribution-use ecosystems for hydrogen within specific regional areas. The clusters enhance technology verification and reduce costs through scale-based economic advantages. The success of hydrogen development will depend on international cooperation between nations. The development of bilateral and multilateral agreements will be important for cross-border hydrogen trading, for advancing mutual safety levels, and for harmonizing certification models such as Guarantees of Origin. These efforts will further develop global hydrogen supply chains and advance the overall energy transition.

### 1.6.3 Challenges and Strategic Priorities

The green hydrogen industry shows promising potential yet it faces multiple significant barriers to reach its full potential. The main problem with green hydrogen production exists because it costs more than gray hydrogen produced from fossil fuels without carbon capture systems. The production cost reduction depends on continuous efficiency advancements and material enhancements and extensive deployment to reach economies of scale.

The storage and transportation of hydrogen presents a major safety challenge. The combination of high flammability and low molecular weight in hydrogen creates significant engineering challenges for storage in high pressure or cryogenic conditions. The embrittling properties of hydrogen in specific materials create difficulties for building strong infrastructure. The management of these challenges demands both strict safety protocols and thorough risk assessments. The development of infrastructure remains a great challenge. The lack of development in pipeline infrastructure and refueling stations and large-scale storage facilities acts as a barrier to hydrogen adoption at scale. The supply chain will not develop properly unless coordinated investment focuses on these essential areas. The production of cost-effective hydrogen transport across long distances requires maximizing efficient logistics networks that include marine, rail, and pipeline transportation systems.

### 1.6.4 Strategic Recommendations

Overcoming these challenges and fostering a green hydrogen future is a collaborative effort of governments, industry, and academia. Governments need to lead the establishment of international agreements that harmonize border controls and safety standards to enable an unimpeded global trade of hydrogen. It is necessary to invest in pilot and demonstration projects that test hydrogen solutions in real-world applications, which will be the key to building public confidence and gathering critical performance information. Furthermore, it is necessary that public money be invested in education and workforce development programs for preparing the future generation of professionals in hydrogen technologies.

The industrial community needs to focus on hastening investment in research and development that will produce affordable electrolyzers, durable storage materials, and high-performance fuel cells. Collaboration with government institutions and academic institutions can promote innovation and hasten the roll-out of infrastructure. Industry stakeholders should embrace circular economy principles by integrating hydrogen production with $CO_2$ capture, wastewater treatment, and biomass valorization processes. Research and academic institutions play a critical role in advancing the science and engineering of green hydrogen. There is a need to work with industry partners to co-develop knowledge, test solutions, and provide insights into commercialization strategies. Investigation efforts should more and more focus on next-generation fields such as photoelectrochemical water splitting, artificial intelligence in optimizing the logistics of hydrogen, and hybrid storage technologies. In addition, life cycle studies and detailed techno-economic evaluations will be essential to informing policy choices and investment plans.

In summary, the future of green hydrogen will be shaped by a combination of technological innovation, prudent policy formulation, international cooperation, and targeted investments. By adopting a cooperative and cross-disciplinary approach, green hydrogen can emerge as a central part of the global transition toward clean energy.

## 1.7 Conclusion

Green hydrogen stands as a fundamental element in the worldwide transition toward sustainable low-carbon energy systems. Green hydrogen operates as a flexible renewable power system which demonstrates potential to decrease carbon emissions in hard-to-decarbonize sectors including industry and transportation and power generation. Multiple elements determine the successful execution of this potential including technological development and robust policy support and international alliances and expanding market potential. The commercial viability and scalability of green hydrogen have become possible through significant advancements in electrolysis efficiency and hydrogen storage systems and fuel cell optimization.

The IPHE and Clean Hydrogen Mission together with national initiatives from Germany, Japan, India, and Australia show growing international consensus about hydrogen's function in energy transitions. The establishment of a hydrogen economy depends on public-private partnerships together with financial support and regulatory modifications. But, major hurdles must still be overcome. Systematic solutions require concentrated research and development together with targeted investment and standard global regulations to address production costs and safety issues and infrastructure deficiencies and logistics challenges. A coordinated multidisciplinary approach that connects governments with industry and universities and civil society will drive the effective advancement of green hydrogen technologies. By having innovation, policy, and collaboration function together well, green hydrogen can be a key component in achieving global climate goals and creating a strong, sustainable energy system for the future.

The chapter illustrates how green hydrogen can be used as a direct substitute for fossil fuels in steel and cement sectors which are hard-to-abate sectors. Green hydrogen is a key element of energy systems as it makes it possible to store surplus renewable energy that enhances grid flexibility and energy reliability. International cooperation and harmonized regulations along with financial incentives will stimulate innovation and deployment, as is shown in the chapter. An enabling policy environment will be necessary to reduce costs and integrate existing infrastructure gaps.

Green hydrogen is not just a technological breakthrough as it is an integral component for constructing an equitable sustainable energy system of tomorrow. National efforts toward net zero emissions need concomitant research coordination with policy formulation and industrial application. The deployment of green hydrogen technology results in an energy transition that delivers cleaner outcomes and builds resilience and inclusivity for generations to come.

# References


[1] Rao, T., Tao, L. and Gonzalez, M., 2018. An overview of biorefinery-derived platform chemicals from a cellulose and hemicellulose biorefinery. Clean Technologies and Environmental Policy, 20(7), pp.1615–1630.

[2] Afdhol, M.K., Lubis, H.Z. and Siregar, C.P., 2019. Bioethanol production from tea waste as a basic ingredient in renewable energy sources. Journal of Earth Energy Engineering, 8(1), pp.21–26.

[3] Liao, Z., 2023. Assessing sustainable impacts of green energy projects for the development of renewable energy technologies: A triple bottom line approach. Processes, 11(8), p.2228.

[4] Shivanna, K., 2022. Climate change and its impact on biodiversity and human welfare. Proceedings of the Indian National Science Academy, 88(2), pp.160–171.

[5] Suzuki, K., Suzuki, Y., Miao, S. and Dowaki, K., 2024. Environmental system analysis on formic acid synthesis from captured $CO_2$ and green $H_2$ using life cycle assessment methodology. IOP Conference Series: Earth and Environmental Science, 1354(1), p.012041.

[6] Nakayama, H., Kameyama, M., Kamiuchi, H. and Dowaki, K., 2021. Optimization of Heat Transfer through Heat Carriers in Bio-H2 Production Using CFD Simulation. Journal of the Japan Institute of Energy, 100(10), pp.194–199.

[7] Pereira, A. and Pereira, R., 2021. Picking our environmental battles: Removal of harmful subsidies or carbon taxation? Journal of Economics and Public Finance, 7(2), p.1.

[8] Elshafei, A.M. and Mansour, R., 2023. Green hydrogen as a potential solution for reducing carbon emissions: a review. Journal of Energy Research and Reviews, 13(2), pp.1–10.

[9] International Energy Agency (IEA), 2022. Global Hydrogen Review, 2022; https://www.iea.org/reports/global-hydrogen-review-2022.

[10] Intergovernmental Panel on Climate Change (IPCC), 2021. Climate Change 2021: The Physical Science Basis. Cambridge University Press, Cambridge, UK and New York, NY, USA, pp.2391, doi:10.1017/9781009157896.

[11] International Renewable Energy Agency (IRENA), 2022. Tracking SDG7, The Energy Progress Report 2022; https://www.irena.org//media/Files/IRENA/Agency/Publication/2022/Jun/SDG7_Tracking_Progress_2022.pdf.

[12] Hydrogen Council, McKinsey & Company, Hydrogen Insite December 2023, https://hydrogencouncil.com/wp-content/uploads/2023/12/Hydrogen-Insights-Dec-2023-Update.pdf.

[13] Renewable Energy Policy Network for the 21st Century (REN21), Renewables 2024 Global Status Report – Energy Systems & Infrastructure; https://www.ren21.net/renewables-2024-global-status-report-energy-supply-module/

[14] Yazdani, M., 2023. Preparation and evaluation of Ir base anode for proton exchange membrane water electrolysis.

[15] Hamed, A.M., Kamaruddin, T.N.A.T., Ramli, N. and Wahab, M.F.A., 2023, December. A review on blue and green hydrogen production process and their life cycle assessments. In IOP Conference Series: Earth and Environmental Science, 1281(1) p.012034. IOP Publishing.

[16] Kang, D., Ma, L., Huang, Y., Qiu, Y., Kang, L. & Huang, S., 2024. The influence of operating parameters on the performance of PEMWE electrolytic cell. Highlights in Science Engineering and Technology, 114, pp.244–252.

[17] Wu, Y., Yao, R., Zhao, Q., Li, J. and Liu, G., 2023. RuO2 nanoparticles anchored on gC3N4 as an efficient bifunctional electrocatalyst for water splitting in acidic media. Dalton Transactions, 52(30), pp.10515–10521.

[18] Noussan, M., Raimondi, P.P., Scita, R. and Hafner, M., 2021. The role of green and blue hydrogen in the energy transition – A technological and geopolitical perspective. Sustainability, 13, p.298.

[19] Zhao, C., Kobayashi, L.Z., Alquaity, A.B.S., Monfort, J.C., Cenker, E., Miralles, N. and Sarathy, S.M., 2024. Solutions for decarbonising urban bus transport: a life cycle case study in Saudi Arabia. Communications Engineering, 3(1), p.95.
[20] Patel, G., Havukainen, J., Horttanainen, M., Soukka, R. and Tuomaala, M., 2024. Climate change performance of hydrogen production based on life cycle assessment. Green Chemistry, 26(2), pp.992–1006.
[21] Yagüe, L., Linares, J.I., Arenas, E. and Romero, J.C., 2024. Levelized cost of biohydrogen from steam reforming of biomethane with carbon capture and storage (golden hydrogen) – application to Spain. Energies, 17(5), p.1134.
[22] Ramos-Escudero, A., Gil-García, I., García-Cascales, M. and Molina-García, Á., 2020. Shallow geothermal potential impact on the energy transition: A case study region of Murcia, Spain. IEEE International Conference on Environment and Electrical Engineering and IEEE Industrial and Commercial Power Systems Europe, pp.1–6.
[23] Meckling, J. and Hughes, L., 2018. Global interdependence in clean energy transitions. Business and Politics, 20(4), pp.467–491.
[24] Jensen, F., Lööf, H. and Stephan, A., 2020. New ventures in Cleantech: Opportunities, capabilities and innovation outcomes. Business Strategy and the Environment, 29(3), pp.902–917.
[25] Yang, J., Li, R. and Solangi, Y.A., 2024. Assessing and prioritizing drivers and strategies for transition to a green energy for sustainable development in China. Heliyon, 10(4).
[26] Theiventhran, G., 2021. Emerging frontiers of energy transition in Sri Lanka. In: Energy Transition in South Asia. Cham: Springer, pp.185–210.
[27] McKinsey 2023, McKinsey & Company – Listen to the article: Global Energy Perspective 2023: Hydrogen outlook. https://www.mckinsey.com/industries/oil-and-gas/our-insights/global-energy-perspective-2023-hydrogen-outlook.
[28] EU Commission 2022, Energy, Climate change, Environment; https://energy.ec.europa.eu/topics/eus-energy-system/hydrogen_en#documents
[29] BloombergNEF 2023, New Energy Outlook, India; https://assets.bbhub.io/professional/sites/24/BNEF_New-Energy-Outlook-India-2023.pdf.
[30] Aghion, P., Dechezleprêtre, A., Hemous, D., Martin, R. and Van Reenen, J., 2016. Carbon taxes, path dependency, and directed technical change: Evidence from the auto industry. Journal of Political Economy, 124(1), pp.1–51.
[31] Bonilla, J., Coria, J., Mohlin, K. and Sterner, T., 2015. Refunded emission payments and diffusion of NOx abatement technologies in Sweden. Ecological Economics, 116, pp.132–145.
[32] Price, J. and Keppo, I., 2017. Modelling to generate alternatives: A technique to explore uncertainty in energy-environment-economy models. Applied Energy, 195, pp.356–369.
[33] Dixit, A.C., Ashok, B.C., SA, M.K. and Harshavardhan, B., 2023. Green hydrogen for Karnataka: regional solutions for a clean energy future. In E3S Web of Conferences (Vol. 455, p.02020). EDP Sciences.
[34] Franco, A. and Giovannini, C., 2023. Routes for hydrogen introduction in the industrial hard-to-abate sectors for promoting energy transition. Energies, 16(16), p.6098.
[35] Staffell, I., Scamman, D., Abad, A., Balcombe, P., Dodds, P., Ekins, P. et al., 2019. The role of hydrogen and fuel cells in the global energy system. Energy & Environmental Science, 12(2), pp.463–491.
[36] Dell'Isola, M., Ficco, G., Moretti, L., Perna, A., Candelaresi, D. and Spazzafumo, G., 2021. Impact of hydrogen injection on thermophysical properties and measurement reliability in natural gas networks.

[37] Dadkhah, A., Van Eetvelde, G. and Vandevelde, L., 2022, June. Optimal investment and flexible operation of power-to-hydrogen systems increasing wind power utilisation. In 2022 IEEE International Conference on Environment and Electrical Engineering and 2022 IEEE Industrial and Commercial Power Systems Europe (EEEIC/I&CPS Bloombergrope) (pp.1–6). IEEE.
[38] Atteya, A.I., Ali, D., Hossain, M. and Sellami, N., 2023. A comprehensive review on the potential of green hydrogen in empowering the low-carbon economy: development status, ongoing trends and key challenges. Green energy and environmental technology, 2.
[39] Yap, J. and McLellan, B., 2023. A historical analysis of hydrogen economy research, development, and expectations, 1972 to 2020. Environments, 10(1), p.11.
[40] Shakya, N., Shrestha, R., Saiju, R. and Thapa, B., 2022. Hydrogen as a fuel for electrifying transportation sector in Nepal: Opportunities and challenges. IOP Conference Series: Earth and Environmental Science, 1037(1), p.012064.
[41] Peng, J., Zheng, Y. and Mao, K., 2021. Heterogeneous impacts of extreme climate risks on global energy consumption transition: An international comparative study. Energies, 14(14), p.4189.
[42] Baum, Z.J., Diaz, L.L., Konovalova, T. and Zhou, Q.A., 2022. Materials research directions toward a green hydrogen economy: a review. ACS omega, 7(37), pp.32908–32935.
[43] Liponi, A., Baccioli, A., Ferrari, L. and Desideri, U., 2022. Techno-economic analysis of hydrogen production from PV plants. In E3S web of conferences (Vol. 334, p. 01001). EDP Sciences.
[44] Yang, X., Nielsen, C.P., Song, S. and McElroy, M.B., 2022. Breaking the hard-to-abate bottleneck in China's path to carbon neutrality with clean hydrogen. Nature Energy, 7(10), pp.955–965.
[45] Di Nardo, A., Calabrese, M., Venezia, V., Portarapillo, M., Turco, M., Di Benedetto, A. and Luciani, G., 2023. Addressing environmental challenges: the role of hydrogen technologies in a sustainable future. Energies, 16(23), p.7908.
[46] Corsi, J.S., Fu, J., Wang, Z., Lee, T., Ng, A.K. and Detsi, E., 2019. Hierarchical bulk nanoporousaluminum for on-site generation of hydrogen by hydrolysis in pure water and combustion of solid fuels. ACS Sustainable Chemistry & Engineering, 7(13), pp.11194–11204.
[47] Martin, P., Ocko, I.B., Esquivel-Elizondo, S., Kupers, R., Cebon, D., Baxter, T. and Hamburg, S.P., 2024. A review of challenges with using the natural gas system for hydrogen. Energy Science & Engineering, 12(10), pp.3995–4009.
[48] Locke, K., 2024. The urgency of hydrogen: environmental issues and the need for change. Future Sustainability, 2(2), pp.46–58.
[49] Sigen, A. and Li, X., 2024. Strategic analysis of past and modern hydrogen storage methods and prospects for its future. Innovation in Science and Technology, 3(2), pp.14–20.
[50] Abbas, A.J., Hassani, H., Burby, M. and John, I.J., 2021. An investigation into the volumetric flow rate requirement of hydrogen transportation in existing natural gas pipelines and its safety implications. Gases, 1(4), pp.156–179.
[51] Ocko, I.B. and Hamburg, S.P., 2022. Climate consequences of hydrogen emissions. Atmospheric Chemistry and Physics, 22(14), pp.9349–9368.
[52] Cheon, K. and Kim, J., 2020. Hydrogen economy in major countries: policies of promotion and lessons learnt from them. Journal of the Korean Society of Mineral and Energy Resources Engineers, 57(6), pp.629–639.
[53] Li, S.N., Li, Y. and Yu, W., 2023. Assessing the Content Quality of Industry Technology Roadmaps. Sustainability, 15(11), p.9058.
[54] Stangarone, T., 2020. South Korean efforts to transition to a hydrogen economy. Clean Technologies and Environmental Policy, 23(2), pp.509–516.
[55] Caillard, A., Yeganyan, R., Cannone, C., Plazas-Niño, F. and Howells, M., 2024. A critical analysis of Morocco's Green Hydrogen Roadmap: a modelling approach to assess country readiness from the energy trilemma perspective. Climate, 12(5), p.61.

[56] Scita, R., Raimondi, P. and Noussan, M., 2020. Green hydrogen: The holy grail of decarbonisation? An analysis of the technical and geopolitical implications of the future hydrogen economy. SSRN Electronic Journal.
[57] Safronova, A. and Barisa, A., 2023. Realizing a green hydrogen economy: An examination of influencing factors. Environmental and Climate Technologies, 27(1), pp.928–949.
[58] Reed, J., Dailey, E., Shaffer, B., Lane, B., Flores, R., Fong, A. et al., 2020. Realizing a self-sustaining renewable hydrogen sector in California. Energy Proceedings.

Neeraj Kumar Yadav, Abhay Pratap Singh, Deoashish Panjiara, Abhay Kumar Choudhary

# 2 Electrolyzer Technologies for Hydrogen Production

**Abstract:** Hydrogen is commonly produced using nonrenewable sources such as fossil fuels, which raises concerns about source depletion, environmental pollution, and low-purity hydrogen production. Hydrogen production using electrolyzer technology is an excellent alternative for a sustainable and carbon-free future. Electrolyzer technology splits water into hydrogen and oxygen using electricity, producing highly pure hydrogen. The potential for the integration of electrolyzer technology with renewable energy sources is gaining popularity. The electrolyzers are mainly comprised of alkaline electrolyzers, proton exchange membrane (PEM) electrolyzers, and solid oxide electrolyzers. The above-mentioned electrolyzer has some advantages and disadvantages. Among the electrolyzers, the alkaline electrolyzer is considered superior in terms of maturity and cost-effectiveness. However, alkaline electrolysis has drawbacks of slower response times and less operational flexibility. PEM electrolyzers have the advantages of improved efficiency, compact design, and quick response time. PEM electrolyzers often employ expensive catalysts such as platinum and iridium, making them less cost-effective. Solid oxide electrolyzers operating at high temperatures are more efficient and offer the potential for co-electrolysis, but additional research into durability and material stability is needed. This chapter critically discussed the electrolyzer technology along with its various advancements and challenges. This chapter also discusses the economic viability of the electrolyzer technology in the current scenario. The application of electrolyzers, along with its sustainability, is discussed at the end of the chapter.

**Keywords:** Electrolyzer technology, hydrogen production, PEM electrolyzer, alkaline electrolyzer, solid oxide electrolyzer, co-electrolysis

## 2.1 Introduction

Hydrogen is the most plentiful element (approx. 75%) in the universe and is seen as a crucial energy carrier for a sustainable future [1, 2]. Hydrogen fuel is a zero-emission fuel that produces only water as a byproduct when combusted and is considered a clean energy carrier [3]. However, this does not imply that hydrogen production is completely free of carbon emissions. Hydrogen can be produced through a variety of processes based on different feedstocks or energy sources [4]. In industrial applications, hydrogen generation can be categorized into two categories: conventional hydrogen forms (gray

 | https://doi.org/10.1515/9783112181072-002

hydrogen and blue hydrogen) and renewable hydrogen forms (green hydrogen and yellow hydrogen) [5]. The gray hydrogen is produced through steam reforming of fossil fuels, resulting in direct $CO_2$ emissions into the atmosphere. Currently, it is the most common, accounting for more than 90% of worldwide $H_2$ production, and is mostly used in the ammonia and petrochemical industries [6]. Blue hydrogen is likewise produced through steam reforming of fossil fuels, but it incorporates carbon capture, utilization, and storage (CCUS) technology to limit $CO_2$ emissions following the hydrogen synthesis process. Yellow hydrogen is also known as solar-driven hydrogen, whose production process utilizes solar energy to split water into hydrogen and oxygen. The most sustainable type is green hydrogen, which is produced from water electrolysis driven by renewable energy sources, maintaining carbon neutrality throughout the whole hydrogen supply chain [7]. Table 2.1 lists the various forms of hydrogen by color, according to their production method, feedstock/energy source, and related $CO_2$ emissions.

**Table 2.1:** Types of hydrogen color based on the production process with $CO_2$ emission [4, 6].

| Type of hydrogen | Feedstock/ energy source | Production process | $CO_2$ emissions per kg of $H_2$ production | Remarks |
|---|---|---|---|---|
| Gray hydrogen | Natural gas and fossil fuels | Steam methane reforming (SMR), autothermal reforming (ATR), or coal gasification | 10–12 kg $CO_2$ | Most common method; most prevalent, >90% of the global $H_2$ production); no carbon capture |
| Blue hydrogen | Natural gas, biomass, and coal | SMR or ATR or sorbent enhanced reforming + carbon capture utilization and storage (CCUS) | 0.6–2 kg $CO_2$ | Capturing $CO_2$ by 60–90% significantly reduces GHG emissions; total $H_2$ costs substantially more than gray hydrogen. |
| Brown and blackhydrogen | Coal, biomass, agricultural waste | Gasification of lignite (brown coal), biomass, and agricultural waste | 22–26 kg $CO_2$ | Considerable amount of $CO_2$ produced; rarely used now due to emissions |
| Turquoisehydrogen | Natural gas or fossil fuels | Methane pyrolysis | 0.4–1 kg $CO_2$ | Emerging technology; solid carbon by-product easier to store; experimental stage |
| Green hydrogen | Water + renewable electricity | Electrolysis using renewable electricity | 0 kg $CO_2$ | Environmentally clean; dependent on renewable energy |

**Table 2.1** (continued)

| Type of hydrogen | Feedstock/ energy source | Production process | $CO_2$ emissions per kg of $H_2$ production | Remarks |
|---|---|---|---|---|
| Pink/purple hydrogen | Water + nuclear energy | Electrolysis using electricity from nuclear power plants | 0 kg $CO_2$ | Clean hydrogen production with stable baseload energy; public perception of nuclear safety |
| Yellow hydrogen | Water + mixed grid electricity or typically solar power | Electrolysis using mixed grid electricity or solar energy | Minimal to zero | Not fully green; emissions depend on the energy source |
| Red hydrogen | Water + nuclear power and heat | High-temperature catalytic splitting of water | Minimal to zero | Emerging/under development in countries with strong nuclear infrastructure |
| White/ goldhydrogen | Rocks, organic waste, water, microorganisms | Serpentinization, biological hydrogen production | Minimal to zero | Early-stage; exploration and trial extraction underway |

Thus, improving green hydrogen production technology is a significant step toward carbon neutrality. Currently, electrolysis accounts for less than 4% of global hydrogen generation, with a considerable part produced as a by-product of the chlor-alkali manufacturing process [8]. The majority of hydrogen is still produced using fossil fuels, primarily by steam reforming of natural gas and coal gasification, and accounts for around 76% of global $H_2$ production. However, these technologies are carbon-intensive, producing 10–26 kg of $CO_2$ for each kilogram of $H_2$ produced, rendering them unsustainable [4]. This dependency on conventional sources stems from the higher costs of electrolysis, which are driven by rising electricity prices and regulatory hurdles. The amount of hydrogen produced from renewable energy must be increased to minimize $CO_2$ emissions and reduce dependency on fossil fuels in the coming years [9]. Water electrolysis is one of the frontrunner techniques that can achieve this since it splits water molecules into gaseous hydrogen ($H_2$) and oxygen ($O_2$) using energy produced from renewable sources. Electrolysis has been investigated for over two centuries. One of the earliest experiments is reported to have taken place in 1785, when Dutch scientist Martinus Van Marum used an electrostatic generator to separate metals such as tin, antimony, and zinc, from their salts – an unintentional attempt that laid the framework for electrolysis. When William Nicholson and Anthony Carlisle in England performed an experiment using a similar setup to break water into hydrogen and oxygen in 1800, they discovered the fundamental process of water electrolysis. Since then, electrolysis has been vital in the discovery of elements that are challenging to extract using conventional chemical processes, like sodium, gallium, potassium, and others like fluorine [10]. However, despite great scientific interest, progress in water electrolysis has been modest throughout the past century. This is due primarily to the lower cost of alternative hydrogen production technologies such as natural gas steam reforming, coal gasification, light hydrocarbon reforming, etc.

## 2.2 Principle of Water Electrolysis

The hydrogen evolution reaction (HER) at the cathode and the oxygen evolution reaction (OER) at the anode are the two main half-reactions involved in water electrolysis (Figure 2.1). Figure 2.1 shows the schematic of the typical water electrolysis using PEM electrolyte. The basic reaction of water electrolysis is given as follows [11]:

$$H_2O + \text{Electricity} + \text{Heat} \rightarrow H_2 + \frac{1}{2}O_2 \tag{2.1}$$

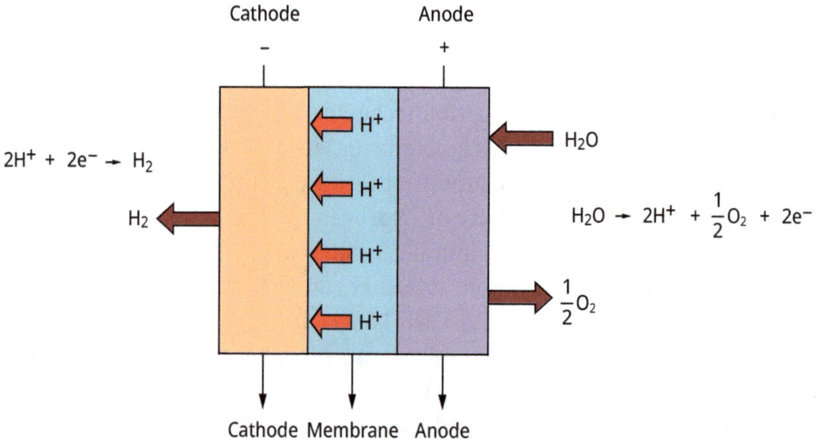


**Figure 2.1:** Schematic of the typical water electrolysis using PEM electrolyte.

Schematic diagram illustrating the process of water electrolysis using a proton exchange membrane (PEM) system.

The OER, which is a lengthy process requiring the transfer of four electrons, has the greatest influence on overall water-splitting efficiency. Once refined, the resulting hydrogen has a purity level of more than 99.9%, making it appropriate for a variety of industrial operations and transportation uses. The most widely used renewable energy sources for producing hydrogen are solar and wind, while base-load energy generation is aided by hydropower, biomass, and geothermal energy [7, 12]. However, one major disadvantage of renewable energy is its intermittent and uneven distribution. Hydrogen production via electrolysis will become more viable as wind and solar power integration improves and fair carbon pricing is implemented. The ability to store surplus electricity as hydrogen is one of the benefits of combining electrolysis and renewable energy. This helps to control variations in energy supply and demand. While metal hydrides can hold modest amounts of hydrogen, larger quantities can be held in salt caverns, storage tanks, or integrated into the gas grid [13].

The study of thermodynamics of water electrolysis is essential for understanding the fundamentals of water electrolysis. The following equation depicts the total energy ($\Delta H$) needed to split water:

$$\Delta H = \Delta S \times T + \Delta G \tag{2.2}$$

In this equation, $\Delta S$ represents the total entropy transition from liquid water to gaseous hydrogen and oxygen, $\Delta G$ represents the reaction's total Gibbs free energy, and $T$ is the temperature in Kelvin. Under typical circumstances of temperature (25°C) and pressure (1 bar), splitting one mole of water by electrolysis requires a minimum theoretical energy input equal to the standard Gibbs free energy of formation for water (237.2 kJ/mol). This energy is delivered in the form of electric power. Furthermore, additional energy is necessary to account for the entropy shift during the reaction, which is measured to be 48.6 kJ/mol. This component is often supplied as thermal energy (heat). The total energy required for accomplishing the water-splitting process is 285.8 kJ/mol [14]. The electrolysis cell potential ($\Delta E$) at thermodynamic equilibrium is given as follows:

$$\Delta E = \frac{-\Delta \mathrm{G}}{n \times F} \tag{2.3}$$

where $n$ is the number of moles and $F$ is the Faradic constant (96,485.3 C/mol). Each electrolyzed water molecule transfers two electrons. The minimum voltage required for electrolysis is approximately 1.23 V, while the thermoneutral voltage for water electrolysis at temperatures below 100 °C is typically 1.48 V [10, 14]. However, in practical applications, the water electrolysis process requires a voltage greater than 1.8 V to overcome activation energy constraints. The high voltage requirement is due to various factors, including the kinetics and ohmic resistance of the electrolyte and electrolyzer cell components, as well as the breaking of strong multiple O–H bonds in water molecules and the re-formation of $H_2$ molecules [4, 15].

## 2.3 Types of Electrolyzers

Electrolyzers are important tools that use electricity to split water molecules into hydrogen and oxygen electrochemically, in the electrochemical process. The electrolyzers are divided into four major types based on the electrolyte and working temperature [1]:

1) Alkaline water electrolyzer (AWE)
2) Anion exchange membrane (AEM) water electrolyzer (AEMWE)
3) PEM water electrolyzer (PEMWE)
4) Solid oxide water electrolyzer (SOWE)

AWE and PEMWE are the low-temperature tools that are well-established with great technological advancement, while SOWE, a high-temperature process, is currently in development. AWE uses a concentrated alkaline solution as an electrolyte, which requires a gas-impermeable separator to prevent hydrogen and oxygen from mixing. Furthermore, AWE technology enables the use of non-noble and cost-effective electrode materials, such as nickel, coated with electrocatalytic substances [16]. On the other hand, PEMWE uses noble metal catalysts such as iridium oxide and platinum, as well as a polymer membrane as the electrolyte that must be humidified [17]. Both AWE and PEMWE operate at low temperatures (30–80 °C) and pressures of up to 30 bar, with a nominal stack efficiency of about 70%. The SOWE operates at high temperatures ranging from 700 to 900 °C, transforming gaseous water into hydrogen and oxygen with higher efficiency (74–86%) [18]. At these high temperatures, favorable thermodynamic phenomena allow theoretical stack efficiency to approach 100%. To be economically effective, the process requires a steady supply of waste heat from industries such as chemical processing, metallurgy, or thermal power generation. Furthermore, the rigorous operating circumstances need the development of superior corrosion-resistant materials. As a result, SOE is currently confined to small stack capacities of less than 10 kW, while AWE can reach 6 MW and PEMWE up to 2 MW [19]. For large-scale applications, investment costs and system longevity are critical factors in deciding whether AWE or PEMWE is the better option [20]. Furthermore, AWE systems typically have a longer operational lifespan and lower yearly maintenance expenses than PEMWE. Nonetheless, PEMWE is frequently selected for applications needing dynamic operation since it provides quick start-up times and a wide range of load flexibility. The constraints of AWE are being addressed gradually by continual improvements. As a result, this chapter focuses on fundamental and current developments in AWE, AEM electrolyzer, PEM electrolyzer, and SOWE using different operating conditions and materials. For safe and efficient functioning, water electrolysis systems must be dynamically optimized. As a result, it is critical to understand how operational dynamics affect system performance and what considerations must be considered when using different water electrolysis systems along with fluctuating renewable energy instead of a stable power supply. Table 2.2 depicts the various advantages and limitations of different water electrolyzer technologies. The next section contains an exhaustive description of all of these electrolyzer technologies.

**Table 2.2:** Advantages and limitations of water electrolyzer technology [1, 15, 21].

| Electrolyzer technology | Advantages | Limitations |
|---|---|---|
| Alkaline water electrolyzer (AWE) | Well-established technology (>100 years), utilized non-noble metal electrocatalyst, low-cost ($300–600/kW) technology, used for industrial scale, stability up to 80,000 h. | Gas crossover (up to 5%), formation of carbonates on the electrode, needs a lot of space, low dynamic operation, corrosive electrolyte (5 M KOH), low $H_2$ purity, low current densities (0.2–0.4 A/cm$^2$). |
| AEM electrolyzer | Utilized solid polymer electrolyte, non-noble metal electrocatalyst, and cheap components. | Under development (TRL: 4–6) and limited stability (up to 10,000). |
| PEM electrolyzer | Produce highly purified gases (99.99%), high current densities (1–2 A/cm$^2$), commercialized technology, quick response and startup, and compact system design. | Higher capital cost ($1,200–2,000/kW), utilized noble metal electrocatalyst, sensitive to water source contaminants, acidic electrolyte. |
| Solid oxide electrolyzer | High efficiency at high working temperature, utilized non-noble metal electrocatalyst, low capital cost, and high purity hydrogen. | Under development (TRL: 4–6), large system design, stability (up to 10,000), safety and sealingissues. |

## 2.3.1 Alkaline Water Electrolyzer (AWE)

The AWE is currently a well-established technology that is utilized industrially to generate hydrogen at multi-megawatt levels around the world. Troostwijk and Diemann initially introduced the electrolysis of water in an alkaline medium in 1789 [14]. The schematic of the operation of the alkaline electrolyzer is depicted in Figure 2.2. AWE is made up of several components, including an electrocatalyst layer on a gas diffusion layer on the anode and cathode sides, a current collector, end plates, and a central part of the electrolyzer consisting of a separator through which diffusion of ions takes place (Figure 2.2). The electrolyzer utilizes non-Pt group metal electrodes, typically nickel-coated stainless steel [15]. The electrolyzer is characterized by the presence of two electrodes submerged in an alkaline electrolyte solution containing KOH or NaOH (20–40%). The alkalinity of the electrolyte facilitates ion conductivity, which is renewed for longevity and efficiency [1]. The separator or diaphragm is typically composed of ceramic oxide materials such as asbestos, potassium titanate, glass frits, and porous pots [15, 22]. A porous $ZrO_2$-stabilized polyphenylene sulfide (PPS) mesh (e.g., Zirfon) is employed as a separator to improve OH- ion conductivity and gas isolation. The separator's primary role is to separate the anode and cathode, hence preventing anolyte and catholyte mixing for safety and purity aspects.

It also serves to provide permeating hydroxyl ions ($OH^-$) and water molecules through the diaphragm's porous structure, allowing the electrochemical reaction to perform effectively [22].

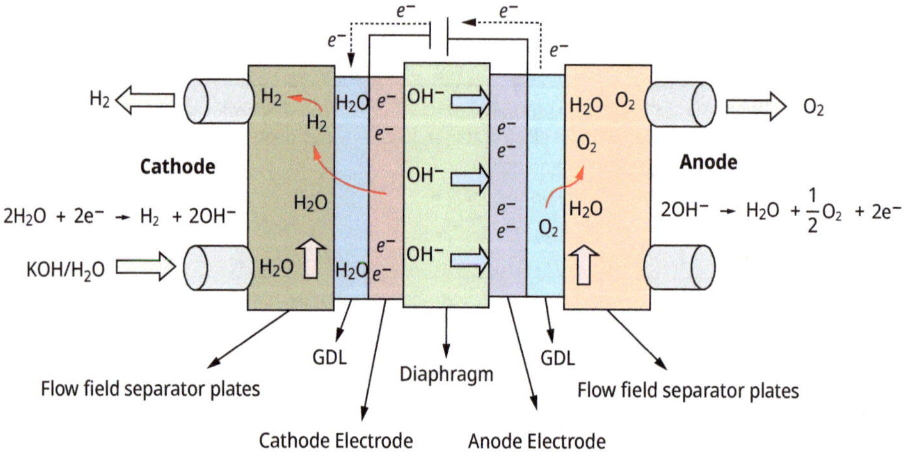


**Figure 2.2:** Schematic of alkaline water electrolyzer and electrode reaction.

The technique utilizes an electrochemical process to convert water molecules in an alkaline solution into hydrogen and oxygen in the presence of electricity. This mechanism includes two separate half-cell reactions: the HER at the cathode and the OER at the anode. Initially, two moles of water molecules in an alkaline solution are reduced by the cathode during electrolysis, yielding two moles of hydroxide ions ($OH^-$) and one mole of hydrogen gas ($H_2$) (Figure 2.2). The produced $H_2$ gas is released from the cathodic surface, and the applied electric field propels the hydroxide ions through a porous separator and toward the anode. One mole of water ($H_2O$), half a mole of oxygen gas ($O_2$), and two electrons are produced by the oxidation of hydroxide ions ($2OH^-$) at the anode. The electrons flow through the external circuit to the cathode, completing the electrical circuit in the electrolyzer and contributing to the process of water splitting. This method ensures that water is constantly split, with hydrogen collected as a clean energy carrier and oxygen released as a by-product [21].

The established technology of alkaline water electrolysis is appropriate for large-scale hydrogen production. In the early nineteenth century, over 400 industrial alkaline electrolyzers were successfully developed and employed for a wide range of industrial uses. Currently, the investment cost for such systems ranges between US $500/kW and $1,000/kW, with an operational lifespan of around 90,000 h [15].

AWE performance is impacted by an array of factors, including electrolyte concentration, diaphragm, anode and cathode material type, and operating conditions. Electrolysis reaction kinetics can be significantly improved by using highly active and stable electrode materials. Furthermore, an optimum electrolyte concentration can improve ionic conductivity, resulting in higher overall efficiency. Operational variables, including temperature and pressure, also play key roles. High-temperature operation can accelerate reaction rates, whereas high-pressure operation can improve $H_2$ gas solubility and reduce downstream compression needs, which could improve electrolyzer performance [1, 16].

One of the major disadvantages of this technology is its limited current density, typically between 0.1 and 0.5 A/cm$^2$. This is due to the moderate transit of $OH^-$ ions and the high impedance of the liquid electrolyte and diaphragm [22]. Additionally, the sensitivity of electrolytes to ambient $CO_2$ causes the production of $K_2CO_3$, lowering the hydroxyl ion concentration and ionic conductivity. Accumulation of $K_2CO_3$ can restrict the gas diffusion layer pores of the anode, reducing ion movement through the separator and decreasing hydrogen production efficiency [15, 22]. Another challenge with alkaline electrolysis is that the gases produced, hydrogen and oxygen, are of low purity (99.9%). This is because gas crossover takes place through the diaphragm between the two half-cells, which reduces overall gas separation efficiency. Furthermore, at low loads (<40%), $H_2$ diffusion into the oxygen evolution chamber can lead to concentrations above the lower explosion limit of >4 mol% $H_2$, posing a serious safety issue. Aside from that, the use of liquid electrolytes inhibits high-pressure operation, necessitating a bulky stack arrangement [14, 22].

The other alkaline electrolyzer, which gained a lot of attention, is the AEMWE. The AEMWE technology incorporates the positive aspects of AWE and PEM electrolyzers. The AEMWE consists of an electrochemical process that splits water into hydrogen and oxygen using an AEM and an external electric field. This method comprises two essential half-cell reactions, similar to AWE, namely HER at the cathode and OER at the anode. Initially, water molecules are reduced by the cathode, which accepts electrons to produce hydroxyl ions ($OH^-$) and $H_2$ gas. The produced $H_2$ gas is released from the cathode surface, and the electrostatic pull of the positively charged anode causes hydroxyl ions to move toward the anode across the AEM. At the anode, hydroxyl ions ($OH^-$) recombine to produce oxygen gas and water molecules and emit electrons. The oxygen generated is then discharged from the anode. The essential working mechanism and accompanying half-cell reactions of AEME are shown in Figure 2.2, where the diaphragm is replaced with an ion exchange membrane.

The AEM electrolyzer is a novel technique for producing sustainable hydrogen that combines the features of AWE and PEM electrolyzers. Various research institutes and organizations have recently been extensively interested in promoting AEM electrolyzers because they use cost-effective non-precious metal (i.e., Ni, Co, and Fe) catalysts, lowering overall costs while providing superior performance than other conventional electrolyzer technologies.

The use of AEMs, which are commonly constructed of quaternary ammonium ion-exchange materials or divinylbenzene (DVB) series, instead of conventional diaphragms (such as asbestos), distinguishes alkaline water electrolysis from AEME. The commonly used AEM membranes include quaternary ammonium ion exchange membranes (Fumatech, Fumasep, and Sustainion®) [15]. This improvement improves efficiency and enables the use of cost-effective transition metal catalysts rather than more expensive noble metal catalysts. Ni, Fe, Co oxides, iridium-based materials ($IrO_2$, $IrO_x$), and transition metal nitrides are commonly used as anode catalysts in AEM electrolyzers, whereas Ni, Ni–Co alloys, and iron-based catalysts (FeNi, FeCo) are used as cathode catalysts [1, 4]. Moreover, AEME can work with a wider range of water sources, such as distilled water or a low-concentration alkaline electrolyte (1 M KOH) or seawater, instead of the highly concentrated KOH solution generally used in alkaline electrolysis [1, 23]. Despite these advantages, further research is needed to increase ionic conductivity and membrane stability, membrane electrode assembly (MEA) stability, and power efficiency to reduce total stack costs, all of which are crucial for large-scale commercialization. According to current reports, the operating stability of AEMWE varies, with Sustainion® membranes obtaining up to 2,000 h, Fumatech membranes (A201 and FAA3-50) reaching 1,000 h, and Enapter's multicore AEM electrolyzer exceeding 35,000 h [15].

#### 2.3.1.1 Current Developments in Alkaline Electrolyzer

AWEs are a well-developed technology that can yet be improved to perform effectively. There are various research and development going on to improve alkaline electrolyzers. The electrode material and catalyst material are the two significant components of the alkaline electrolyzer and are the primary focus of research and development. Efforts to improve efficiency, cost-effectiveness, and sustainability have resulted in significant advances in electrode materials for alkaline water electrolysis (AWE) systems throughout the years. The first anode materials were primarily noble metals, particularly platinum (Pt) [24]. Anode materials in AWE were initially developed in the mid-twentieth century when scientists sought to produce hydrogen efficiently and sustainably. During this age, platinum became a popular anode material. The excellent electrochemical characteristics of platinum made it a popular choice in electrocatalysis. It demonstrated exceptional catalytic activity during the OER, a critical stage in water electrolysis that splits water molecules into hydrogen and oxygen gases at the anode. However, it is widely known that the high cost of platinum is the biggest hindrance to its commercial use. Growing environmental awareness and economic concerns have drawn attention to transition metal oxides (TMOs) as highly desirable materials due to their widespread availability and low cost [25]. Compounds like ruthenium dioxide ($RuO_2$), iridium dioxide ($IrO_2$), and manganese dioxide ($MnO_2$) show interesting electrocatalytic activity [26–28]. $RuO_2$ and $IrO_2$ exhibit high activity in the OER, making them ideal for anode development. To overcome durability and cost limits, research efforts have been made on improving crystal structures and developing composite materials.

Metal nitrides, particularly titanium nitride (TiN) and molybdenum nitride (MoN), have recently gained popularity due to their superior electrocatalytic characteristics, corrosion resistance, and cost-effectiveness. Researchers investigated several compositions, including manganese-based oxides like $MnO_2$ and cobalt-based oxides like $Co_3O_4$ and $Co_2O_3$ [29, 30]. These materials were extremely versatile, allowing for precise modifications to improve their electrocatalytic properties.

The emphasis on developing electrocatalysts for the AWE is significant for achieving high efficiency. A significant goal in water splitting research is to synchronize the kinetics of the related processes to eliminate energy inefficiencies. A significant achievement in this field has been the utilization of transition metal-based compounds as highly effective bifunctional electrocatalysts. Alloys such as NiFe, NiMo, and NiCo have demonstrated exceptional performance in driving hydrogen and oxygen evolution processes. Scientists have been able to increase catalytic efficiency while maintaining long-term operational durability by carefully modifying their elemental composition and structural properties. Qazi et al. [31] developed a highly effective bifunctional nano-heterostructured electrocatalyst, NiCo-$NiCoO_2$@$Cu_2O$@CF, for water splitting. The electrocatalyst was tested in a 1 M KOH alkaline electrolyte and showed increased activity for both HER and the OER. To obtain 10 $mA/cm^2$ of current density, overpotentials of 133 mV for HER and 327 mV for OER were sufficient. The catalyst had good kinetics, with Tafel slopes of 119 mV/dec for HER and 118 mV/dec for OER, indicating efficient catalytic action in alkaline circumstances.

Platinum (Pt) is widely recognized as one of the most effective catalysts for HER due to its high electrocatalytic performance, allowing for efficient hydrogen synthesis. Despite its excellent efficiency, the high cost and restricted availability of platinum have created significant impediments to its widespread usage in hydrogen-generating systems. To overcome these issues, researchers developed platinum-based alloy catalysts such as Pt–Ir and Pt–Ni, which offer a balance of high catalytic activity and low cost [16]. These alloys not only maintain high performance but also help to reduce the overall costs of hydrogen generation. In addition to these alloys, nickel (Ni) and its compounds have emerged as effective HER catalysts. Nickel is far less expensive than platinum and has good electrocatalytic properties, making it suitable for commercial-scale uses. Nickel-based materials, such as hydroxides and oxides, have strong electrochemical activity, which contributes to the excellent performance of alkaline water electrolysis systems. Researchers have investigated other catalysts beyond noble and base metals, including TMOs and hydroxides like cobalt oxide ($Co_3O_4$), iron oxide ($Fe_2O_3$), and manganese oxide ($MnO_2$) [16, 32, 33]. These non-noble metal compounds are not only cheap and abundant, but they also provide significant HER performance, reducing dependency on limited noble metals. Furthermore, novel research into non-metal catalysts, particularly nitrogen-doped carbon materials, has opened up new opportunities for developing efficient and cost-effective HER catalysts that do not rely on traditional metal-based systems.

### 2.3.2 Proton Exchange Membrane Water Electrolyzer (PEMWE)

The fundamental principle of the PEM electrolyzer is similar to the PEM fuel cell, which utilizes a sulfonated polymer as the membrane electrolyte to conduct protons. In this configuration, protons ($H^+$) operate as ionic charge carriers and deionized water permeates through the membrane, enabling electrochemical operations. The working temperature of this technology is (30–80 °C) and can handle large current densities of 1–2 A/cm$^2$. The hydrogen and oxygen gases produced are also very pure, with purity levels of up to 99.999% [21]. It is reported that the stability of PEME exhibits excellent stability for 60,000 h and demonstrates very little performance degradation, while the target stability is 1,000,000 h [34]. One major benefit of PEME over AWE is the faster HER kinetics. This is primarily because platinum electrodes have a high catalytic surface activity, and the membrane electrolyte provides an intrinsically low pH environment.

In the PEM electrolyzer, water molecules are broken into hydrogen and oxygen by electrochemical reactions at the electrodes – hydrogen is produced at the cathode and oxygen at the anode. Water is introduced at the anode and oxidized to produce oxygen gas ($O_2$), protons ($H^+$), and electrons ($e^-$). Protons are transferred from the membrane electrolyte to the cathode. Meanwhile, electrons generated at the anode move through an external circuit, providing the required cell voltage to power the operation. At the cathode, these electrons mix with protons to produce hydrogen gas. The anode and cathode reaction in PEMWE is shown in Figure 2.3.

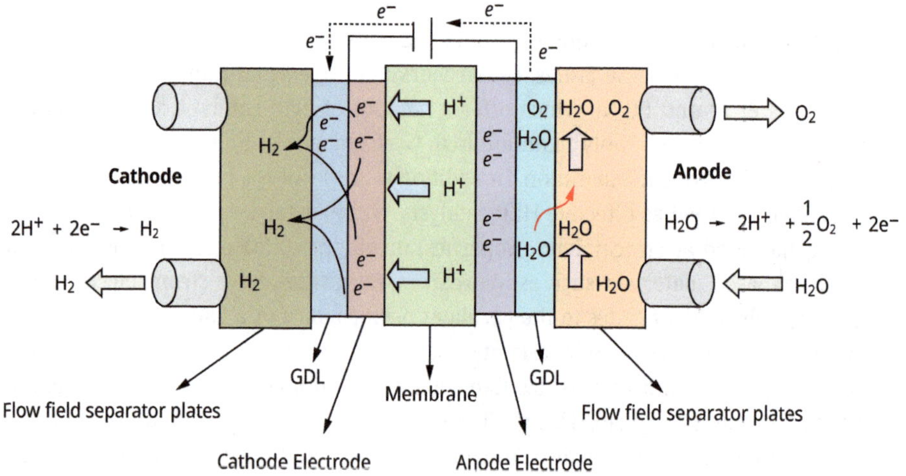


**Figure 2.3:** Schematic of the PEM water electrolyzer and electrode reactions.

The major components of the PEMWE consist of a membrane, an electrocatalyst painted on GDL, a current collector, and separator plates. The central part of the

PEMWE consists of a membrane through which transportation of protons takes place. It is the heart of any kind of PEM water electrolyzer. Perfluorosulfonic acid (PFSA) polymer membranes, such as Nafion, Fumapem, Flemion, and Aciplex, are among the most extensively used membranes due to their superior properties [21, 35]. These include exceptional mechanical strength, high efficiency, great oxidative stability, and consistent dimensional performance throughout temperature ranges. Nafion membranes (particularly Nafion 115, 117, and 212) are the most often used materials in PEMWE. This preference is partly owing to their capacity to function at greater current densities (up to 2 A/cm$^2$), paired with remarkable longevity, strong proton conductivity, and consistent mechanical integrity [14]. The membrane is sandwiched between two electrodes (electrocatalyst painted on the GDL), namely, an anode and a cathode, and this assembly is termed a membrane electrode assembly (MEA). The MEA is prepared by different methods. One method is an electrocatalyst slurry painted on the GDL, and the membrane is sandwiched between electrodes, which is known as the clamping method [36, 37]. The other method involves the coating of the catalyst on the membrane surface and the hot-pressed [38]. Current collectors are crucial to the overall performance and efficiency of the PEM electrolysis system. They must have strong corrosion resistance and electrical conductivity to endure the acidic environment, extreme overpotentials, and oxygen exposure. In addition to electrical performance, current collectors should provide adequate mechanical support for the membrane. Optimized porosity and pore volume in current collectors are critical for efficient gas removal and continuous water access to the electrode's catalytic sites. Properly built current collectors are very essential. During operation, protons travel from the anode electrode surface to the cathode via the proton-conducting membrane. Meanwhile, electrons flow through the current collectors and separator plates to the cathode. Once electrons reach the cathode electrode surface, they mix with protons to generate hydrogen gas, which departs the cell through the cathode-side current collector.

In PEMWE, noble metal-based electrocatalysts are widely used. Typically, platinum or palladium catalysts are utilized at the cathode to drive the hydrogen evolution process, while iridium oxide ($IrO_2$) or ruthenium oxide ($RuO_2$) serve as anode catalysts for the OER. Although this technology has various advantages, including high gas purity, exceptional efficiency, and the ability to operate at high current densities, the high cost of noble metals significantly raises the entire cost of the system. To address these cost-related difficulties, extensive research has been conducted to produce cheaper options. Numerous studies have introduced low-cost electrocatalysts with increased activity and durability, particularly those based on Ru-containing multimetallic oxides. These materials seek to retain performance while decreasing dependency on costly elements such as Ir and Pt. Furthermore, the utilization of alternative, cost-effective carbon-based support materials has been investigated and successfully proved to cut costs while maintaining efficiency.

#### 2.3.2.1 Current Development in PEM Water Electrolyzer

PEMWE is a mature technology that is commercially employed for industrial applications on megawatt (MW) scales. Several prominent firms have already started building large-scale PEM electrolyzers. Nonetheless, additional developments are necessary to reduce the overall cost of $H_2$ gas production. One of the key difficulties in PEM water electrolysis is to improve the HER at the cathode by finding more efficient electrocatalysts. Platinum-based materials have long been recognized and used as the standard for HER due to their outstanding catalytic performance and extraordinary stability in acidic conditions. Currently, Pt supported on high-surface-area carbon materials is regarded as the standard for HER in PEM electrolyzers. However, the high cost of platinum is a significant disadvantage. As a result, extensive research has been conducted to reduce total costs by improving the performance and endurance of electrocatalysts while minimizing the use of expensive materials. To improve economic feasibility, measures such as reducing Pt loading or developing other, less expensive materials with equal catalytic activity have been actively investigated. Many researchers used very low platinum catalyst loading to minimize its cost [39, 40]. Carbon nanoparticles are commonly used to improve the electrochemically active surface area while reducing the amount of platinum required [41]. These electrocatalysts are painted on the gas diffusion electrodes, which use low-cost, high-surface-area conductive materials such as carbon black or carbon nanotubes. As a result, several types of carbon black (CB) have been widely embraced as support materials for platinum-based catalysts, which are commonly used in the HER [39]. However, even with lower platinum loading, carbon-supported catalysts still account for a significant portion of the system's cost, owing to performance losses caused by carbon-support corrosion or degradation.

Xie et al. [42] synthesized a unique electrode design by integrating platinum nanowires (PtNWs) onto ultrathin titanium-based gas diffusion layers for use in the HER of PEME. This PtNW/Ti electrode had excellent electrochemical characteristics, with a low overpotential of 63 mV at a high current density of 100 $mA/cm^2$ and a Tafel slope of just 35 mV/dec. When used as a cathode in a PEM electrolyzer, the electrode reduced cell voltage to 1.643 V at 1 $A/cm^2$ and achieved a high efficiency of 90.08%. Notably, this performance was accomplished with 15 times less catalyst than is commonly utilized in standard membrane electrode assemblies (MEAs). Xie et al. [43] created ionomer-free integrated electrodes by growing $MoS_2$ nanosheets ($MoS_2$NSs) directly on carbon fiber paper, resulting in a 1 T-2 H heterophase structure (1 T-2 H $MoS_2$NS/CFP). The electrode, loaded with only 0.14 $mg/cm^2$ $MoS_2$, was tested for the HER in 0.5 M $H_2SO_4$ using a PEM water electrolyzer system. The electrode demonstrated exceptional catalytic performance, with a low overpotential of 192 mV and a Tafel slope of 44 mV/dec. Qayum et al. [44] synthesized Ni-FeOx/FeNi for HER and OER. The Ni-FeOx/$FeNi_3$ electrocatalyst is a prospective alternative to currently available noble metal-based electrocatalysts for large-current water electrolysis due to its superior electrocatalytic characteristics and long-term durability.

The OER occurs at the anode side of the PEM water electrolyzer. The electrocatalyst used for the OER is the metal oxides. Among the various metal oxides, $IrO_2$ and $RuO_2$ are widely used electrocatalysts for OER in PEMWE. $IrO_2$ and $RuO_2$ have strong electrical conductivity (about $10^4$ S/cm). This is due to the advantageous overlap of inner d-orbitals caused by short metal-metal distances and comparable cationic radii, which allows for efficient electron transport through their d-bands [21]. $RuO_2$ has a higher activity for the OER than other metal oxides [39]. $RuO_2$ has excellent catalytic activity but has disadvantages of lack of durability under extremely acidic conditions of perfluorosulfonic membranes and high anodic potentials during OER. In contrast, $IrO_2$ provides great chemical stability in acidic conditions, but with decreased catalytic effectiveness [45]. To address these limitations, recent research efforts have focused on the development of new catalyst materials for efficient and stable OER operation, thereby improving both performance and durability while lowering material costs. Jiang et al. [46] synthesized $WO_x$ nanorods decorated with an Ir film (Ir@$WO_x$NRs) using an electrodeposition approach to improve OER performance in acidic conditions. The Ir@$WO_x$NRs electrode had a low noble metal loading of 0.14 mg/cm$^2$ and demonstrated excellent electrochemical activity and endurance. It obtained a high current density of 2.2 A/cm$^2$ at 2 V, and during a long-term test over 1,030 h at a constant current density of 0.5 mA/cm$^2$, it demonstrated a moderate deterioration rate of 49.7 µV/h. Jiang et al. [47] expanded on this by introducing a $WO_3$ nano-array electrode coated with a heterogeneous IrRu alloy layer in changing Ir/Ru ratios, also by electrodeposition. The $Ir_{3.2}Ru_{1.6}$@$WO_3$ electrode outperformed commercial $IrO_2$, with a low overpotential of 245 mV at 10 mA/cm$^2$. Min et al. [48] developed a high-efficiency IrOx/W-$TiO_2$ electrocatalyst to reduce the use of noble metals in PEM-based water electrolysis. Researchers tested its OER activity in half-cell and single-cell configurations, comparing it to undoped $IrO_2$. In half-cell experiments, the IrOx/W-$TiO_2$ catalyst considerably increased Ir mass activity compared to undoped and commercial $IrO_2$. Single-cell studies demonstrated higher electrochemical performance and endurance compared to unsupported IrOx and commercial $IrO_2$, with a current density of 1 A/cm$^2$ at 1.602 V and a low precious metal loading of less than 0.2 mg/cm$^2$.

The electrocatalysts for the HER and OER are widely developed to obtain a highly efficient PEM water electrolyzer, as discussed above. Efforts have also been made to improve the membrane material for PEM water electrolysis [49, 50]. The efficiency of the electrolyzer also highly depends on the efficient ion transport through the membrane. The integrated system of water electrolysis is also developed to obtain a sustainable electrolyzer. Electrolysis and photovoltaic (PV) solar technology can be combined together to represents a possible improvement in renewable energy systems. This method uses electricity generated by PV panels to power the electrolysis process, which converts water into hydrogen and oxygen. The resulting hydrogen is a clean, storable energy source, or industrial feedstock, while oxygen can also be used for good purposes. This integrated approach promotes both sustainable energy generation and efficient storage technologies [51].

### 2.3.3 Solid Oxide Water Electrolyzer (SOWE)

SOWEs are developing as a promising method for producing sustainable hydrogen on a large scale, utilizing renewable energy. The SOWE has been thoroughly investigated in recent studies [3, 52, 53]. SOWEs produce more hydrogen and convert energy more efficiently in comparison to PEMWEs and AWEs. The ability of SOWE to operate at high temperature (500–850 °C) is responsible for it high performance, which results in faster ion conductivity and reduced voltage needs, hence lowering the consumption of energy. The reported energy requirements for electrolyzing water and carbon dioxide are about 3.1 kWh per normal cubic meter of hydrogen and 3.4 kWh per normal cubic meter of carbon monoxide, respectively [54]. SOWEs can achieve electrical efficiencies of up to 95%, outperforming low-temperature systems by 25–30% in the temperature range of 750–850°C. This performance is due to the increased ionic conductivity of solid electrolytes and lower electrode polarization [55]. SOWEs may directly produce synthesis gas (a mixture of $H_2$ and CO), a useful raw ingredient in the commercial synthesis of compounds like ethanol and ethylene [18].

Despite the tremendous promise of SOWE technology, various obstacles prevent widespread use. Notably, the high costs of both manufacturing and maintenance, along with enormous energy consumption, provide significant hurdles. Currently, the cost of producing hydrogen from solid oxide electrolysis cells (SOECs) ranges from $2.99/kg to $7.02/kg, which is higher than the production costs of PEM electrolysis cells ($2.50–6.80/kg) and alkaline electrolysis cells (AECs) ($2.36–6.98/kg) [56, 57]. Figure 2.4 shows the components, electrode reaction, and the operation of the solid oxide electrolyzer. The working principle of a solid oxide electrolyzer involves the reduction of water molecules (steam) into hydrogen and oxide ions ($O^{2-}$) at the cathode side. The oxide ion produced at the cathode side is transported through the membrane to the anode side, where it is converted into oxygen and generates two moles of electrons.

The electrons generated pass through the external circuit to the cathode side. Oxygen ions migrate faster and have better transport properties than protons. Additionally, oxide ion electrolytes operate better in high-temperature and corrosive situations. The components of the solid oxide electrolyzer also consist of flow filed separator plates, gas diffusion layer, electrocatalyst layer painted on the GDL, current collector, and the membrane. The three main components of a typical SOWE are: two porous electrodes (anode and cathode) and a dense ceramic electrolyte of dense ceramic material that conducts oxide ions ($O^{2-}$). The most often utilized membrane electrolyte material is yttria-stabilized zirconia (YSZ). This material is commonly made up of zirconium oxide doped with 8 mol% yttria, which helps to stabilize the ceramic's cubic phase. YSZ is recommended for its stability and performance at high temperatures (700–850 °C). This

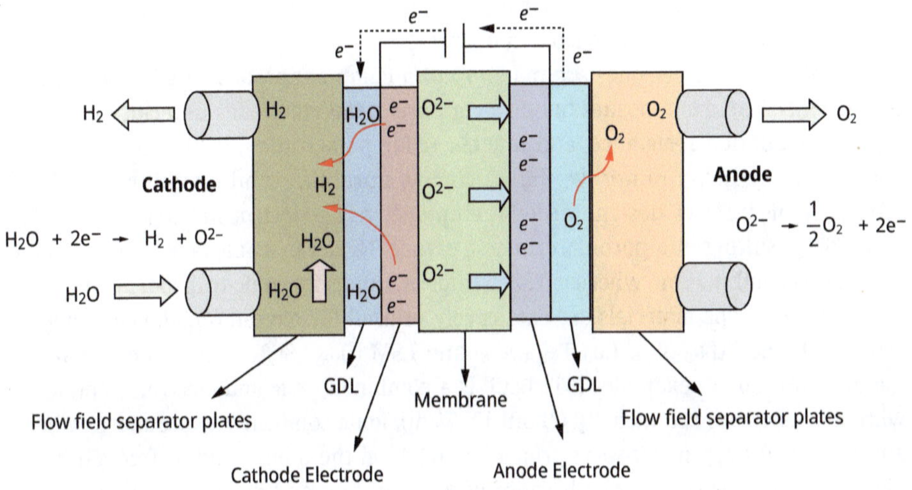


**Figure 2.4:** Schematic of solid oxide water electrolyzer and electrode reactions.

electrolyte material is beneficial for high-temperature applications due to its high oxide ion conductivity (ranging from $10^{-2}$ to $10^{-1}$ S/cm) and great chemical and thermal stability [58]. Furthermore, adding cerium dioxide into the electrolyte helps to increase the electrode performance and stability of the SOWE, particularly at lower temperatures. Cerium dioxide has superior ionic transport characteristics, unlike YSZ, which can improve electrolyte efficiency and increase the functional lifespan of the cathode [59]. A lot of research has been conducted on structural properties, surface qualities, and transport mechanisms of cerium dioxide [60–62]. Studies show that SOEs with cerium-based electrolytes function poorly at 700 °C due to high tensile stress at the interface with the oxygen electrode [63]. Thin-film cerium dioxide electrolytes, on the other hand, are more resistant to stress caused by electrochemical processes when supported, making them better suited for stable operation [64].

Choosing appropriate electrode materials can enhance gas movement and electrochemical reaction activity, resulting in high conversion efficiency. Metal and metal-ceramic are commonly utilized materials for SOE cathode electrodes. Metal-based electrodes in SOEs contain both precious metals like platinum (Pt) and more common metals like nickel (Ni), cobalt (Co), and iron (Fe). Nickel-based electrodes are particularly popular because of their superior electrochemical performance and affordable cost. Copper (Cu) has also been studied as an electrode material due to its ability to reduce carbon deposition; nevertheless, its overall electrochemical performance is unsatisfactory. Platinum electrodes outperform iron and cobalt electrodes but fall short of nickel electrodes. Furthermore, interactions between platinum particles and the electrolyte can result in reduced efficiency. The high price of platinum, along with the structural breakdown of its porous framework at increasing temperatures, precludes its practical use in SOWE technology.

Metal-ceramic electrodes are made up of a combination of metallic and ceramic components. These composites outperform pure metal electrodes in terms of thermal stability, oxidation resistance, and low electrode polarization at high temperatures. These features help to improve the efficiency, durability, and overall reliability of SOWEs. In most SOWE designs, nickel is employed as an electrocatalyst in conjunction with YSZ, resulting in a porous cermet structure. Nickel promotes electronic conduction in this combination, whereas YSZ creates channels for ionic transport [58].

Perovskite-type materials are extensively utilized for oxygen (anode) electrodes in SOEs, with LSCF ($La_{0.58}Sr_{0.4}Co_{0.2}Fe_{0.8}O_{3-\delta}$) and LSM ($(La_{1-x}Sr_x)_{1-y}MnO_{3-\delta}$) being among the most prominent examples [65]. LSCF is a combined ionic and electronic conductor with high electrical conductivity (about $10^2$ S/cm), ionic conductivity (around $10^{-2}$ S/cm), and improved oxygen transport properties. LSM, on the other hand, is frequently employed as a reference material because of its consistent and dependable performance in high-temperature electrochemical applications [66].

#### 2.3.3.1 Current Development in SOWE

The development of SOWE is not yet advanced as AWE and PEMWE. It is currently in development stage and yet to commercialized. The efficiency of SOWE is quite high due to its operation at high temperatures and the use of non-noble metal electrocatalysts. However, there is a lot of technological advancement required and economic difficulties remain a concern. One main difficulty is long-term operating stability; present systems typically work for roughly 20,000 h, while the aim is 100,000 h. The cost of producing hydrogen remains high, estimated at US $2,000/kW of $H_2$ in 2020. The goal is to lower this to US $200/kW by 2050 [15]. To achieve these objectives, electrode materials must be improved in terms of durability, either through electrochemical optimization of existing materials or the introduction of more stable alternatives such as advanced perovskites. Furthermore, scaling up solid oxide electrolyzer systems and integrating them with renewable energy sources is crucial for significantly lowering costs. In this regard, Li et al. [67] investigated and synthesized a composite oxide material, $Ni_{1-x}Cd_xO$-SDC, which was evaluated as a hydrogen electrode for use in reversible solid oxide cells, showing promising potential for improved performance. Zhao et al. [68] used an infiltration approach to build a high-efficiency oxygen electrode made of LSM-YSZ-$Ce_{0.9}Co_{0.1}O_{2-\delta}$. They then evaluated its use in hydrogen production using a solid oxide electrolyzer (SOE). This electrode exceeded expectations, achieving a current density of 1.26 A/cm$^2$ and a hydrogen emission rate of 873 mL/cm$^2$ h. In a separate study, Zheng et al. [69] developed an innovative oxygen electrode material based on Mn-doped Ruddlesden-Popper oxide, specifically $La_{1.5}Sr_{0.5}NiO_{4+\delta}$ (LSNMx), and evaluated its electrochemical performance in SOE systems, highlighting its potential to enhance oxygen electrode activity. Roy and Samanta

[3] investigated a solar-assisted power-to-hydrogen system based on proton-conducting solid oxide electrolyzer cells to fulfill the objective of Net-Zero 2050 because it can be produced utilizing renewable energy sources that emit no greenhouse gases. The study conducts a full thermoeconomic evaluation of the system, including optimization utilizing a genetic algorithm technique. The research determines the ideal values for energy efficiency and hydrogen production costs, which are 25.15% and $1.021/kg, respectively. Exergy research shows that the solar PV thermal unit has the highest losses, accounting for 67.83%, followed by the parabolic trough solar collector at 13.31%. Furthermore, the performance of the solar-driven power-to-hydrogen arrangement is compared to other hydrogen generation technologies to determine its relative effectiveness.

## 2.4 Comparative Analysis of Electrolyzer Technologies

Each kind of electrolyzer has unique operational characteristics, benefits, and limits that affect its appropriateness for diverse hydrogen production applications.

It is well known that the AWEs are the most widely used technology for hydrogen production and also the oldest technology. AWE uses a liquid alkaline solution as the electrolyte, such as potassium hydroxide. The operating temperature of PEMWE is in low to moderate range and also known for its affordability and long-term reliability due to its maturity status and simple design. However, due to some limitations like low efficiency and slow reaction to changing power inputs, they make them unsuitable for renewable energy sources [61]. The operation of the AWE for the longer duration is crucial, especially while operating under variable conditions. The anticipated lifetime of an alkaline electrolyzer based on acceptable efficiency is 60,000–90,000 h [61].

The drawback of the AWE is minimized to some extent by the development of PEMWE, which uses an acidic solid polymer membrane electrolyte to conduct protons. The performance of the PEMWE is comparatively high, which is capable of functioning at large current densities. Apart from this, it is compact in nature and produces hydrogen with high purity and pressure. Furthermore, their other advantages, such as quick response times and excellent efficiency, make them suited for use in situations with variable energy demands. Their major issues include sensitivity to water contaminants, as well as the high cost of electrode materials, particularly noble metal catalysts such as platinum and iridium, which limit their feasibility for large-scale applications [70]. The PEM water electrolyzer can work up to 20,000–60,000 h [61].

In contrast, AEM water electrolyzers are a newer type of promising technology that utilizes an anion-conducting polymer membrane and combines the benefits of alkaline with PEM technology. They are being studied to combine the low material costs of alkaline systems with the structural advantages of PEM technology, as well as to offer high efficiency with the potential use of inexpensive Ni-based catalysts instead of high-cost, scarce noble metal catalysts. AEMWE is still in development, but it shows promise for lowering dependency on precious metals while maintaining high performance at low temperatures [15].

SOWEs run at high temperatures using a solid ceramic electrolyte. Their high-temperature operation increases energy efficiency, particularly when combined with waste heat or thermal sources. However, the materials utilized must be resistant to heat stress and chemical deterioration over time, which presents issues for long-term durability and system complexity [18]. The solid oxide electrolyzer has the problem of durability; present systems typically work for roughly 20,000 h, while the aim is 100,000 h. Despite the tremendous promise of SOE technology, various obstacles prevent widespread use. Notably, the high costs of both manufacture and maintenance, along with enormous energy consumption, provide significant hurdles. Currently, the cost of producing hydrogen from SOEs ranges from \$2.99/kg to \$7.02/kg, which is higher than the production costs of PEMWEs (\$2.50–6.80/kg) and AWEs (\$2.36–6.98/kg) [56, 57]. Table 2.3 outlines the operational aspects and essential components of the four types of water electrolyzers.

**Table 2.3:** Comparison between water electrolyzer technologies [1, 4, 15].

| Specification | AWE | AEM electrolyzer | PEM electrolyzer | SOE electrolyzer |
|---|---|---|---|---|
| Technology maturity | Mature (TRL: 10) | R&D stage (TRL:4–6) | Commercial (TRL: 9) | R&D stage (TRL:4–6) |
| Electrolyte | Concentrated alkaline liquor (20–40%KOH/NaOH) | Sustainion®, Fumatech, Fumasep, and DVB membrane | PFSA membranes (Nafion®) | YSZ and ceria-based electrolytes |
| Working fluid (feedstock) | High concentrationsolution (30 wt% KOH) | 1 M KOH and DI water | DI water | Steam |
| Anode material (OER electrode) | Ni-coated stainless steel (SS) | Ni-based materials, $IrO_2$ | $IrO_2$ and Ir-based materials | Perovskite type (e.g., LSCF andLSM) |
| Cathode material (HER electrode) | Ni-coated SS | Co-based materials, Ni-based materials | Pt-based catalysts and Ni-Mo | Ni/YSZ |
| Bipolar plate (anode) | Ni-coated SS | Ni-coated SS | Pt-coated Ti | None |
| Bipolar plate (anode) | Ni-coated SS | Ni-coated SS | Gold-coated Ti | Co-coated SS |
| GDL | Ni mesh | Ni foam/carbon cloth | Ti mesh/carbon cloth | Ni mesh/foam |

**Table 2.3** (continued)

| Specification | AWE | AEM electrolyzer | PEM electrolyzer | SOE electrolyzer |
|---|---|---|---|---|
| Operating temperature (°C) | 70–90 | 40–80 | 50–80 | 500–850 |
| Operation pressure (bar) | 1–30 | <35 | <70 | 1 |
| Efficiency (%) | 50–78 | 57–59 | 70–80 | 80–90 |
| Current density (A/cm$^2$) | 0.1–0.5 | 0.2–0.5 | Up to 2 | 0.3–1 |
| Operating voltage (V) | 1.8–2.4 | 1.6–2.4 | 1.8–2.2 | 0.9–1.2 |
| $H_2$ purity | 99.5–99.9998% | 99.9–99.999% | 99.9–99.999% | 99.90–99.99% |
| Compatibility with renewable energy | Medium | Potentially high | Excellent, superior tolerance for variable renewable energy sources | Low |
| Environmental impact | Moderate | Potentially low | Lowest | Highest |
| Applications | $NH_3$ synthesis and industrial $H_2$ | Small-scale applications, energy storage, and chemical synthesis | Transportation and backup power | Integrationwith waste heat, large-scale hydrogen, anddecarbonization |

TRL, technology readiness level; DVB, divinylbenzene; PFSA, perfluorosulfonic acid; YSZ, yttria-stabilized zirconia; LSM, $(La_{1-x}Sr_x)_{1-y}MnO_3$; LSCF, $La_{0.58}Sr_{0.4}Co_{0.2}Fe_{0.8}O_{3-\delta}$; GDL, gas diffusion layer; OER, oxygen evolution reaction; HER, hydrogen evolution reaction.

## 2.5 Challenges and Research Directions

Although extensive research has been done on the development of various types of electrolyzers to obtain green hydrogen, there are still various challenges in the commercial production of hydrogen through electrolysis. The challenges involve the performance for the long term and cell durability, cell operating temperature, and stack design optimization for achieving MW-scale commercial system. To overcome these challenges, various improvements are required in the different electrolyzers. The AWE has many challenges, such as cell efficiency, crossover through diaphragms, and lower operating current density. Significant inventions and improvements are needed to boost current density. The diaphragm used in the alkaline electrolyzer should have the minimum possible thickness to decrease the resistance and hence improve the current density. The diaphragm should improve in such a way that minimal crossover of gases occurs through it.

Similar to the alkaline electrolyzer, an electrolyzer is called an AEM water electrolyzer, which uses a membrane electrolyte instead of a diaphragm, as discussed earlier. The membrane's limited lifespan is a significant hurdle to scaling up this technology. Currently, membrane durability is typically limited to roughly 30,000 h, owing to chemical instability that causes the breakdown of the polymer backbone. This degradation endangers long-term performance and reliability. As a result, considerable advances and creative ways are required to enhance membrane stability and extend its operating lifetime.

The development of PEMWE is one of the great steps toward the production of green hydrogen as it has many advantages such as:
- Capacity to operate at higher current densities
- Produce high-purity hydrogen and oxygen
- Transport gases at elevated pressures
- Needs less space due to its compact form

The above benefits of PEMWE help to produce high performance and produce green hydrogen, but apart from these benefits, it also suffers from some disadvantages. The major drawback of PEMWE, which creates the main hurdle in its commercialization, is the high cost of its components. As a result, lowering prices and making this technology more economically feasible are the essential focus of ongoing research and development activities. The high cost for the development of PEMWE is mainly due to the electrocatalyst and membrane material. Advances in electrocatalyst design are critical, as the use of expensive noble metals such as platinum and iridium raises the cost of PEMWE and limits their scalability. The high cost of the issue of the PEMWE is solved by using low-cost materials and easily available electrocatalysts. Large-scale hydrogen generation and economic viability can be improved by using non-precious, earth-abundant materials instead of rare and noble metal resources. Membranes are critical to the operation of PEM water electrolyzers, and there is an urgent need for improvements in their durability, efficiency, and cost-effectiveness. One option could be to reduce the membrane thickness while increasing its mechanical strength. The system's durability and overall efficiency can be improved by using thinner membranes that have robust structural integrity, resulting in decreased operational energy consumption.

In comparison to other technologies, solid oxide water electrolysis is considered a relatively new technology that is also noted for its great efficiency. However, low durability is one of its key drawbacks. Significant improvements in long-term stability are necessary to make this technology more viable. This could be achieved by regulating the mechanical strength and chemical composition of system components while also increasing the ionic conductivity of the electrolyte employed. Another issue is that SOWE components are prone to oxidation and corrosion when exposed to high temperatures for an extended period of time. Furthermore, the high working temperatures make it difficult to ensure that the electrolysis stack is properly sealed.

## 2.6 Conclusions

The fundamental concepts of water electrolysis and water electrolyzer technology, as well as their most recent developments, have been covered in this chapter. Hydrogen ($H_2$) is seen as a promising energy carrier for the future because it produces no carbon emissions when used as a fuel source, with the exception of water as a byproduct. Green hydrogen production technologies, such as water electrolysis, when coupled with renewable energy sources, represent a sustainable hydrogen generation pathway for achieving a decarbonized energy landscape. The water electrolysis process is a hope that can fulfill the aim of producing a low-carbon energy system if all the constraints are eliminated or minimized. The water electrolysis process is done by four different electrolyzers, namely AWE, PEMWE, AEMWE, and solid oxide electrolyzer (SOE). Each type of electrolyzer operates on different principles and materials and produces various performance characteristics. AWE is the oldest water electrolyzer, which is a well-established technology, but it suffers from the problem of comparably low efficiency and reactivity. Another major breakthrough in water electrolyzer technology is the PEMWE. The performance of the PEMWE is quite high and is appropriate for the dynamic applications. The major drawback of this technology is its high cost due to the costly materials used in the manufacturing, particularly the membrane and electrode. AEMWE technology is similar to the AWE, but it uses a solid AEM instead of a liquid electrolyte. AEMWE is not fully developed yet, but it aims to develop an electrolyzer of low cost and high performance using low-cost materials in a low temperature range. The above-discussed electrolyzers are the low-temperature range electrolyzers. A high-temperature range electrolyzer is developed, which is known as SOE. SOE electrolyzer operates at high temperature and it exhibit high efficiency as compared to the low temperature electrolyzer. But due to the operation at high temperature, it has issues of material stability and durability. The continuous research and development are going on for the development of a highly efficient and low-cost electrolyzer that can be produced on a commercial scale. Water electrolysis technologies will become increasingly important as the need for green hydrogen rises, owing to the increase in renewable energy and decarbonization activities.

## References

[1] Akyüz ES, Telli E, Farsak M. Hydrogen generation electrolyzers: paving the way for sustainable energy. International Journal of Hydrogen Energy. 2024 Sep 4;81:1338–62.

[2] Pinsky R, Sabharwall P, Hartvigsen J, O'Brien J. Comparative review of hydrogen production technologies for nuclear hybrid energy systems. Progress in Nuclear Energy. 2020 May 1;123:103317.

[3] Roy D, Samanta S. A solar-assisted power-to-hydrogen system based on proton-conducting solid oxide electrolyzer cells. Renewable Energy. 2024 Jan 1;220:119562.

[4] Anand C, Chandraja B, Nithiya P, Akshaya M, Tamizhdurai P, Shoba G, Subramani A, Kumaran R, Yadav KK, Gacem A, Bhutto JK. Green hydrogen for a sustainable future: A review of production methods, innovations, and applications. International Journal of Hydrogen Energy. 2025 Mar 20;111:319–41.

[5] Algburi S, Munther H, Al-Dulaimi O, Fakhruldeen HF, Sapaev IB, Al Seedi KF, Khalaf DH, Jabbar FI, Hassan Q, Khudhair A, Nakimera G. Green Hydrogen Role in Sustainable Energy Transformations: A Review. Results in Engineering. 2025 Apr 26:105109.

[6] Ajanovic A, Sayer M, Haas R. The economics and the environmental benignity of different colors of hydrogen. International Journal of Hydrogen Energy. 2022 Jul 5;47(57), pp.24136–54.

[7] Guo X, Zhu H, Zhang S. Overview of electrolyser and hydrogen production power supply from industrial perspective. International Journal of Hydrogen Energy. 2024 Jan 2;49:1048–59.

[8] Paulu A, Matuštík J, Trecáková T, Kočí V. Overlooked source of hydrogen: The environmental potential of chlor-alkali by-product. International Journal of Hydrogen Energy. 2024 Jan 2;49:1437–43.

[9] Oliveira AM, Beswick RR, Yan Y. A green hydrogen economy for a renewable energy society. Current Opinion in Chemical Engineering. 2021 Sep 1;33:100701.

[10] Risbud M, Kaur G, Dhawale DS, Zhu H, Haque N, Giddey S. Electrolyzer technologies for hydrogen economy. In Hydrogen economy 2023 Jan 1 (pp. 459–485). Academic Press: Elsevier, London, UK.

[11] Li W, Tian H, Ma L, Wang Y, Liu X, Gao X. Low-temperature water electrolysis: fundamentals, progress, and new strategies. Materials Advances. 2022;3(14):5598–644.

[12] Mahmoud M, Ramadan M, Olabi AG, Pullen K, Naher S. A review of mechanical energy storage systems combined with wind and solar applications. Energy Conversion and Management. 2020 Apr 15;210:112670.

[13] Usman MR. Hydrogen storage methods: Review and current status. Renewable and Sustainable Energy Reviews. 2022 Oct 1;167:112743.

[14] Carmo M, Fritz DL, Mergel J, Stolten D. A comprehensive review on PEM water electrolysis. International journal of hydrogen energy. 2013 Apr 22;38(12):4901–34.

[15] Kumar SS, Lim H. An overview of water electrolysis technologies for green hydrogen production. Energy reports. 2022 Nov 1;8:13793–813.

[16] Emam AS, Hamdan MO, Abu-Nabah BA, Elnajjar E. A review on recent trends, challenges, and innovations in alkaline water electrolysis. International Journal of Hydrogen Energy. 2024 Apr 25;64:599–625.

[17] Bodkhe RG, Shrivastava RL, Soni VK, Chadge RB. A review of renewable hydrogen generation and proton exchange membrane fuel cell technology for sustainable energy development. International Journal of Electrochemical Science. 2023 May 1;18(5):100108.

[18] Xu Y, Cai S, Chi B, Tu Z. Technological limitations and recent developments in a solid oxide electrolyzer cell: A review. International Journal of Hydrogen Energy. 2024 Jan 2;50:548–91.

[19] Buttler A, Spliethoff H. Current status of water electrolysis for energy storage, grid balancing and sector coupling via power-to-gas and power-to-liquids: A review. Renewable and Sustainable Energy Reviews. 2018 Feb 1;82:2440–54.

[20] Brauns J, Turek T. Alkaline water electrolysis powered by renewable energy: A review. Processes. 2020 Feb 21;8(2):248.
[21] Kumar SS, Himabindu V. Hydrogen production by PEM water electrolysis–A review. Materials Science for Energy Technologies. 2019 Dec 1;2(3):442–54.
[22] Li C, Baek JB. The promise of hydrogen production from alkaline anion exchange membrane electrolyzers. Nano Energy. 2021 Sep 1;87:106162.
[23] Henkensmeier D, Najibah M, Harms C, Žitka J, Hnát J, Bouzek K. Overview: State-of-the art commercial membranes for anion exchange membrane water electrolysis. Journal of Electrochemical Energy Conversion and Storage. 2021 May 1;18(2):024001.
[24] Santos DM, Sequeira CA, Macciò D, Saccone A, Figueiredo JL. Platinum–rare earth electrodes for hydrogen evolution in alkaline water electrolysis. International journal of hydrogen energy. 2013 Mar 19;38(8):3137–45.
[25] Kim JE, Kim TH, Park CS, Jung K, Yoon J, Lee KB, Kang KS. Probing the (de) activation of Raney nickel–iron anodes during alkaline water electrolysis by accelerated deactivation testing. Electrochemistry Communications. 2023 Dec 1;157:107601.
[26] Zhou Q, Liao L, Zhou H, Li D, Tang D, Yu F. Innovative strategies in design of transition metal-based catalysts for large-current-density alkaline water/seawater electrolysis. Materials Today Physics. 2022 Sep 1;26:100727.
[27] Capozzoli L, Caprì A, Baglio V, Berretti E, Evangelisti C, Filippi J, Gatto I, Lavacchi A, Pagliaro M, Vizza F. Ruthenium-loaded titania nanotube arrays as catalysts for the hydrogen evolution reaction in alkaline membrane electrolysis. Journal of Power Sources. 2023 Apr 1;562:232747.
[28] Du X, Liao Y, Ma M, Wang Y, Zhang J, He G. Advance of manganese dioxide-based electrocatalyst for water splitting. Journal of Solid State Chemistry. 2024 Jan 1;329:124369.
[29] Vidales AG, Choi K, Omanovic S. Nickel-cobalt-oxide cathodes for hydrogen production by water electrolysis in acidic and alkaline media. international journal of hydrogen energy. 2018 Jul 19;43 (29):12917–28.
[30] Guo M, Deng R, Wang C, Zhang Q. Recent progress of advanced manganese oxide-based materials for acidic oxygen evolution reaction: Fundamentals, performance optimization, and prospects. Journal of Energy Chemistry. 2023 Mar 1;78:537–53.
[31] Qazi UY, Javaid R, Zahid M, Tahir N, Afzal A, Lin XM. Bimetallic NiCo–NiCoO2 nano-heterostructures embedded on copper foam as a self-supported bifunctional electrode for water oxidation and hydrogen production in alkaline media. International Journal of Hydrogen Energy. 2021 May 25;46 (36):18936–48.
[32] El-Deab MS, Awad MI, Mohammad AM, Ohsaka T. Enhanced water electrolysis: electrocatalytic generation of oxygen gas at manganese oxide nanorods modified electrodes. Electrochemistry communications. 2007 Aug 1;9(8):2082–7.
[33] Zhang S, Ma X, He Y, Zhu Y, Wang Z. Amorphous mixed Ir–Mn oxide catalysts for the oxygen evolution reaction in PEM water electrolysis for $H_2$ production. International Journal of Hydrogen Energy. 2023 Apr 1;48(28):10532–44.
[34] Weiß A, Siebel A, Bernt M, Shen TH, Tileli V, Gasteiger HA. Impact of intermittent operation on lifetime and performance of a PEM water electrolyzer. Journal of the electrochemical society. 2019;166(8):F487–97.
[35] Mališ J, Mazúr P, Paidar M, Bystron T, Bouzek K. Nafion 117 stability under conditions of PEM water electrolysis at elevated temperature and pressure. International Journal of Hydrogen Energy. 2016 Jan 30;41(4):2177–88.
[36] Yadav NK, Pramanik H. Performance enhancement of NaOH doped physically crosslinked alkaline membrane electrolyte by addition of TEOS to PVA solution for direct sodium borohydride fuel cell application. International Journal of Hydrogen Energy. 2024 Jan 2;50:1373–94.

[37] Reddy PB, Pramanik H. Process Parameter Optimization via RSM of a PEM based Water Electrolysis Cell for the Production of Green Hydrogen. Journal of Electrochemical Science and Technology. 2024;15(3):388–404.

[38] Audichon T, Mayousse E, Morisset S, Morais C, Comminges C, Napporn TW, Kokoh KB. Electroactivity of $RuO_2$–$IrO_2$ mixed nanocatalysts toward the oxygen evolution reaction in a water electrolyzer supplied by a solar profile. International journal of hydrogen energy. 2014 Oct 13;39(30):16785–96.

[39] Xu W, Scott K. The effects of ionomer content on PEM water electrolyser membrane electrode assembly performance. International Journal of Hydrogen Energy. 2010 Nov 1;35(21):12029–37.

[40] Ayers KE, Renner JN, Danilovic N, Wang JX, Zhang Y, Maric R, Yu H. Pathways to ultra-low platinum group metal catalyst loading in proton exchange membrane electrolyzers. Catalysis Today. 2016 Mar 15;262:121–32.

[41] Lee BS, Park HY, Choi I, Cho MK, Kim HJ, Yoo SJ, Henkensmeier D, Kim JY, Nam SW, Park S, Lee KY. Polarization characteristics of a low catalyst loading PEM water electrolyzer operating at elevated temperature. Journal of Power Sources. 2016 Mar 31;309:127–34.

[42] Xie Z, Yu S, Yang G, Li K, Ding L, Wang W, Cullen DA, Meyer III HM, Retterer ST, Wu Z, Sun J. Ultrathin platinum nanowire based electrodes for high-efficiency hydrogen generation in practical electrolyzer cells. Chemical Engineering Journal. 2021 Apr 15;410:128333.

[43] Xie Z, Yu S, Ma X, Li K, Ding L, Wang W, Cullen DA, Meyer III HM, Yu H, Tong J, Wu Z. MoS2 nanosheet integrated electrodes with engineered 1T-2H phases and defects for efficient hydrogen production in practical PEM electrolysis. Applied Catalysis B: Environmental. 2022 Sep 15;313:121458.

[44] Qayum A, Peng X, Yuan J, Qu Y, Zhou J, Huang Z, Xia H, Liu Z, Tan DQ, Chu PK, Lu F. Highly Durable and Efficient Ni-FeO x/$FeNi_3$ Electrocatalysts Synthesized by a Facile In Situ Combustion-Based Method for Overall Water Splitting with Large Current Densities. ACS Applied Materials & Interfaces. 2022 Jun 10;14(24):27842–53.

[45] Sun X, Xu K, Fleischer C, Liu X, Grandcolas M, Strandbakke R, Bjørheim TS, Norby T, Chatzitakis A. Earth-abundant electrocatalysts in proton exchange membrane electrolyzers. Catalysts. 2018 Dec 13;8(12):657.

[46] Jiang G, Yu H, Li Y, Yao D, Chi J, Sun S, Shao Z. Low-loading and highly stable membrane electrode based on an Ir@ WO x NR ordered array for PEM water electrolysis. ACS Applied Materials & Interfaces. 2021 Mar 24;13(13):15073–82.

[47] Jiang G, Yu H, Yao D, Li Y, Chi J, Zhang H, Shao Z. Boosting the oxygen evolution stability and activity of a heterogeneous IrRu bimetallic coating on a WO 3 nano-array electrode for PEM water electrolysis. Journal of Materials Chemistry A. 2022;10(22):11893–903.

[48] Min X, Shi Y, Lu Z, Shen L, Ogundipe TO, Gupta P, Wang C, Guo C, Wang Z, Tan H, Mukerjee S. High performance and cost-effective supported IrOx catalyst for proton exchange membrane water electrolysis. Electrochimica acta. 2021 Jul 20;385:138391.

[49] Xu X, Li R, Tang C, Wang H, Zhuang X, Liu Y, Kang W, Shi L. Cellulose nanofiber-embedded sulfonated poly (ether sulfone) membranes for proton exchange membrane fuel cells. Carbohydrate Polymers. 2018 Mar 15;184:299–306.

[50] Sachs M, Sprick RS, Pearce D, Hillman SA, Monti A, Guilbert AA, Brownbill NJ, Dimitrov S, Shi X, Blanc F, Zwijnenburg MA. Understanding structure-activity relationships in linear polymer photocatalysts for hydrogen evolution. Nature communications. 2018 Nov 23;9(1):4968.

[51] Ouabi H, Lajouad R, Kissaoui M, El Magri A. Hydrogen production by water electrolysis driven by a photovoltaic source: A review. e-Prime-Advances in Electrical Engineering, Electronics and Energy. 2024 Jun 1;8:100608.

[52] Liu C, Dang Z, Xi G. Numerical study on thermal stress of solid oxide electrolyzer cell with various flow configurations. Applied Energy. 2024 Jan 1;353:122041.

[53] Rowberg AJ, Slomski HS, Kim N, Strange NA, Gorman BP, Shulda S, Ginley DS, Kweon KE, Wood BC. Impact of Sr-Containing Secondary Phases on Oxide Conductivity in Solid-Oxide Electrolyzer Cells. Chemistry of Materials. 2024 Jun 15;36(13):6464–74.

[54] Küngas R, Blennow P, Heiredal-Clausen T, Nørby TH, Rass-Hansen J, Hansen JB, Moses PG. Progress in SOEC development activities at Haldor Topsøe. ECS Transactions. 2019 Jul 10;91(1):215.

[55] Liu Y, Tian Y, Wang Y, Li Y, Pu J, Ciucci F, Chi B. Nano film Pr2Ni0. 8Cu0. 2O4+ δ decorated La0. 6Sr0. 4Co0. 2Fe0. 8O3-δ oxygen electrode for highly efficient and stable reversible solid oxide cells. Electrochimica Acta. 2022 Oct 20;430:141032.

[56] Khouya A. Levelized costs of energy and hydrogen of wind farms and concentrated photovoltaic thermal systems. A case study in Morocco. International Journal of Hydrogen Energy. 2020 Nov 13;45(56):31632–50.

[57] Mohammadi A, Mehrpooya M. Thermodynamic and economic analyses of hydrogen production system using high temperature solid oxide electrolyzer integrated with parabolic trough collector. Journal of Cleaner Production. 2019 Mar 1;212:713–26.

[58] Nechache A, Hody S. Alternative and innovative solid oxide electrolysis cell materials: A short review. Renewable and Sustainable Energy Reviews. 2021 Oct 1;149:111322.

[59] Kim J, Jun A, Gwon O, Yoo S, Liu M, Shin J, Lim TH, Kim G. Hybrid-solid oxide electrolysis cell: A new strategy for efficient hydrogen production. Nano Energy. 2018 Feb 1;44:121–6.

[60] Mullins DR. The surface chemistry of cerium oxide. Surface Science Reports. 2015 Mar 1;70(1):42–85.

[61] Schmitt R, Nenning A, Kraynis O, Korobko R, Frenkel AI, Lubomirsky I, Haile SM, Rupp JL. A review of defect structure and chemistry in ceria and its solid solutions. Chemical Society Reviews. 2020;49(2):554–92.

[62] Chatzichristodoulou C, Norby P, Hendriksen PV, Mogensen MB. Size of oxide vacancies in fluorite and perovskite structured oxides. Journal of Electroceramics. 2015 Feb;34:100–7.

[63] Zhu S, Wang Y, Rao Y, Zhan Z, Xia C. Chemically-induced mechanical unstability of samaria-doped ceria electrolyte for solid oxide electrolysis cells. International journal of hydrogen energy. 2014 Aug 13;39(24):12440–7.

[64] Lenser C, Zhang J, Russner N, Weber A, Guillon O, Menzler NH. Electro-chemo-mechanical analysis of a solid oxide cell based on doped ceria. Journal of Power Sources. 2022 Sep 1;541:231505.

[65] Tietz F, Sebold D, Brisse A, Schefold J. Degradation phenomena in a solid oxide electrolysis cell after 9000 h of operation. Journal of Power Sources. 2013 Feb 1;223:129–35.

[66] Wang Y, Li W, Ma L, Li W, Liu X. Degradation of solid oxide electrolysis cells: Phenomena, mechanisms, and emerging mitigation strategies – A review. Journal of Materials Science & Technology. 2020 Oct 15;55:35–55.

[67] Li P, Dong R, Yang P, Ma X, Yan F, Zhang P, Fu D. Performance enhanced of NiCe0. 8Sm0. 2O1. 9 hydrogen electrode for reversible solid oxide cells with cadmium substitution. Journal of Electroanalytical Chemistry. 2021 Feb 1;882:115018.

[68] Zhao Z, Wang X, Tang S, Cheng M, Shao Z. High-performance oxygen electrode Ce0. 9Co0. 1O2-δ-LSM-YSZ for hydrogen production by solid oxide electrolysis cells. International Journal of Hydrogen Energy. 2021 Jul 21;46(50):25332–40.

[69] Zheng Y, Jiang H, Wang S, Qian B, Li Q, Ge L, Chen H. Mn-doped Ruddlesden-Popper oxide La1. 5Sr0. 5NiO4+ δ as a novel air electrode material for solid oxide electrolysis cells. Ceramics International. 2021 Jan 1;47(1):1208–17.

[70] Xie Y, Luo F, Yang Z. Strategies for the enhancements in catalytic performance and stability of anodic electrocatalyst in PEM water splitting. Energy Reviews. 2024 Jul 15:100103.

Mohammed Irfan, Rahul, Shiv Charan Prajapati, Ritesh Patidar

# 3 Hydrogen Production Using Hybrid Renewable Systems

**Abstract:** The rise in global greenhouse gas emissions and the depletion of fossil fuel energy resources prompted a paradigm shift to alternative energy sources. To reduce carbon emissions and toxic effluents from traditional sources, it became important to find a carbon-free and clean energy source. Hybrid renewable energy systems are essential to the production of hydrogen, which is becoming a major energy carrier for a sustainable future. This study investigates the use of cutting-edge advanced methods like electrolysis, gasification, and thermochemical processes to generate hydrogen from renewable sources like sun, wind, and biomass. The intermittent nature of independent renewable energy sources is addressed by hybrid systems, which provide a steady and uninterrupted power supply. Electrolysis, which separates water into hydrogen and oxygen, is powered by electricity generated by photovoltaic panels and wind turbines. To ensure the best possible use of resources, excess energy is either stored or diverted to the generation of hydrogen. By generating syngas that is rich in hydrogen, biomass gasification enhances the system's overall performance. This is a brief discussion of the advanced processes used to produce hydrogen. Techniques for producing hydrogen using hybrid mechanisms can be broadly categorized according to the source and process, including: thermochemical, photocatalytic, biomass gasification, solid oxide electrolysis cells, anion exchange membrane electrolysis, etc. As hybridization and integration of different pathways for hydrogen generation increases the scalability, cost-effectiveness, and energy reliability and hydrogen production becomes more feasible for use in transportation and industrial purposes. This strategy minimizes carbon emissions, lessens reliance on fossil fuels, and minimizes energy waste. Innovation in hydrogen production methods is being propelled by sustained research and policy assistance, notwithstanding technological and financial obstacles.

**Keywords:** Greenhouse, gas emissions, electrolyzer, hydrocarbon reforming, carbon-neutral energy, decarbonization, photovoltaic, energy conversion

## 3.1 Introduction

Hydrogen, the most basic and prevalent element in the universe, has long captivated scientists and engineers. Because of its exceptional fuel qualities, hydrogen is a vital component in the worldwide shift to greener energy systems because of its capacity to generate energy without emitting hazardous pollutants. The history of hydrogen,

 | https://doi.org/10.1515/9783112181072-003

from its discovery centuries ago to its present uses and potential, represents humanity's pursuit of sustainability and creativity.

### 3.1.1 Historical Perspective

When metals dissolved in acids in the sixteenth century, Paracelsus made the first observation of hydrogen. Henry Cavendish extracted "inflammable air," which was eventually determined to be hydrogen, in 1766, and showed how it exploded when combined with oxygen to make water [1]. The basis for comprehending the chemical characteristics of hydrogen was established by this finding.

Jacques Charles's use of hydrogen gas in balloon flights in the late eighteenth century gave hydrogen notoriety and signaled the beginning of its practical uses. Industrial developments like coal gasification, which created hydrogen-containing mixes for heating and lighting, occurred in the nineteenth century [2]. By the twentieth century, large-scale hydrogen generation was made possible by technologies like steam methane reforming (SMR), which made hydrogen essential to sectors like petroleum refining and ammonia synthesis.

### 3.1.2 Present Status

Hydrogen ($H_2$) is now acknowledged as a flexible energy carrier with uses in manufacturing, energy storage, and transportation. Vehicle hydrogen utilization has been transformed by fuel cell technology. Onboard, fuel cell electric cars, like the Hyundai Nexo and Toyota Mirai, turn hydrogen into power while only releasing water vapor. They are thus perfect for cutting greenhouse gas emissions in industries like shipping and heavy-duty transportation [3].

$H_2$ is essential to industry for the production of steel, ammonia for fertilizers, and oil refinement. It is also being utilized more and more to store excess renewable energy. Green hydrogen may be produced by electrolysis, a process driven by renewable energy sources, using excess electricity from solar or wind farms [4].

Space exploration is also powered by hydrogen. Liquid hydrogen is a desirable rocket fuel due to its high energy density, allowing for missions that push the limits of human exploration.

### 3.1.3 Future Aspects

Hydrogen's capacity to decarbonize global energy systems is what will determine its future. Green hydrogen, which is created by electrolysis using renewable energy, is leading this shift. Infrastructure is being substantially invested in by governments; for example, Europe intends to add 40 GW of electrolyzer capacity by 2030. Innovations such as solid-state hydrogen storage systems are lowering prices while improving efficiency and safety. New technologies like methane pyrolysis provide encouraging avenues for the generation of clean hydrogen [5]. Through the separation of methane into solid carbon and hydrogen, this process produces "turquoise hydrogen" without emitting $CO_2$. The capacity of solar-driven processes, such as photocatalysis, to directly create hydrogen from sunlight is also drawing interest.

According to the International Energy Agency (IEA) [6] in 2023, the world's demand for hydrogen increased by 2.5% over 2022 to a record high of over 97 Mt. Until the Covid-19 outbreak caused an economic downturn in 2020, demand had been steadily increasing for many decades. We predict that growth will continue in the upcoming years, and the demand for hydrogen worldwide may approach 100 Mt. In order to accomplish this demand, $H_2$ production methods, along with present techniques, require modern and hybrid approaches as well.

The IEA estimates that if sectors embrace hydrogen for decarbonization, the world's need for it might double by 2050. Because of its adaptability, it is essential for reaching net-zero emissions goals.

## 3.2 Hydrogen as a Fuel and Hydrogen Chemistry

Hydrogen is a one-of-a-kind element with numerous captivating features. Here are some of hydrogen's most notable properties [7]:

- With an atomic weight of 1.008, hydrogen is a very light element in the periodic table. At normal temperature and pressure, it is the lightest of all gases and has the lowest density of any gas.
- Colorless and odorless: Hydrogen is a gas that has no color, odor, or taste. It is also highly combustible and can combine with air to generate explosive combinations.
- Hydrogen has the largest energy content per unit mass of any fuel, making it an appealing source of energy. It is also a clean and sustainable fuel source since, when burned in air or in a fuel cell, it creates only water as a by-product.

Hydrogen is a versatile fuel that can be used in a variety of applications in the energy sector. There are various forms of hydrogen, as shown in Figure 3.1, utilized in energy applications, each with its own set of benefits and drawbacks.

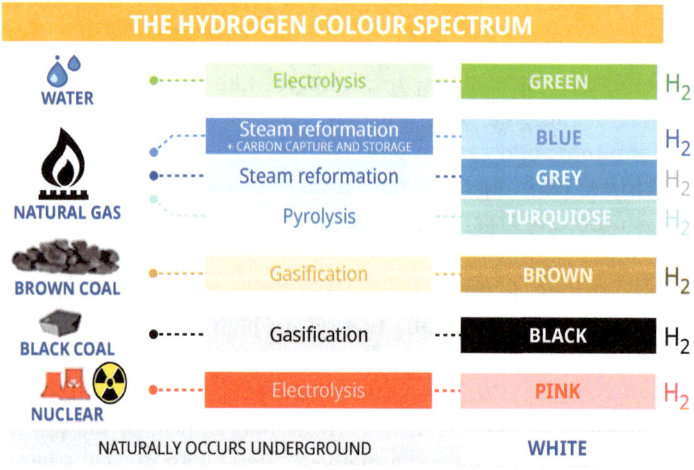


**Figure 3.1:** Source and color spectrum of hydrogen.

The following are some of the most prominent forms of hydrogen used in the energy sector [8]:

- Gray hydrogen is the most often used kind of hydrogen in the energy industry. It is created from fossil fuels such as natural gas or coal via a process known as steam methane reforming. Because this process emits carbon dioxide ($CO_2$) into the atmosphere, gray hydrogen contributes significantly to greenhouse gas emissions.
- Blue hydrogen is created in the same way as gray hydrogen is, but with the addition of carbon capture and storage (CCS) technology. CCS collects and stores $CO_2$ produced during the SMR process, lowering greenhouse gas emissions. Blue hydrogen is a low-carbon substitute for gray hydrogen.
- Green hydrogen is created by using sustainable energy sources, such as solar or wind power, to power water electrolysis. The hydrogen and oxygen in water molecules are separated in this process, yielding pure hydrogen and oxygen. Green hydrogen is a completely renewable and sustainable energy source that emits no greenhouse gases.
- Yellow hydrogen is created by using nuclear power to power the electrolysis of water. This procedure generates pure hydrogen and oxygen while emitting no greenhouse gases. Yellow hydrogen is a low-carbon alternative to gray hydrogen, but its applications are limited due to the scarcity of nuclear power.
- Turquoise hydrogen is created by combining natural gas with renewable energy sources like solar or wind power. This technology collects and stores the $CO_2$ created during the SMR process, making it a low-carbon alternative to gray hydrogen [9].
- Brown and black hydrogen that has been produced from either brown (lignite) or black (bituminous) coal. This is accomplished by gasification, a process that employs heat to convert carbon-rich coal into gas. This releases the hydrogen inside,

but it also emits carbon dioxide. These emissions, like gray hydrogen, are not caught or stored and are instead discharged into the atmosphere [10].

- Pink hydrogen is also known as purple hydrogen and is created by the electrolysis process, although in this case, the electrolyzer is fueled by nuclear energy [11].
- White hydrogen is supposed to be a naturally occurring variant of pure hydrogen that can be discovered underground. However, extracting it is not now feasible, and nothing is known about this form of hydrogen.

Because of its clean combustion and high energy density (lower heating value (LHV) of 120 MJ/kg) [12], hydrogen is essential for facilitating the shift to sustainable energy systems. The simplest element in the periodic table, hydrogen, is essential to sustainable energy systems and chemical reactions. It is a focus of study targeted at decarbonizing industry and moving toward sustainable energy solutions because of its distinct chemistry and fuel potential. The chemical characteristics, combustion processes, manufacturing processes, and pertinent facts for the scientific study of hydrogen are highlighted in this chapter.

Under normal circumstances, hydrogen primarily occurs as a diatomic molecule ($H_2$). Because it only has one electron, it is very reactive and easily forms connections with other elements. Hydrogen and oxygen combine in combustion processes to create water and release energy according to the following equation:

$$2H_2 + O_2 \rightarrow 2H_2O \qquad (\Delta H = -286\,\mathrm{kJ/mol}) \tag{3.1}$$

Because it doesn't produce greenhouse gases like carbon dioxide, this exothermic process is essential to using hydrogen as fuel. Nevertheless, hydroxyl radicals ($OH^-$) are essential to the intricate branched-chain processes involved in hydrogen combustion. These reactions include intermediate steps like the production of radicals and chain-ending recombination events.

The characteristics of hydrogen also make it perfect for fuel cells. Protons ($H^+$) and electrons are created when hydrogen is oxidized at the anode of proton-exchange membrane (PEM) fuel cells. While electrons go via an external circuit to produce electricity, protons move through the electrolyte. Water is created at the cathode when oxygen joins protons and electrons.

Yet, current hydrogen production relies heavily on SMR, which emits 10–12 kg $CO_2$ per kg $H_2$ and dominates 95% of global production. SMR's reaction is:

$$CH_4 + H_2O \rightarrow CO + 3H_2 \tag{3.2}$$

Electrolysis, powered by renewables, offers a greener alternative. Despite its potential, electrolysis only makes about 5% of output because of its high price and energy requirements. Decarbonizing production is the goal of new techniques like solar-driven photocatalysis and methane pyrolysis, which produces solid carbon.

## 3.3 Hydrogen Production Techniques

There are several methods for producing hydrogen, as depicted in Figure 3.2 flowchart, and each has its own feedstock, technology, cost, and environmental effect. In order to manufacture hydrogen more effectively, sustainably, or economically than conventional standalone methods, hybrid hydrogen production techniques integrate many energy sources or processes, usually thermal, electrical, chemical, or biological approaches. By overcoming the drawbacks of single-mode systems, these methods hope to facilitate the shift to clean, large-scale hydrogen generation.

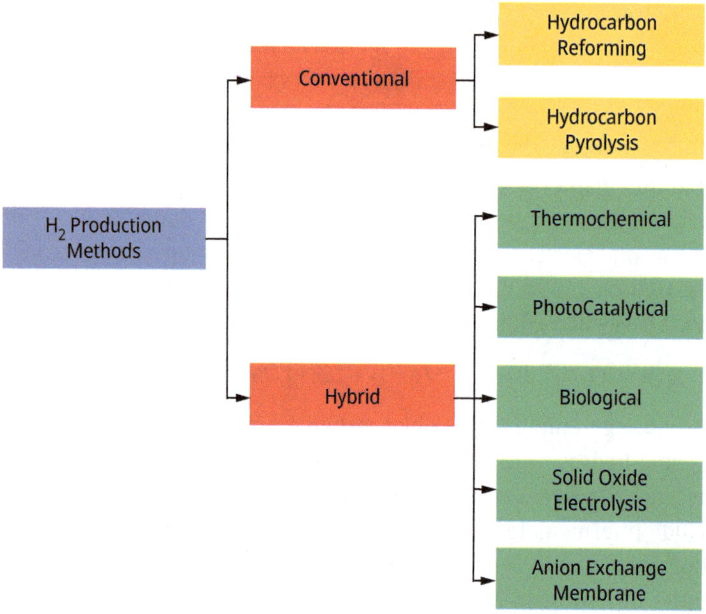


**Figure 3.2:** Flowchart of various hydrogen production techniques.

Numerous techniques are used to make hydrogen: biomass gasification transforms organic material; electrolysis divides water using electricity; photobiological and photoelectrochemical processes employ sunlight with bacteria or semiconductors; and SMR uses natural gas. High heat, often from nuclear sources, is used in thermochemical water splitting to produce hydrogen without emitting carbon.

This chapter is a brief description of the advanced methods for producing hydrogen. Based on the process and source, hydrogen generation techniques using hybrid mechanisms might be broadly categorized as: thermochemical, photocatalytic water splitting, biological hydrogen production, solid oxide electrolysis cells (SOECs), anion-exchange membrane (AEM) electrolysis.

### 3.3.1 Thermochemical Method

Thermochemical processes, which are usually carried out at high temperatures, use heat and chemical reactions to divide water ($H_2O$) into hydrogen ($H_2$) and oxygen ($O_2$). Unlike electrolysis, these processes depend on thermal energy, often from nuclear or solar sources, rather than direct electrical consumption.

Here's a detailed look into the steps of thermochemical hydrogen production:

I. Use high-temperature heat (typically >500 °C, sometimes over 1,000 °C).
II. Involve cyclic chemical reactions to split water.
III. Ideal for integrating with nuclear or concentrated solar power sources.
IV. Can achieve high efficiency with no direct $CO_2$ emissions if clean heat is used.

Thermochemical cycles for hydrogen production are multi-step chemical processes that use heat (often from solar or nuclear sources) to split water ($H_2O$) into hydrogen ($H_2$) and oxygen ($O_2$) without electricity. They rely on heat-driven chemical reactions where intermediate compounds are recycled, so only water goes in and hydrogen and oxygen come out. Following are the pathways of producing hydrogen via thermochemical mechanism.

#### 3.3.1.1 Sulfur–Iodine (S–I) Cycle

The sulfur–iodine (S–I) cycle, Figure 3.3, is a high-temperature thermochemical process that splits water into hydrogen and oxygen using heat, typically from nuclear or solar sources. It involves three key steps: sulfuric acid decomposition, Bunsen reaction, and hydrogen iodide decomposition, producing hydrogen efficiently without direct electricity input or carbon emissions as represented in Eqs. (3.3), (3.4), and (3.5), respectively.

Key Reactions:

i. Decomposition of $H_2SO_4$ (high temperature ~850 °C):

$$H_2SO_4 \rightarrow SO_2 + H_2O + \tfrac{1}{2}O_2 \tag{3.3}$$

ii. Bunsen reaction (low temperature ~120 °C):

$$SO_2 + 2H_2O + I_2 \rightarrow 2HI + H_2SO_4 \tag{3.4}$$

iii. Decomposition of Hydrogen Iodide (medium-temp ~450 °C):

$$2HI \rightarrow I_2 + H_2 \tag{3.5}$$

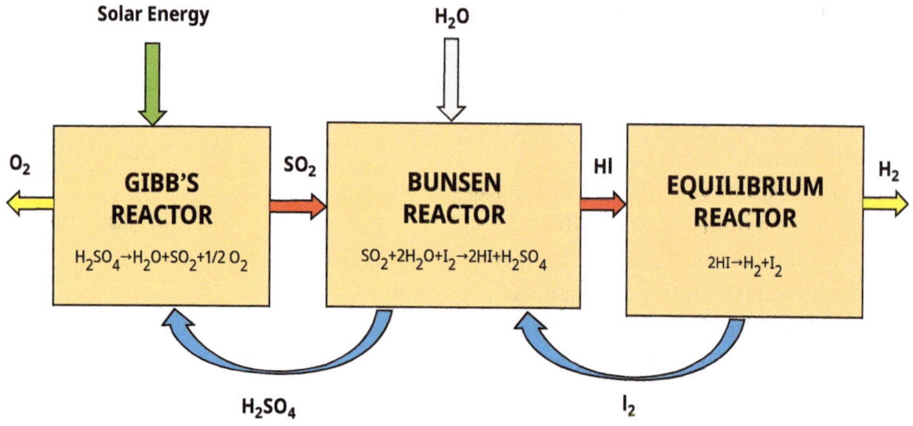


**Figure 3.3:** Hydrogen production by thermochemical via sulfur–iodine cycle.

#### 3.3.1.2 Hybrid Sulfur Cycle

The hybrid sulfur (HyS) cycle is a two-step thermochemical process for hydrogen production using heat (~800–850 °C) and electricity. Sulfur dioxide reacts with water to produce sulfuric acid and hydrogen gas electrochemically. The sulfuric acid is then thermally decomposed into sulfur dioxide and oxygen. This closed-loop cycle efficiently splits water into hydrogen and oxygen with low emissions, as shown in Eqs. (3.6) and (3.7).

Key Reactions:

i. Thermal decomposition (high temperature ~850 °C):

$$H_2SO_4 \rightarrow SO_2 + H_2O + {}^1\!/_2O_2 \tag{3.6}$$

ii. Electrochemical step (low temperature):

$$SO_2 + 2H_2O \rightarrow H_2SO_4 + H_2(\text{using electricity}) \tag{3.7}$$

#### 3.3.1.3 Copper–Chlorine (Cu–Cl) Cycle

The mechanism of this cycle begins with the breakdown of dichlorocopper in solid state [13], to produce gas chlorine and copper in solid state within the temperature range of 400–500 °C, as shown in Eq. (3.8), which is the highest of the cycle. In the second step, half a mole of oxygen and two moles of hydrochloric acid are obtained using water and chlorine gas; both products are in gaseous state; this reaction is called the reverse Deacon reaction as shown in Eq. (3.9). In the third step, Cu reacts with the hydrochloric acid (HCl) to form solid copper(II) chloride ($CuCl_2$) that can be used again in the first reaction as shown in Eq. (3.10). This reaction is exothermic. Hence, one of the potential methods for creating hydrogen at temperatures between 400 and 500 °C without emitting greenhouse gases is the Cu–Cl-based thermochemical cycle [14].

Key Reactions:

i. Production of Cu and $Cl_2$ (400–500 °C):

$$CuCl_2(s) \rightarrow Cu(s) + Cl_2(g) \tag{3.8}$$

ii. Production of HCl and $O_2$ (350–400 °C):

$$Cl_2 + H_2O + Q \rightarrow 2HCl + \tfrac{1}{2}O_2 \tag{3.9}$$

iii. Production of $H_2$ (<50 °C):

$$Cu(s) + 2HCl(g) \rightarrow CuCl_2(s) + H_2(g) + Q \tag{3.10}$$

Thermochemical cycle heat sources include high-temperature nuclear reactors, such as the High-Temperature Gas-cooled Reactor (HTGR), systems that use mirrors or lenses to concentrate solar electricity (CSP), and heat from industrial waste. Thermochemical hydrogen production via different cycles with certain parameters, as summarized in Table 3.1, has the advantage of using no fossil fuels and emitting no $CO_2$ when driven by renewable energy, effective on a large scale, particularly with solar and nuclear power, water is the only feedstock. Challenges include high material and capital expenses, corrosion brought on by substances like halogens or sulfuric acid, intricate safety and engineering requirements, R&D and pilot project phases are still ongoing.

**Table 3.1:** Summarizing thermochemical hydrogen production method.

| Process name | Max. temperature | Chemical | Advantages | Disadvantages |
|---|---|---|---|---|
| **Sulfur–iodine** | ~850 °C | Sulfur | High potential efficiency (~50%) | Complex process with corrosive chemicals |
| | | | Can be driven by solar/nuclear heat | Challenging separation steps |
| **Hybrid sulfur** | ~850 °C | Sulfuric acid | Combines thermal and electrochemical methods | Corrosive acid handling |
| | | | Fewer steps than S–I | Requires efficient $SO_2$ management |

Finally, thermochemical water-splitting cycles, such as the sulfur–iodine, hybrid sulfur, and copper–chlorine cycles, offer efficient, carbon-free hydrogen production using heat from nuclear or solar sources. These closed-loop processes convert water into hydrogen and oxygen without direct electricity input, making them promising for large-scale applications. Despite their high theoretical efficiency and sustainability, technical challenges such as high operating temperatures, material corrosion, and system integration remain. Continued advancements in materials science, reactor design, and process optimization are essential for transitioning these methods from laboratory research to industrial-scale hydrogen production.

### 3.3.2 Photocatalytic Water Splitting (Artificial Photosynthesis) Method

A possible method for directly converting solar energy into hydrogen fuel that mimics natural photosynthesis is photocatalytic water splitting, often known as artificial photosynthesis, as represented in Figure 3.4. A sustainable route to a hydrogen economy that is carbon neutral is provided by the method, which uses semiconductor materials to collect sunlight and power the redox processes that split water molecules into hydrogen ($H_2$) and oxygen ($O_2$).

#### 3.3.2.1 Essentials of Water Splitting via Photocatalysis

The process of photocatalysis depends on photons with energy equal to or higher than a semiconductor's bandgap, exciting it. When exposed to light, holes are left behind when electrons are stimulated from the valence band (VB) to the conduction band (CB). If the bandgap matches the water redox potentials, these charge carriers may drive the oxygen evolution reaction (OER) and hydrogen evolution reaction (HER) as shown in Eq. (3.11). A number of variables, characteristics, and properties of photocatalyst required, including light absorption, charge separation and transfer, redox activity, and material stability, affect how efficient this process is, discussed in Table 3.2.

Hisatomi et al. [15] concluded that the goal of artificial photosynthesis is to replicate natural photosynthesis, which uses water as the only raw material to transform solar energy into chemical fuels, especially hydrogen. One intriguing method for sustainably producing green hydrogen is photocatalytic water splitting [15].

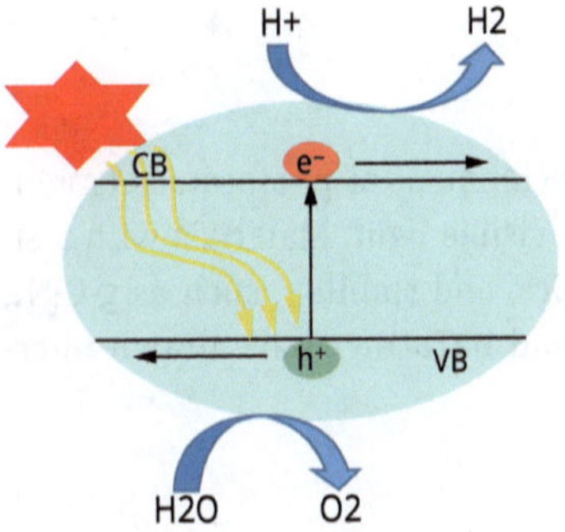


**Figure 3.4:** $H_2$ produced by photocatalytic water splitting (artificial photosynthesis) method (adapted after [16]).

Overall reaction involves

$$2H_2O + h\nu \rightarrow 2H_2 + O_2 \quad (3.11)$$

Mechanism:

1. Light Absorption: Semiconductor absorbs photons (hν ≥ bandgap).
2. Charge Separation: Generation of electron-hole pairs.
3. Surface Reaction:
   - Reduction Reaction (Cathode):

$$2H^+ + 2e^- \rightarrow H_2 \quad (3.12)$$

   - Oxidation Reaction (Anode):

$$H_2O \rightarrow \tfrac{1}{2}O_2 + 2H^+ + 2e^- \quad (3.13)$$

**Table 3.2:** Characteristics and properties of photocatalyst required [15].

| Characteristics | Remarks |
|---|---|
| **Bandgap efficiency** | 1.6–2.4 eV for visible light activation |
| **Charge separation efficiency** | High mobility and low recombination rates |
| **Redox potential** | Conduction and valence band alignment with $H^+/H_2$ and $O_2/H_2O$ redox levels |

For effective photocatalytic hydrogen production, a photocatalyst should efficiently absorb light, facilitate charge separation, and minimize electron-hole recombination. Additionally, it should be stable, nontoxic, cost-effective, and scalable.

### 3.3.2.2 Current Developments and Trends

#### 3.3.2.2.1 Innovative Photocatalyst Materials

In recent years, studies have been made in the development of photocatalysts with optimal bandgaps (1.6–2.4 eV) for the absorption of visible light. Materials with visible-light activity, adjustable electronic characteristics, and stability, such as g-$C_3N_4$, $BiVO_4$, $ZnIn_2S_4$, and perovskite oxides (e.g., $SrTiO_3$ and $BaTaO_2N$), have drawn interest [15].

#### 3.3.2.2.2 Cocatalyst Engineering

By offering hydrogen evolution reaction (HER) and oxygen evolution reaction (OER) sites, the deposition of cocatalysts, such as Pt, $RuO_2$, Ni, or Co-based materials, a good photocatalyst improves charge separation and catalytic activity. In order to concurrently improve both processes, recent research has also focused on dual cocatalysts [17].

#### 3.3.2.2.3 Z-Scheme Systems

Inspired by natural photosynthesis, Z-scheme heterojunctions have been developed to combine two different photocatalysts, each optimized for either HER or OER. This configuration enables more efficient charge separation and utilization of a broader light spectrum [18].

#### 3.3.2.2.4 Single-Atom Catalysts (SACs)

SACs, where isolated metal atoms are anchored on supports like graphene or g-$C_3N_4$, offer superior catalytic activity, atom efficiency, and capable active sites. SAC-based photocatalysts are a cutting-edge development in this field [19].

#### 3.3.2.2.5 Computational Screening and Machine Learning

High-throughput DFT calculations and machine learning models are increasingly used to predict and design new photocatalysts, accelerating discovery and optimization [20].

### 3.3.2.3 Challenges

Despite encouraging advancements, a number of obstacles prevent photocatalytic water splitting from being used commercially.

#### 3.3.2.3.1 Low Solar-to-Hydrogen (STH) Efficiency

The majority of photocatalysts continue to have low STH efficiencies (<1%) and quantum yields, which restricts their practicality. Enhancing charge separation and light harvesting is still essential.

#### 3.3.2.3.2 Stability and Deterioration

Under extended light or in aquatic conditions, many materials experience photo corrosion or deterioration, particularly in the presence of actual sunshine and pH variations.

#### 3.3.2.3.3 Catalyst Cost and Scalability

Using noble metals as co-catalysts, such as Pt and Rh, increases expenses. Development of scalable and earth-abundant materials is still a top goal.

#### 3.3.2.3.4 Half-Reactions Against Full Water Splitting

The simultaneous complete water splitting without sacrificial agents remains a major barrier, even if many systems are able to perform efficient hydrogen or oxygen evolution independently.

#### 3.3.2.3.5 Integration and Reactor Design

Integration into scalable and stable photoreactors that have long-term operation, gas separation, and light management, all of which are currently in the research stage, is necessary for real-world application.

Finally, photocatalytic water splitting, inspired by natural photosynthesis, offers a sustainable and environmentally friendly approach to hydrogen production using sunlight and water. While the method holds great promise due to its simplicity and zero carbon emissions, current challenges, such as low solar-to-hydrogen efficiency, catalyst stability, and scalability, limit its commercial viability. Continued research on advanced photocatalysts, system design, and solar energy utilization is essential to make artificial photosynthesis a feasible, large-scale hydrogen generation technology in the future.

### 3.3.3 Biological Hydrogen Production (Biomass Gasification) Method

A carbon-containing feedstock, in this case biomass, is heated to high temperatures under sub-stoichiometric circumstances with an oxidant (oxygen, air, and/or steam) to prevent full combustion. This process is known as gasification.

Biomass gasification is the process of mixing lignocellulosic residues, energy crops, or organic wastes with a gasifying agent (air, $O_2$, steam, or supercritical water) at 600–1,000 °C to create a combustible gas combination (syngas) rich in $H_2$, CO, and $CO_2$ as represented in Figure 3.5. Because the carbon comes from recent photosynthesis and the feedstocks are renewable, the whole pathway is regarded as "biological" in the bioenergy literature, even if the main reactions are thermochemical [21]. Gasification is one of the most developed bio-hydrogen routes, with pilot plants currently routinely producing hydrogen yields of 80–120 g $H_2$/kg dry biomass.

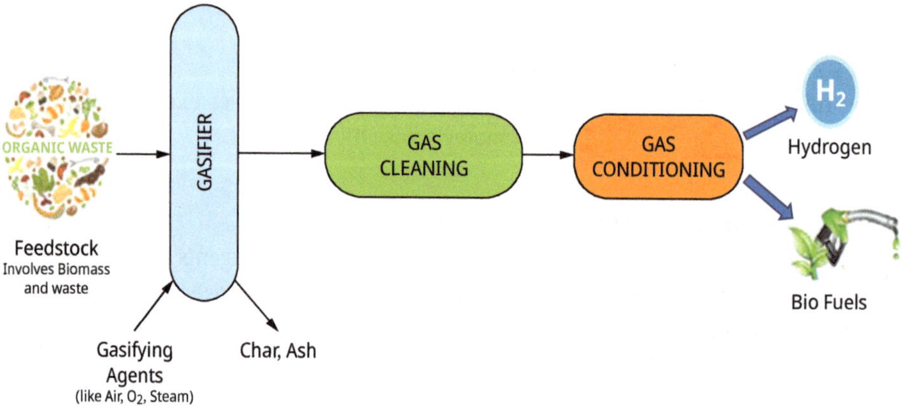


**Figure 3.5:** General process steps in gasification.

#### 3.3.3.1 Fundamental

Important reactions include the water-gas shift ($CO + H_2O \leftrightarrow CO_2 + H_2$) and steam reforming of volatiles ($C_nH_m + n\,H_2O \rightarrow (n + m/2)\,H_2 + n\,CO$). Tar cracking and syngas composition are controlled by operating temperature, equivalency ratio, and residence duration. While sorption-enhanced gasification (e.g., CaO loops) in situ eliminates $CO_2$, moving the equilibrium toward $H_2$, catalytic steam gasification using Ni/Fe catalysts improves $H_2$ selectivity [22].

#### 3.3.3.2 New Developments

##### 3.3.3.2.1 Reactors with Dual Fluidized Beds

Allothermal operation, i.e., a process where heat for the gasification reaction is provided by an external source, rather than being generated by the combustion of the feedstock itself, with indirect heat transfer, is made possible by dual-bed systems, which divide the gasification and combustion zones. Pure steam gasification and tar-free syngas rich in $H_2$ (>55 vol%) without $N_2$ dilution are made possible by this decoupling [23].

#### 3.3.3.2.2 Super Critical Water Gasification (SCWG)

SCWG at 374–650 °C and 22–30 MPa removes the energy cost of drying for wet slurries (>70%). According to [24], recent stainless-steel continuous reactors have achieved 50–60 mol% $H_2$ with almost full carbon conversion and little char. Ru/C catalysts and corrosion-resistant Ni-based alloys have been used for more than 1,000 h.

#### 3.3.3.2.3 Nanostructuring Catalysts

Higher resistance to sulfur and chlorine found in agricultural wastes is shown by nickel supported on magnesium-promoted ceria or mesoporous biochar (pore size 2–50 nm). Ni–Fe bimetallic nanoparticles produced by atomic layer deposition (ALD) inhibit the production of coke and remain active for more than 500 h [25].

#### 3.3.3.2.4 Integrated Gasification with Solid Oxide Fuel Cell (IG-SOFC)

When biomass gasifiers are coupled with SOFC stacks, $H_2$ and CO are electrochemically converted, increasing total electrical efficiency to above 55% LHV. Stable operation on forest wastes with less than 1 ppm tar after a hot-gas filter and reformer was confirmed by demonstrations at 250 kW (VTT, Finland) in 2023 [26].

#### 3.3.3.2.5 Negative Configurations of $CO_2$

Bioenergy with CCS (BECCS) produces green $H_2$ and net-negative emissions (−0.4 to −1.0 t $CO_2$ e MW/h, i.e., negative tons of $CO_2$ equivalent per megawatt-hour of energy produced that implies net removal of $CO_2$ from the atmosphere) when gasification is coupled with biochar capture or amine $CO_2$ scrubbing and geological storage [27].

### 3.3.3.3 Trends

Diversification of feedstock: Credits to circular economy laws, clean wood chips are being replaced by low-cost wastes, including rice husk, sewage sludge, and algal cake. According to Zhang et al. [28], in digital twin modeling, real-time CFD-based optimization lowers tar by 30% and increases cold-gas efficiency by 5%. Modular, skid-mounted devices (1–5 MWth, megawatt thermal) are intended for distant microgrids and distributed hydrogen for agroindustries. Integration of hydrogen refueling initiatives in the EU and Japan includes using pressure swing adsorption (PSA)-purified $H_2$ (>99.97%) for heavy-duty fuel-cell vehicles and placing gasifiers next to HRS (hydrogen refueling stations).

The three primary kinds of gasifiers are entrained flow, fluidized bed, and fixed bed. Less sophisticated fixed-bed gasifiers are appropriate for small-scale CHP (heat and power) generation. Compared to fixed-bed gasifiers, fluidized-bed gasifiers are more effective and capable of processing a wider range of feedstocks. Entrained-flow gasifiers can handle both liquid and solid feedstocks and are appropriate for wider capacity ranges. They provide complete carbon conversion and the creation of syngas free of tar and phenol by operating at high temperatures above ash slagging points [29]. To produce the desired gas product (in this case, hydrogen), a gasifier typically requires pretreatment of the feedstock, gas conditioning, and end product purification. This study provides a short description of emerging gasification techniques, such as catalytic reformers and plasma gasifiers. Figure 3.6 represents the classification of different approaches for $H_2$ production using biomass gasification.

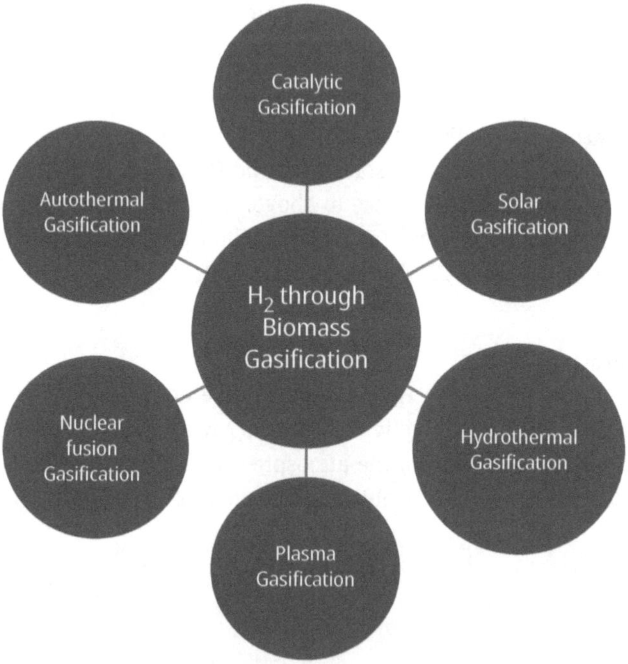


**Figure 3.6:** Different approaches for $H_2$ production using biomass gasification.

A producer gas with different proportions of $H_2$, CO, $CO_2$, $CH_4$, $C_xH_y$, and tars is created based on operation parameters, including temperature, catalysts, and oxidant. The average production gas compositions from various biomass gasifier types are shown in Table 3.3. Different gas conditioning units and pre- and post-treatment procedures are used to create high concentrated hydrogen since the concentration of hydrogen produced in gasifier reactors is often less than 50%.

**Table 3.3:** Gas compositions obtained from different types of gasifiers (main components only) [30].

| Process | Gas composition (vol%) | | | | | $H_2$:CO ratio |
|---|---|---|---|---|---|---|
| | CO | $H_2$ | $CO_2$ | $CH_4$ | $C_xH_y$ | |
| **Entrained flow** | 60 | 35 | 5 | – | – | 0.58 |
| **Bubbling fluidized bed** | 26 | 19 | 37 | 17 | 1 | 0.73 |
| **Circulating fluidized bed** | 33 | 30 | 31 | 6 | <1 | 0.91 |
| **Dual fluidized bed** | 23 | 39 | 25 | 10 | 3 | 1.70 |

#### 3.3.3.4 Challenges

**Tar Formation and Removal**: Downstream catalysts are soiled by poly-aromatic hydrocarbons (>1 g/m$^3$). Although tars are reduced to less than 50 mg/m$^3$ by catalytic hot-gas cleaning, catalyst deactivation by K, Cl, and P is still an issue [31].

**Variability of Feedstock:** Ash and moisture heterogeneity cause bed agglomeration and uneven syngas quality. Cost is increased by advanced pre-processing (pelletization, the process of transforming fine powders or granules into small, spherical or semispherical pellets. It's a technique used to agglomerate materials, making them easier to handle, store, and use in various applications, torrefaction, a thermal process where biomass is heated in a low-oxygen environment, typically between 200 and 300 °C, to enhance its fuel properties and make it resemble coal in appearance and characteristics. This process reduces moisture content, increases energy density, and improves ignitability, resulting in a more stable and easily handled fuel.

**Catalyst Poisoning:** Without guard beds, alkali metals, $H_2S$, and HCl from biomass reduce the life of Ni catalysts to less than 200 h.

**Competitiveness in the Economy**: In areas with inexpensive natural gas, the levelized cost of $H_2$ from biomass gasification is more than SMR with carbon capture and storage (CCS). Competitiveness is enhanced by waste-tipping fees and carbon price [32].

**Scale-Up and Financing**: Limited long-term performance hinders bankability; few plants surpass 20 MWth (megawatt thermal).

Approaching to eliminate CO and $CO_2$, as well as trace elements like $H_2S$, $NH_3$, HCl, and tar, the gas must be cleaned and upgraded. After the generated gas has been improved and conditioned, it is sent to a water gas shift (WGS) reactor, where steam and CO combine with a catalyst to create $H_2$ and $CO_2$. A high-quality hydrogen stream may be produced using PSA, which can further purify the 65–70 vol% hydrogen that normally makes up the gas exiting the WGS. As a crucial step in the process, the $CO_2$ is separated into a concentrated residual stream, which creates a favorable environment for $CO_2$ collection and storage and, as a result, reduces $CO_2$ emissions.

### 3.3.4 Solid Oxide Electrolysis Cell method

A potential technique for producing hydrogen with great efficiency is SOECs, especially when combined with renewable energy sources. SOECs use solid oxide or ceramic electrolytes to electrolyze steam into hydrogen and oxygen while operating at high temperatures (600–1,000 °C) [33]. In addition to having better efficiency than low-temperature electrolysis techniques, this process provides possible avenues for the co-electrolysis of $CO_2$ and $H_2O$ to generate syngas. Figure 3.7 shows $H_2$ production by steam electrolysis in SOEC mode.

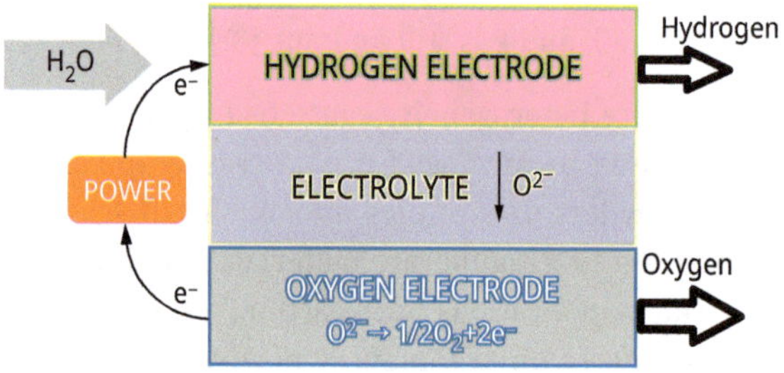


**Figure 3.7:** $H_2$ production by steam electrolysis in solid oxide electrolysis cell (SOEC).

#### 3.3.4.1 Operating Principles of SOECs

SOECs function by applying an external voltage to drive the electrolysis of steam at high temperatures. The core reactions are shown as Eqs. (3.14) and (3.15):

- Cathode (Hydrogen Electrode):

$$H_2O + 2e^- \rightarrow H_2 + O^{2-} \tag{3.14}$$

- Electrolyte: $O^{2-}$ ions migrate through the solid oxide electrolyte.
- Anode (Oxygen Electrode):

$$2O^{2-} \rightarrow O_2 + 4e^- \tag{3.15}$$

This high-temperature operation enhances reaction kinetics and allows for the utilization of thermal energy, reducing the electrical energy required for electrolysis.

#### 3.3.4.2 Advantages of SOEC Technology

- **High Efficiency:** SOECs can achieve electrical efficiencies up to 90% when integrated with heat recovery systems.
- **Fuel Flexibility:** SOECs can co-electrolyze steam and $CO_2$ to produce syngas, enabling the synthesis of various fuels and chemicals.
- **Integration with Renewable Energy:** Their high efficiency and ability to utilize waste heat make SOECs suitable for coupling with renewable energy sources, enhancing overall system sustainability.

#### 3.3.4.3 Materials and Cell Components

SOECs share structural similarities with solid oxide fuel cells (SOFCs), comprising:
- Electrolyte: Yttria-stabilized zirconia (YSZ) is commonly used for its high ionic conductivity and stability at elevated temperatures.
- Cathode (Hydrogen Electrode): Typically composed of nickel-YSZ cermet, offering good catalytic activity and electronic conductivity.
- Anode (Oxygen Electrode): Materials like lanthanum strontium cobaltite or lanthanum strontium manganite are employed for their mixed ionic and electronic conductivity.

By tackling issues like chemical deterioration and thermal expansion mismatches, material advancements seek to improve performance and durability. SOEC is poised to play a significant role in the hydrogen economy, i.e., green hydrogen production. Their high efficiency makes them ideal for producing green hydrogen when powered by renewable energy.

### 3.3.5 Anion-Exchange Membrane Method

AEM electrolysis is a new method for producing hydrogen. It provides the possibility of producing hydrogen with alkaline conditions and non-precious metal catalysts with great efficiency and at an inexpensive cost [34].

**Working:** In order for water to split into hydrogen and oxygen, AEM electrolysis works by moving hydroxide ions ($OH^-$) from the cathode to the anode via an AEM. A polymeric membrane functionalized with quaternary ammonium groups that promote $OH^-$ transport is a common feature of AEM systems. Water is delivered to the cathode, where it is converted to hydroxide ions and hydrogen gas, which move to the anode via the membrane. Hydroxide ions undergo oxidation at the anode to produce water and oxygen gas.

In AEM electrolysis, water is split into hydrogen and oxygen using an AEM that conducts hydroxide ions ($OH^-$) from the cathode to the anode. The half-reactions are as shown in Eqs. (3.16), (3.17), and overall reaction in Eq. (3.18):

- Anode (Oxygen Evolution Reaction):

$$4OH^- \rightarrow O_2 + 2H_2O + 4e^- \tag{3.16}$$

- Cathode (Hydrogen Evolution Reaction):

$$4H_2O + 4e^- \rightarrow 2H_2 + 4OH^- \tag{3.17}$$

- The overall reaction is:

$$2H_2O \rightarrow 2H_2 + O_2 \tag{3.18}$$

The hydroxide ions generated at the cathode migrate through the AEM to the anode, where they are oxidized to produce oxygen and hydrogen gas.

#### 3.3.5.1 Materials and Components

a. **AEM:** The membrane must exhibit high hydroxide ion conductivity, chemical stability in alkaline environments, and mechanical robustness. Examples include hydroxide AEMs like direct methanol fuel cells and direct-ethanol fuel cells. Other AEMs like poly(arylene ether), polyolefin, and polyphenylene backbones are used to create AEMs for different applications.
b. **Catalysts**
   - **Anode:** Typically utilizes nickel-based catalysts (e.g., $NiFe_2O_4$) for the oxygen evolution reaction.
   - **Cathode:** Often employs nickel or Raney nickel for the hydrogen evolution reaction.
c. **Electrolyte:** Aqueous solutions of potassium hydroxide (KOH) are commonly used to enhance ionic conductivity and support the electrochemical reactions.

#### 3.3.5.2 Performance Metrics

Recent studies have demonstrated significant advancements in AEM electrolysis performance [35]:

- **Current Density:** Achieving current densities up to 10 $A/cm^2$ with cell voltages around 2.3 V, indicating high reaction rates and efficiency.
- **Durability:** Long-term operation exceeding 12,000 h at 1 $A/cm^2$ current density with minimal voltage degradation, showcasing the system's stability.

- **Hydrogen Purity:** Producing hydrogen with purity levels comparable to those from PEM electrolysis, suitable for various applications.

#### 3.3.5.3 Recent Advancements and Trends

##### 3.3.5.3.1 Improved Membrane Materials

The creation of highly conductive and chemically stable AEMs is one of the most active research topics. Membranes based on poly (aryl piperidinium) and poly (phenylene oxide) with tethered quaternary ammonium groups are recent inventions that have shown enhanced conductivity and stability in alkaline environments.

##### 3.3.5.3.2 Catalyst Development

Developments in non-PGM catalysts are essential, particularly for the oxygen evolution reaction (OER). In alkaline conditions, transition metal oxides such $NiFeO_x$, $CoO_4$, and Mn-based materials have shown encouraging performance, providing strong electrochemical activity and endurance [35].

##### 3.3.5.3.3 Integration with Renewable Energy

Direct connection of AEM electrolyzers with intermittent renewable sources, such as wind and sun, is being investigated more and more. They are appropriate for decentralized hydrogen generation, especially in distant or off-grid applications, due to their versatility and low operating voltage.

##### 3.3.5.3.4 System Scaling and Commercialization

These days, businesses and research teams are concentrating on scaling AEM electrolysis for pilot and commercial operations. As an illustration of the technology's usefulness, Enapter has marketed small AEM electrolyzers for modular green hydrogen systems.

#### 3.3.5.4 Challenges and Limitations

Despite significant progress, AEM electrolysis faces several technical and durability challenges.

##### 3.3.5.4.1 Membrane Stability

Through nucleophilic substitution or Hofmann elimination, alkaline conditions encourage the breakdown of cationic functional groups, including quaternary ammonium. One significant obstacle is maintaining stability for more than 2,000–5,000 h [36].

#### 3.3.5.4.2 Reduced Efficiency

AEM systems often exhibit lower current densities and efficiency when compared to the other electrolyzers. One of the primary goals of the study is to achieve comparable performance while continuing to employ non-PGM catalysts.

#### 3.3.5.4.3 Water Management

Especially when the power supply is dynamic or sporadic, controlling water crossing, drying the membrane, and preserving hydration balance without flooding continue to be significant technical challenges.

#### 3.3.5.4.4 Scalability and Durability

There is little testing done on durability in real-world scenarios with temperature, pressure, and fluctuating loads. Commercialization requires ensuring mechanical and electrochemical resilience for large-scale applications.

It is anticipated that developments in AEM electrolysis technology will concentrate on material innovation, the process of creating novel membrane materials with improved stability and conductivity. Developing long-lasting, highly active non-precious metal catalysts is known as catalyst development. System optimization is the process of enhancing cell integration and design to increase scalability and performance. The goal of these advancements is to make AEM electrolysis a viable and competitive process for producing hydrogen on a wide scale [37]. Table 3.4 demonstrates a parallel analysis between the SOEC technique and the AEM method [25].

**Table 3.4:** Comparative study among the SOEC and AEM methods.

| Factor | Anion-exchange membrane electrolysis | Solid oxide electrolysis cell |
|---|---|---|
| **Temperature range** | Low to moderate: 40–70 °C | High: 700–850 °C |
| **Electrolyte material** | Polymer with quaternary ammonium groups transporting $OH^-$ ions | Ceramic solid electrolyte (typically yttria-stabilized zirconia, YSZ) conducting $O^{2-}$ ions |
| **Conducting species** | $OH^-$ ions | $O^{2-}$ ions |
| **Feed** | Water (liquid phase) | Steam or steam + $CO_2$ (co-electrolysis) |
| **Electrode catalysts** | Non-noble metals (e.g., NiFe, NiMo) due to alkaline conditions | Ni–$ZrO_2$ cermet for cathode, perovskite oxides for anode |
| **Startup time** | Fast; suitable for intermittent renewables | Slow; requires thermal cycling management for variable renewable energy compatibility |

**Table 3.4** (continued)

| Factor | Anion-exchange membrane electrolysis | Solid oxide electrolysis cell |
|---|---|---|
| **Carbon intensity** | Zero if powered by renewables; low if powered by low-carbon grid | Zero if heat/electricity is renewable or nuclear |
| **System complexity** | Low to medium | High; thermal management, sealing, and material compatibility are critical |
| **Technology readiness level (TRL)** | TRL 5–7 (pilot scale, improving) | TRL 6–8 (pilot to demonstration stage) |
| **Co-product potential** | Oxygen only | Syngas ($H_2$ + CO) via co-electrolysis with $CO_2$ |
| **Use in grid integration** | Suitable for variable renewable electricity (solar/wind) | Challenging; control theory needed for dynamic operation |

## 3.4 Future Aspects

A thorough strategy that includes state policies, industry involvement, national missions, and regulatory frameworks in order to become a worldwide leader in the production and use of hydrogen. The full potential of hydrogen as a sustainable fuel will need sustained attention to infrastructure development, global partnerships, and innovation.

Green hydrogen is at the heart of the world and India's hydrogen future, which aims to decarbonize electricity, transportation, and industry. With financing of ₹19,744 crore, the National Green Hydrogen Mission (2023) seeks to create 5 MMT annually by 2030. Under PLI initiatives, India takes advantage of its cheap renewable cost and push for electrolyzer manufacture. Policies that require refineries and fertilizers to purchase hydrogen increase demand. Pilot projects are moving forward in regions like Gujarat and Rajasthan that are known for their steel, mobility, and hydrogen. India views hydrogen as a key component of its net-zero routes, with export aspirations and public-private research and development in catalysts and storage [38].

## 3.5 Conclusion

High-temperature reactions (such as SMR, pyrolysis, and water-splitting cycles) are used in thermochemical hydrogen generation to transform thermal energy into chemical fuel. Advanced techniques such as methane pyrolysis produce $H_2$ without $CO_2$, and multi-step cycles (such as S–I and ceria loops) are powered by nuclear or solar heat. Integrating variable heat sources and material durability in harsh environments is a major issue. Entropy control, heat integration, and fusing solid carbon valorization with $CO_2$ reuse are the main solutions. Modular reactors, carbon separation, and thermal storage are combined in emerging designs; nonetheless, large-scale viability is contingent upon established economics and appropriate governmental frameworks.

The exact band-structure design is essential for photocatalytic water splitting; the semiconductor's valence band must be below the $O_2/H_2O$ level and its conduction band must be above the $H^+/H_2$ redox level, while recombination is suppressed by long carrier lifetimes and quick surface kinetics. With DFT (density functional theory) + GW (Green's function) and time-dependent DFT mapping defect-tolerant oxynitrides, perovskites, and carbon nitrides, the focus of current theory is on Z- and S-scheme heterojunctions that resemble real photosynthesis. The search for metastable, highly absorbing phases is accelerated by machine-learning potentials. The kinetic-thermodynamic interface presents difficulties; efficiency is limited by photo-corrosion, surface traps, and photon attenuation in particle slurries. Tandem or multilayer photonic absorbers and bifunctional cocatalysts that reduce OER (oxygen evolution rate)/HER (hydrogen evolution rate) overpotentials below 250 mV are required for >10% solar-to-hydrogen performance, according to models; single-atom Ni or Co on pyridinic-N carbon nitride is a prime candidate. While automated synthesis and operando spectroscopy promise to shorten discovery cycles and accelerate commercial viability, suggested methods combine atom-economic catalysts with waveguide or photonic-crystal reactors and immobilized photo-electrodes.

Through tar reforming, steam-char reactions, and devolatilization, biomass gasification transforms lignocellulosic feedstocks into $H_2$-rich syngas, with equivalency ratios of 0.2–0.3 and maximal $H_2$ outputs at 850–950 °C. Dual-fluidized-bed and sorption-enhanced systems that isolate combustion from gasification and use in-bed CaO carbonation to shift the equilibrium toward hydrogen are preferred by advanced theory. Multiphase CFD demonstrates that tar aerosols nucleate hot spot fouling downstream, whereas Temkin isotherms quantify catalyst poisoning by K, Cl, and S. Kinetic Monte Carlo modeling relates Ni- catalyzed tar cracking to particle size and basicity. In the absence of tar concentrations below 5 mg and catalyst life beyond 4,000 h, levelized hydrogen prices remain uncompetitive. When combined with BECCS, gasification produces −0.4 to 1 t $CO_2$ eq MW/h. When combined with SOFCs, it increases electrical efficiency to above 55% LHV. In order to move biomass hydrogen toward economic feasibility, solutions include coke-resistant ALD bimetallic catalysts and "digital twin" control, which anticipates tar spikes and optimizes operating conditions. This may possibly quintuple catalyst life and reduce operating expenditure (OPEX) by 20%.

**SOECs** use oxide-ion conduction in a YSZ electrolyte to split steam at around 800 °C. Since the Gibbs free energy for $H_2O$ dissociation drops to around 200 kJ/mol at this temperature, the electrical demand may be reduced to 3 kWh/Nm$^3$ $H_2$ using recovered process heat. Perovskite anodes, where $O^{2-}$ ions generate $O_2$, and Ni–$ZrO_2$ ceramic cathodes, which decrease steam, control electrokinetics. According to modeling, nuclear or other high-temperature heat integration may propel SOEC plants to 85–90% HHV efficiency and cheap hydrogen cost if cell lifespan surpasses 40,000 h. Simultaneous co-electrolysis of $H_2O + CO_2$ produces syngas with a 15% lower energy penalty than separate processes. Ni-free perovskite electrodes, compliant glass-ceramic sealing, and roll-to-roll metal supports that match thermal expansion are important durability solutions. While control-optimized start-stop procedures reduce thermomechanical fatigue by half while preserving flexibility for variable renewable energy operation, aluminized coatings and Mn–Co spinel getters prevent chromium volatilization.

**AEM** electrolyzers work in alkaline environments by transporting hydroxide ions via membranes functionalized with quaternary ammonium. They provide a PEM-like design employing inexpensive non-noble catalysts (NiFe layered double hydroxides, LDH for OER, oxygen evolution reaction, and NiMo for HER, hydrogen evolution reaction) at moderate temperatures (≤70 °C). Chemical durability, low gas crossover, and high $OH^-$ conductivity are essential. Performance relies on controlling losses and maximizing ion transport, which calls for tiny electrode designs and ultra-thin membranes. Advanced control systems for renewable integration, carbonate removal, and stabilizing membrane chemistry are some of the solutions. Scaling up toward affordable, gigawatt-level manufacturing is being accelerated by open-source technologies and design iteration.

## References


[1] Cavendish, H., 1766. XIX. Three papers, containing experiments on factitious air. Philosophical Transactions of the Royal Society of London, (56), pp.141–184.

[2] Smil, V., 2017. Energy and civilization: a history. MIT press: Cambridge, MA, USA.

[3] Staffell, I., Scamman, D., Abad, A.V., Balcombe, P., Dodds, P.E., Ekins, P., Shah, N. and Ward, K.R., 2019. The role of hydrogen and fuel cells in the global energy system. Energy & Environmental Science, 12(2), pp.463–491.

[4] Balta-Ozkan, N., Yildirim, J., and Connor, P. M. (2015). Regional distribution of hydrogen energy potential: A review. Renewable and Sustainable Energy Reviews, 42, pp.1015–1033.

[5] Broom, D.P., 2011. Hydrogen storage materials: the characterisation of their storage properties. Green energy and technology'(Springer, London, 2011).

[6] IEA (2023) Global Hydrogen Review 2023. Paris: International Energy Agency.

[7] Atkins, P.W., De Paula, J. and Keeler, J., 2023. Atkins' physical chemistry. Oxford University Press: Oxford, UK.

[8] Dincer, I. and Acar, C., 2015. Review and evaluation of hydrogen production methods for better sustainability. International journal of hydrogen energy, 40(34), pp.11094–11111.

[9] Das, S et al. 2021, 'Renewable-assisted methane pyrolysis for turquoise hydrogen: An Indian case study', Journal of Cleaner Production, 328, art. 129692.

[10] Chandra, P and Dutta, A 2021, 'Life-cycle emissions of hydrogen from Indian coal gasifiers', International Journal of Hydrogen Energy, 46(60), pp.30708–30720.
[11] Sharma, R, Singh, G and Dhar, S 2023, 'Techno-economic assessment of nuclear-driven electrolysers for hydrogen production in India', Nuclear Engineering and Design, 403, art. 110003.
[12] Balta-Ozkan, N., Yildirim, J., and Connor, P. M. (2015). Regional distribution of hydrogen energy potential: A review. Renewable and Sustainable Energy Reviews, 42, 1015–1033.
[13] Ozbilen, A., Dincer, I. and Rosen, M.A., 2011. A comparative life cycle analysis of hydrogen production via thermochemical water splitting using a Cu–Cl cycle. International Journal of Hydrogen Energy, 36(17), pp.11321–11327.
[14] Orhan, M.F., Dincer, I. and Rosen, M.A., 2010. An exergy–cost–energy–mass analysis of a hybrid copper–chlorine thermochemical cycle for hydrogen production. International Journal of Hydrogen Energy, 35(10), pp.4831–4838.
[15] Hisatomi, T., Kubota, J. and Domen, K., 2014. Recent advances in semiconductors for photocatalytic and photoelectrochemical water splitting. Chemical Society Reviews, 43(22), pp.7520–7535.
[16] Marschall, R., 2021. 50 years of materials research for photocatalytic water splitting. European Journal of Inorganic Chemistry, 2021(25), pp.2435–2441.
[17] Chen, Y., Zhao, X. and Li, H. (2022) 'Progress in biomass gasification for hydrogen production: A review', Renewable and Sustainable Energy Reviews, 158, 112134.
[18] Maeda, K. and Domen, K., 2010. Photocatalytic water splitting: recent progress and future challenges. The Journal of Physical Chemistry Letters, 1(18), pp.2655–2661.
[19] Zhou, P., Wang, Y., and Zhang, H. (2022). Single-atom photocatalysts for solar water splitting. Advanced Functional Materials, **32**(1), 2106730.
[20] Tang, L., Zeng, G., Wang, L., Wang, J. and Tang, X., 2021. Joint connection of experiment and simulation for photocatalytic hydrogen evolution: strength, weakness, validation and complementarity. Journal of Materials Chemistry A, 9(11), pp.6749–6774.
[21] Chen, X., Shen, S., Guo, L. and Mao, S.S., 2010. Semiconductor-based photocatalytic hydrogen generation. Chemical reviews, 110(11), pp.6503–6570.
[22] Dutta, A. and Barrett, S. (2023) 'Sorption enhanced steam gasification of lignocellulosic biomass for hydrogen', Energy Conversion and Management, 276, 116516.
[23] Thunman, H., Seemann, M. and Berdugo Vilches, T. (2021) 'Advanced biomass gasification in dual fluidised beds', Energy Science & Engineering, 9(3), pp.312–329.
[24] Kumar, R., Patel, M. and Watanabe, M. (2024) 'Continuous supercritical water gasification of sewage sludge: Catalyst stability and hydrogen yield', Journal of Supercritical Fluids, 200, 106045.
[25] Singh, A. et al., 2022. Hydrogen production from AEM electrolysers: A techno-economic review, International Journal of Hydrogen Energy, 47(12), pp. 7769–7784.
[26] Rautanen, T., Kivisaari, T. and Kauppinen, J. (2023) 'Demonstration of a 250 kW integrated biomass gasification SOFC plant', International Journal of Hydrogen Energy, 48(5), pp. 3672–3684.
[27] IEA Bioenergy (2022) Bio-hydrogen and negative emissions: BECCS pathways. Task 33 Report.
[28] Zhang, L., Wei, H. and Tang, X. (2023) 'Digital-twin optimisation of a 5 MW biomass gasifier for hydrogen', Bioresource Technology, 368, 128244.
[29] Wagner, N.J., Coertzen, M., Matjie, R.H. and van Dyk, J.C., 2008. Coal gasification. In Applied coal petrology (pp. 119–144). Elsevier: Amsterdam, The Netherlands.
[30] Moioli, E. and Schildhauer, T., 2022. Negative $CO_2$ emissions from flexible biofuel synthesis: Concepts, potentials, technologies. Renewable and Sustainable Energy Reviews, 158, p.112120.
[31] Milne, T., Evans, R. and Abatzoglou, N. (2022) 'Tar formation during biomass gasification: Mechanisms and mitigation', Progress in Energy and Combustion Science, 89, 101020.
[32] IEA (2023) Global Hydrogen Review 2023. Paris: International Energy Agency.
[33] Laguna-Bercero, M.A., 2012. Recent advances in high temperature electrolysis using solid oxide fuel cells: A review. Journal of Power sources, 203, pp.4–16.

[34] Park, J.E., Kang, S.Y., Oh, S.H., Kim, J.K., Lim, M.S., Ahn, C.Y., Cho, Y.H. and Sung, Y.E., 2019. High-performance anion-exchange membrane water electrolysis. Electrochimica Acta, 295, pp.99–106.
[35] Zheng, Zhou, D., Li, P., Xu, W., Jawaid, S., Mohammed-Ibrahim, J., Liu, W., Kuang, Y. and Sun, X., 2020. Recent advances in non-precious metal-based electrodes for alkaline water electrolysis. ChemNanoMat, 6(3), pp.336–355.
[36] Maric and Maynard Peng, S., 2021. Anion Exchange Membrane Water Electrolysis. In Electrochemical Hydrogen Production from Water Splitting: Basic, Materials and Progress (pp. 99–146). Singapore: Springer Nature Singapore.
[37] Pushkareva, I.V., Pushkarev, A.S., Grigoriev, S.A., Modisha, P. and Bessarabov, D.G., 2020. Comparative study of anion exchange membranes for low-cost water electrolysis. International Journal of Hydrogen Energy, 45(49), pp.26070–26079.
[38] MNRE (2023). National Green Hydrogen Mission. Ministry of New and Renewable Energy. Available at: https://mnre.gov.in (Accessed: 21 May 2025).

Ambneesh Mishra, Samanshi Rohilla, Dan Bahadur Pal,
Shraddha Awasthi

# 4 Cryogenic Hydrogen Storage and Transportation Technologies

**Abstract:** Hydrogen is gaining prominence as a clean energy vector capable of mitigating the environmental impact of fossil fuels. Yet, its widespread deployment depends on the development of efficient, safe, and scalable storage and transport technologies. This chapter explores two promising liquid-phase hydrogen storage technologies: cryogenic hydrogen storage, a physical method reliant on ultra-low temperatures, and liquid organic hydrogen carriers (LOHCs), which store hydrogen chemically in organic compounds. It discusses their operating principles, technological advancements, safety and infrastructure needs, energy requirements, and suitability for small- and large-scale applications. A comparative analysis highlights that both technologies, despite their challenges, contribute to the hydrogen value chain and hold transformative potential for future clean energy systems.

**Keywords:** Hydrogen storage, cryogenic storage, liquid organic, hydrogen carriers, clean energy, energy infrastructure, environmental impact

## 4.1 Introduction

There is a global drive to shift to clean energy to achieve sustainable development. Green hydrogen offers optimism to decarbonize sectors such as industry, transport, and power generation [30]. Today, we regard green hydrogen as an upcoming energy source. However, rather than a source, it would be more appropriate to consider it a vector or carrier of energy that can potentially store and carry huge amounts of energy derived from renewable sources, and release it for use as needed [16]. In this way, while it replaces nonrenewable fossil fuels, it also helps overcome the limitations of renewables, such as intermittency and the requirement of large and relatively incapacitated infrastructure for storage and transport [27]. With these factors combined with the facilitation of storage, hydrogen can effectively contribute to the circular economy [30]. In the global shift toward renewables, a key consideration is that not all countries have equal access to resources like wind and solar energy, while others have them in abundance. Hydrogen can be produced in such renewable-rich regions, stored, and shipped worldwide, fostering a hydrogen economy, enabling equitable energy distribution, and opening up new trade opportunities [42]. The term "hydrogen economy" was coined by John Bockris in 1970 while he was delivering a speech at the General Motors (GM) Technical Centre. Even though the usage of hydrogen as an

 | https://doi.org/10.1515/9783112181072-004

energy carrier is a concept two centuries old, it has now gained momentum in research and application in response to the global energy crisis post-1970s. With the urgent need to shift from fossil fuels to a clean and sustainable source of energy, the hydrogen economy seems to have reached a turning point, where scientists and governments around the world are finding ways to make the best use of a clean fuel having a lower heating value and higher heating value as 33.3 kW/kg and 120 MJ/kg, respectively [30]. The hydrogen industry is divided into three broad segments: upstream, concerned with the production of hydrogen; midstream, concerned with its storage and transport; and downstream, concerned with its usage [42]. India's National Green Hydrogen Mission proposes incentives for manufacturing electrolyzers and the production of green hydrogen [21]. Thus, we may expect a boost in the upstream segment of its industrial chain, that is, production. Hydrogen's potential as a clean energy carrier depends heavily on efficient storage and transportation methods; hence, to make improvements in green hydrogen as a prospective sustainable fuel, it is also imperative to optimize the midstream segment of the hydrogen industry. While green hydrogen is a promising fuel, in its elemental gaseous state, it is a light gas with a low ignition point, requiring extensive infrastructure to safely maintain it. In the pursuit of addressing these concerns, several alternatives have emerged, each having its advantages and disadvantages. Given the stable chemistry of hydrogen, it may be stored for extended periods. Hydrogen is mostly stored in either gaseous or liquid form, using two main types of storage systems: physical storage, like compressed gas or cryogenic liquid, and material-based storage, where hydrogen is absorbed or chemically bound, such as in metal hydrides or liquid organic hydrogen carriers (LOHCs) [26]. This chapter focuses on two liquid-phase methods: cryogenic hydrogen, a physical storage system, and LOHCs, a material-based approach. It is necessary for humankind to be wise in the choice of technologies and also to use science for their development in terms of their capacity, cost, safety, efficiency, and sustainability. With this view in consideration, the chapter examines upcoming hydrogen storage systems, providing an overview and discussing their potential and challenges for the future.

## 4.2 Challenges in the Midstream Segment of the Hydrogen Industry: Storage and Transport

Compared to other fuels, hydrogen has the theoretical capacity to store the maximum energy per mass [26]. However, research and design are needed to incorporate technologies and materials for efficient hydrogen capture and regulated release, as well as to improve the overall efficiency of the process. As discussed earlier, hydrogen is not the primary source of energy. It uses energy produced by other sources. For hydrogen to produce clean energy in the true sense, the primary sources should ensure minimum emissions in the production process. This dependency has been hindering hydrogen's penetration into the energy market [27]. Moreover, at present, hydrogen is primarily derived from fossil fuels such as natural gas and coal, which contribute to rising atmospheric $CO_2$ levels. To lessen environmental damage, producing hydrogen through renewable electricity sources is essential.

While hydrogen is clean and nontoxic, the low volumetric energy density at normal atmospheric pressure (0.0899 kg/m$^3$) coupled with its flammability, makes its management, especially storage and transport, difficult and raises safety concerns. Hydrogen is usually compressed at high pressures or liquefied at very low temperatures to increase its energy-carrying capacity for a given volume [16]. However, these techniques are expensive and have complex and large design and material needs.

As Calabrese et al. [7] rightfully emphasize the importance of safety as a foundational element across the hydrogen value chain, in their proposed hydrogen pyramid model, it emerges as a critical factor in hydrogen storage and transport due to its unique physical and chemical properties.

One such major safety issue in liquid hydrogen storage is boil-off, the phenomenon in which liquid hydrogen vaporizes due to reasons such as heat entry, thermal stratification, and spin isomer alterations. Not only does this create economic losses, but it also poses explosion hazards if not controlled. Hydrogen's high flammability, extremely low minimum ignition energy (0.02 mJ), and invisible flame, together with high laminar flame speed (up to 3 m/s), render it extremely susceptible to unauthorized ignition and challenging to detect in initial stages [7]. It is very dangerous when blended with gases such as fluorine, chlorine, oxygen, carbon monoxide, or even air, requiring detailed safety precautions from start to finish [6].

Hydrogen permeates through materials like metals and plastics with embrittlement or cracking, leading to leaks. This requires cautious material selection in hydrogen systems. Moreover, hydrogen possesses high diffusion rates in gases, liquids, as well as solids [27]. Negative Joule-Thompson coefficient contributes to cooling requirements so as to suppress expansion and resulting heating [27].

For the realization of a prosperous hydrogen economy, it is important that there are stationary and mobile hydrogen storage systems in place. Yet the limitations that come with the weight and volume of hydrogen are less severe for stationary applications than for mobile ones [16]. The limited size and weight capacity of light-duty vehicles, and customer demand for more than 300 miles of driving range, pose a challenge to bring the technology into these vehicles. Owing to its low volumetric energy density, hydrogen can presently be stored in tanks of greater size and higher pressure than other gaseous fuels. Medium- and heavy-duty vehicles have sufficient room to fit more extensive tanks, but they are likely to be under weight restrictions and constraints [30]. Fixed hydrogen storage units can spare more space, handle higher temperatures and pressures, and include extra capacity for accommodating slower reaction kinetics. Nevertheless, the absence of adequate materials to use in storage tanks is a major technical hurdle in their development [1].

For ships and automobiles running on hydrogen, leak or accident risks are troubling. But based on studies, a safer way has been proposed to run on hydrogen fuel.

The necessity for specialized infrastructures has promoted reluctance and hindrance in the utilization of elemental hydrogen. But recent technological and scientific advances provide optimism that the future hydrogen economy will operate without the necessity to deal with substantial amounts of elemental hydrogen. Material-based storage is a new storage system that has higher energy density. Yet, they experience complications in chemical kinetics and reactions. Each method is accompanied by its relative merits and difficulties. Continuous research and development are ongoing to resolve the loopholes and shortcomings. For existing uses, it is crucial to choose the most appropriate method with respect to the contextual parameters, including transport distance, available infrastructure, cost factor, safety requirement, end use application, in balancing performance, feasibility, environmental, and economic effects.

## 4.3 Cryogenic Hydrogen

Cryogenic hydrogen is a physical storage technology. In this technique, $H_2$ gas is stored below its boiling point −252.87 °C [16]. It can be stored for longer periods by cooling it to temperatures that change its state to liquid and increasing its density [16]. The hydrogen needs to undergo quick and consistent stages of compression and expansion while maintaining cryogenic temperatures [26]. Equipment used in the process includes compressors, heat exchangers, and expansion turbines [26]. Cryogenic hydrogen technique involves liquefication along with refrigerants, starting with pre-cooling substances like liquid nitrogen, followed by further cooling using helium or mixed refrigerants in cryogenic systems [2]. The process is combined with the use of a cryocooler or a liquefier [16]. Thereafter, the liquefied hydrogen is stored in cryogenic tanks and transported in specialized pipelines, road tankers, or ship containers, as per the suitability, designed to maintain cryogenic conditions [26]. Liquefying nitrogen decreases its volume by approximately 848 times, making storage more efficient [4].

### 4.3.1 Storage System

Aluminium alloys, which are favored for space applications, have benefits like reduced weight and improved weldability, while stainless steel is frequently used for liquid hydrogen storage because of its resistance to hydrogen embrittlement and good low-temperature and corrosion performance [29]. The inner vessel, the insulating layer, and the outer shell are the three primary components of a liquid hydrogen storage tank. The liquid hydrogen is kept in the inner vessel, which has to tolerate very low temperatures. While multilayer insulation and the vacuum between the two reduce heat transfer through convection, conduction, and radiation, the outer shell offers structural protection [20]. Better systems employ superinsulation in several layers of insulating materials, like fiber glass paper and aluminium foil [40].

The ability to prevent boil-off is an important factor considered in the selection of insulation materials [34]. Cryogenic valves provide support in maintaining structural integrity and safety. They are useful in regulating flow into and out of the tank [35]. Hydrogen cooling systems are used in hydrogen-fueled cars to facilitate quick and consistent refueling.

### 4.3.2 Performance and Efficiency

Its density of 70.8 kg/m$^3$ is twice that of compressed hydrogen at 70 MPa [33], and its storage capacity in cryogenic tankers increases to five times that of compressed hydrogen in tube trailers [13]. Liquefying hydrogen leads to energy losses of up to 40%, compared to just 10% in compressed storage [13], due to its high energy demand (12.5–15 kWh/kg) and associated costs [33]. Expensive materials for advanced tanks further add to the cost [28]. It also results in the highest $CO_2$ emissions per kg stored [9]. Persistent boil-off losses, insulation degradation, and vacuum maintenance challenges reduce long-term efficiency.

The thermal performance of storage tanks is significantly influenced by their physical design. One of the main reasons for boil-off is heat leakage, which is influenced by the size, shape, and insulation of the tank. Although they are costly and difficult to manufacture, spherical tanks are perfect because of their uniform stress distribution and low surface-to-volume ratio [7].

### 4.3.3 Applications and Limitations

Static storage in controlled environments, such as hydrogen production plants and aviation and aerospace facilities, is the primary application for this technology. As cryogenic technologies have advanced, tanks and systems have emerged that enable the large-scale transportation of liquid hydrogen, including intercontinental shipping and truck deliveries, where the high cost and technical complexity can be justified [13]. However, there are still obstacles to overcome in order to implement this technology on a small and medium scale [19, 31]. It takes a lot of energy to cool hydrogen to its cryogenic state, and in smaller systems, it can be challenging to maintain these extremely low temperatures without experiencing excessive boil-off (evaporation losses). These systems are also more expensive and harder to maintain, limiting their feasibility for smaller refueling stations or local-scale hydrogen applications. Therefore, even though cryogenic storage is practical and becoming more and more popular for transporting hydrogen on a large scale, technical and financial constraints at smaller scales continue to limit its wider use. Because of its high energy requirements, high costs, and technical complexity, cryogenic hydrogen storage is currently not widely used at smaller scales, despite being a tried-and-true, high-capacity solution perfect for large-scale applications. To increase its viability throughout the hydrogen economy, more innovation is required.

## 4.4 Liquid Organic Hydrogen Carrier (LOHC)

One promising development in hydrogen storage and transportation technology is the liquid organic hydrogen carrier (LOHC). According to Preuster et al. [27], these systems make use of organic compounds, usually cyclic aromatic compounds, that change from hydrogenated to dehydrogenated forms as they absorb and release energy [27]. In an endothermic process, electrolysis uses energy from renewable resources to produce gaseous hydrogen [10]. As the hydrogen gas and the organic compound undergo a process known as hydrogenation, this energy is locked. Despite being an exothermic process, the energy stored for later use is retained during hydrogenation [10]. The organic compounds hydrogen now contains energy in a liquid form that makes storage and transportation possible. Heat causes the hydrogen-rich compound to release hydrogen atoms, which then recombine to form hydrogen gas at the energy generation point. We refer to this as dehydrogenation. A cyclic process could be initiated by dehydrogenating the dehydrogenated compound that was obtained. When a catalyst is present, both hydrogenation and dehydrogenation take place. However, this efficiency may differ based on the catalyst and LOHC. Hydrogen can be embedded into a liquid through this cyclic process, making storage and transportation more effective [27]. Catalysts are essential for regulating kinetics and thermodynamics in both processes. Although the reactions may require a lot of energy, their viability is being greatly enhanced by developments in catalyst design and carrier selection [12].

### 4.4.1 Types and Performance of LOHCs

Different categories of LOHCs have been explored and are used. Examples of LOHCs include cycloalkanes like cyclohexane, methylcyclohexane, and decalin; heterocycles with N and B; and formic acid. Because of their strong attraction to hydrogen atoms, cycloalkanes have a 6–8% by weight capacity. They are safe to use because of their liquid state, low boiling point, and low toxicity during the dehydrogenation stage. Because of their abundance and low cost, toluene and benzene are common cycloalkanes. However, during dehydrogenation, cycloalkanes need high energy inputs between 63 and 69 kJ/mol $H_2$, which means temperatures above 300 °C are required. Cycloalkanes also have the drawback of coking catalysts, which lowers their effectiveness [12].

Toluene/methylcyclohexane, which stores roughly 6.1% hydrogen by weight, was one of the first LOHC systems ever studied. However, under severe conditions, this system has problems like coking and side-product formation, and it needs high dehydrogenation temperatures (about 320 °C). Chiyoda Corporation, for instance, has commercialized it for hydrogen logistics in spite of these obstacles. Newer systems, such as dibenzyltoluene/perhydro-dibenzyltoluene (H0-DBT/H18-DBT), have drawn attention in an effort to get around these restrictions. These substances have been used in industry as heat transfer oils because they are low in toxicity, thermally stable, and liquid over a wide temperature range. Utilizing low-grade waste heat is challenging because even these systems need a large amount of heat input (about 65 kJ/mol $H_2$) for dehydrogenation [27].

In the past, one of the first LOHC systems to be examined was based on N-heteroaromatic compounds for low-energy hydrogen storage and release. In order to store and release hydrogen with less energy (i.e., a lower enthalpy of hydrogenation/dehydrogenation), researchers have been investigating LOHC systems, which introduce *N*-heteroaromatics. Researchers at Air Products and Chemicals made significant progress in this field by identifying LOHCs with a lower enthalpy of hydrogenation and dehydrogenation through systematic theoretical screening. With a low dehydrogenation enthalpy of 50 kJ/mol $H_2$ and a hydrogen storage capacity of 5.8% by mass, *N*-ethylcarbazole ($H_0$-NEC) and its hydrogenated form, dodecahydro-*N*-ethylcarbazole ($H_{12}$-NEC), are among the most studied systems. Other noteworthy *N*-heteroaromatic systems that exhibit low dehydrogenation enthalpies (~52 kJ/mol $H_2$) and allow for hydrogen release at much lower temperatures (100–140 °C) are indole/indoline and 1,2,3,4-tetrahydroquinoline/quinoline. Lowering melting points and increasing stability have been made possible by initiatives such as the creation of eutectic mixtures. The potential for thermal degradation, specifically the breaking off of N-alkyl groups at relatively low temperatures, even before the aromatic ring structure starts to decompose, is a known disadvantage of these compounds despite these advancements [27].

The addition of nitrogen improves the dehydrogenation kinetics and thermodynamics of all these N-containing compounds, increasing the efficiency of the hydrogen release process at lower temperatures. Appropriate catalysts, basic supports, and electron-donating or conjugated groups can all help to improve this. But nitrogen also adds some weaknesses: it increases the molecules' thermal liability and raises the possibility of adverse reactions, which can result in undesirable degradation products that could impair long-term performance and reversibility. Other issues include cost, maintaining the liquid state throughout, and catalyst poisoning that inhibits hydrogen binding to LOHC. Overall, adding nitrogen to these LOHCs enhances the kinetics of hydrogen release, but it also calls for careful molecular design to reduce degradation [12, 27].

The hydrogenation and dehydrogenation enthalpies of boron-containing LOHC systems are notably low; for example, 1,2-dihydro-1,2-azaborine has values as low as 36 kJ/mol $H_2$ [22]. This makes it possible for hydrogen to be released at comparatively low temperatures, sometimes lower than 100 °C. Their practical application is hampered by a number of issues, though. For example, many of these compounds are solids at room temperature, which makes handling and storage more difficult, particularly in damp environments or at temperatures as high as 60 °C [14]. These systems are limited to one-time use due to their poor reversibility, which prevents them from being recharged with molecular hydrogen once hydrogen is released [27]. Furthermore, the restricted worldwide supply of boron raises questions regarding the scalability of these LOHC systems, even if technical difficulties are fixed [8].

The hydrogen content of formic acid is comparatively high at 4.4% wt. The corrosive chemical formic acid can cause severe burns to the eyes and skin [7]. Although adding bases or solvents, such as water and ammonia, changes the reaction equilibrium and increases the yield of FA, it drastically lowers its energy density from 1.8 kWh/L in pure form [25]. As a byproduct of dehydrogenation, they emit $CO_2$, which raises entropy and reduces the energy required for the process. A carbon-neutral process could be produced if the $CO_2$ and hydrogen were reacted to regenerate formic acid. Additionally, the mechanism might follow an unfavorable dehydration pathway, release CO, and poison the fuel's catalyst.

In order to suppress the formation of CO and favor the dehydrogenation pathway, selective catalysts are essential [24]. According to Mekonnin et al. [20], formic acids are deemed appropriate for mobile applications [20].

Because LOHCs are liquid at room temperature, they are easily handled, stored, and transported using standard infrastructure. They don't need to be kept under stress. They are more secure. Fuel can be recycled and dehydrogenated after dehydrogenation. By investigating and creating more effective techniques, research is being done to enhance LOHC systems and increase their efficiency. Improvements to catalysts and supports for all LOHC are still being researched.

### 4.4.2 Catalysts in LOHC Systems

Both hydrogenation and dehydrogenation require catalysts. Because of their scalability, durability, and ease of separation, heterogeneous catalysts are preferred for industrial applications, even though homogeneous catalysts have been used extensively in lab settings [27, 36]. Activity and stability are strongly influenced by the support material and catalyst shape. The significance of reaction conditions has been emphasized by studies employing model surfaces such as Pt(111). In practice, reactions that work well in the lab frequently need higher temperatures, resulting in a "temperature gap" [27].

### 4.4.3 Advantages and Limitations

Without needing high pressure or cryogenic conditions, LOHCs can be stored and transported using current infrastructure while remaining liquid under ambient conditions. In terms of safety and public acceptance, this offers them a significant advantage over compressed or liquefied hydrogen [8, 27]. Recyclability of the system is ensured by the ability to rehydrate and reuse the dehydrogenated carrier following hydrogen release. However, there are still issues with reversibility, side-product formation, coking, and catalyst degradation, especially in heteroatomic systems. Furthermore, cost-efficiency is still a work in progress, and certain systems demand a lot of energy [12, 22].

## 4.5 Overcoming Hydrogen Storage Challenges

The biggest obstacles to creating a sustainable hydrogen economy are still storing and transporting hydrogen. Low volumetric energy density, high flammability, material compatibility, high energy and infrastructure requirements, and safety concerns are some of the main problems with hydrogen. LOHCs and cryogenic hydrogen storage are two well-known technologies that provide unique answers to these problems.

### 4.5.1 Volumetric Energy Density

According to Markiewicz et al. [18], liquid hydrogen storage can reach a volumetric hydrogen density of up to 71 g/L [18]. Methanol has a volumetric energy density that is much higher than that of liquid hydrogen, while many other LOHCs have densities that are similar or marginally higher [36].

### 4.5.2 Boil Off

The inevitable heat infiltration that causes boil-off gas (BOG) to gradually evaporate up to 0.4% per day in small tanks and 0.06% in larger systems is a significant drawback for cryogenic storage. Boil-off rates are influenced by tank design, insulation type, and materials such as composites or stainless steel. Techniques like integrated refrigeration, baffles to prevent sloshing, top-fill transfer, and onboard boil-off recovery (like that of Kawasaki's SUISO FRONTIER ship) lower losses and increase safety. A major explosion will only occur if the gas ignites in the event of a leak or tanker rupture [2, 13]. By chemically binding hydrogen in organic liquids, LOHCs, on the other hand, completely eliminate the risk of boil-off and the losses that come with it. This allows for storage and transportation at room temperature and pressure [20, 36].

### 4.5.3 Infrastructure Requirements

To maintain temperatures close to −253 °C, cryogenic hydrogen storage requires specialized infrastructure such as liquefaction plants, double-walled vacuum-insulated tanks, and specially designed cryogenic transport vessels [23]. Because of their stability in ambient conditions, LOHCs can take advantage of the current fuel infrastructure, which includes road tankers, pipelines, and standard tanks. They do, however, need specialized hydrogenation and dehydrogenation facilities with sophisticated reactors and heat control mechanisms [25].

### 4.5.4 Energy Requirements

The storage of liquid hydrogen ($LH_2$) requires the extremely energy-intensive process of cooling hydrogen to cryogenic temperatures. Excluding the extra energy required to control boil-off during storage and transportation, the liquefaction alone uses approximately 15.2 kWh/kg of hydrogen [25], or about 11% of its lower heating value (120 MJ/kg) [38]. Liquid organic hydrogen carrier (LOHC) systems, on the other hand, store hydrogen in chemical compounds at room temperature, avoiding the need for liquefaction. The hydrogenation and dehydrogenation processes in LOHC require thermal energy. According to Niemann et al. [25], the energy needed varies greatly based on the LOHC compound used, ranging from 16.5 kJ/mol $H_2$ (high efficiency) to 68.1 kJ/mol $H_2$ (low efficiency). Therefore, LOHCs transfer energy use to the chemical conversion stages, whereas cryogenic storage concentrates energy use up front.

### 4.5.5 Small-Scale Versus Large-Scale Transport

In general, medium- to large-scale transportation and storage, like international shipping and truck-based delivery systems, are better suited for cryogenic hydrogen storage. It is essential for applications requiring the storage or transportation of large amounts of hydrogen due to its high gravimetric hydrogen density. Because of their maturity and capacity, cryogenic systems are regarded as crucial in industries such as energy shipping and space exploration [30].

Cryogenic hydrogen presents difficulties for small-scale applications, though. Only in situations where hydrogen consumption is rapid and cost is not a limiting factor is the liquefaction process economically feasible due to its high energy intensity and ongoing boil-off losses [12, 30]. Furthermore, the complexity of cryogenic use in decentralized or low-demand settings is increased by the high insulation and safety infrastructure requirements.

Conversely, LOHC systems are showing promise as large-scale and small-scale hydrogen transport options. They are safer and more compatible with the current liquid fuel infrastructure because they can store hydrogen at room temperature and pressure [12, 36]. However, the cost and availability of carrier liquids, particularly in stationary applications, determine their economic viability [25]. Although pilot projects show promise, cost uncertainties and the requirement for additional data validation continue to be barriers to large-scale deployment [17]. LOHCs exhibit competitive potential for large-scale transportation. Aromatic compounds are already hydrogenated on an industrial scale, and research into effective dehydrogenation is progressing, particularly for on-board applications in automobiles that call for reactors with high power densities and quick response times [36]. The full-scale commercial viability of LOHCs is still being assessed, despite the fact that they are feasible at the pilot level [17]. Table 4.1 compares the ways in which LOHC and cryogenic hydrogen storage systems have helped to address the main issues with long-term use and transportation of hydrogen:

**Table 4.1:** Comparison of cryogenic hydrogen storage and LOHC systems.

| Challenge | Cryogenic hydrogen storage | LOHC systems |
|---|---|---|
| Volumetric energy density | High (~71 g/L); allows compact storage for large-scale applications [18]. | Comparable or higher in some carriers (e.g., methanol); effective for bulk hydrogen [36]. |
| Boil-off losses | Advanced insulation, top-fill transfer, baffles, and BOG recovery systems reduce losses (e.g., SUISO FRONTIER). | No boil-off; hydrogen is chemically bound, enabling zero-loss storage even over long durations [20]. |
| Infrastructure compatibility | Requires specialized cryogenic infrastructure; high capital and safety requirements [23]. | Uses existing fuel infrastructure; only hydrogenation/dehydrogenation units need special setup [25]. |
| Energy requirements | Liquefaction is energy-intensive (~15.2 kWh/kg $H_2$); additional energy for boil-off management [38]. | Avoids liquefaction; energy is needed for chemical conversion (16.5–68.1 kJ/mol $H_2$), depending on LOHC used [25]. |
| Small-scale transport suitability | Less suitable due to high boil-off losses and cost; viable only with fast consumption [30]. | Suitable due to ambient storage; safer and more cost-effective for decentralized systems [12]. |
| Large-scale transport suitability | Essential for sectors like shipping and aerospace; mature and high-capacity solution [30]. | Competitive potential; hydrogenation is industrially established; dehydrogenation tech is improving [36]. |
| Safety considerations | Boil-off risks and flammability are managed with safety systems; explosion only if ignited [2]. | Chemically stable at ambient conditions; inherently safer in transport and storage [36]. |
| Commercial readiness | Mature for large-scale use; significant progress in insulation, BOG control, and vessel design. | Proven in pilot projects; large-scale use still under cost and performance assessment [17]. |

For large-scale transportation, LOHCs show promise as a competitive option. Research into efficient dehydrogenation is advancing, especially for on-board applications in cars that require reactors with high power densities and fast response times. Aromatic compounds are already hydrogenated on an industrial scale [36]. Although LOHCs are feasible at the pilot level, their full-scale commercial viability is still being evaluated [17]. The ways that LOHC and cryogenic hydrogen storage systems have contributed to resolving the primary concerns with long-term hydrogen use and transportation are contrasted in the following table.

## 4.6 Other Methods

### 4.6.1 Compressed Hydrogen Storage

In order to increase the volumetric energy density of hydrogen gas, it is physically compressed into high-pressure tanks at 350 bar (35 MPa) or 700 bar (70 MPa). This approach is well-established and facilitates quick refueling [26]. But compressing hydrogen uses a lot of energy, which lowers efficiency overall [26]. For example, it takes about 20.48 MJ/kg of energy to compress hydrogen gas to 600 bar, but only about 1.25 MJ/kg to compress methane to the same pressure. This highlights a significant difference, with hydrogen demanding over 16 times more energy than methane [16].

At 350 and 700 bar, the volumetric hydrogen density is roughly 24 and 40 g/L, respectively. Type I (all-steel, heavy, low pressure), Type II (steel liner with composite wrap), Type III (aluminium liner with composite wrap, lighter but with low thermal conductivity), and Type IV (plastic liner with full composite wrap, lightest but most expensive) are the four primary types of tanks [26]. For broad adoption, constructing substantial refueling infrastructure continues to be a major obstacle [11].

### 4.6.2 Cryo-compressed Hydrogen Storage

Combining cryogenic and high-pressure storage techniques, cryo-compressed hydrogen storage overcomes the drawbacks of each approach when applied separately, including low energy density in compressed gas storage and boil-off losses in cryogenic storage [3, 26]. This method achieves a higher volumetric capacity of about 80 g/L compared to about 70 g/L in liquid hydrogen by storing hydrogen at cryogenic temperatures (20–70 K) and elevated pressures (up to 35 MPa) [37, 41]. Tanks are usually constructed from sophisticated composite materials like carbon fiber to withstand extremely low temperatures as well as high pressure [5].

Because hydrogen is less reactive at low temperatures, cryo-compressed storage provides better volumetric and gravimetric energy densities along with increased safety. High system complexity, the requirement for sophisticated insulation and cooling, and inadequate infrastructure are some of its drawbacks. Cost-wise, it is in the middle of compressed and cryogenic storage, with continuous research and development aimed at increasing effectiveness and cutting costs [26].

### 4.6.3 Material-Based Hydrogen Storage

In addition to LOHCs, a number of other material-based hydrogen storage techniques have been thoroughly investigated due to their potential to offer more secure, effective, and portable storage options. Chemical storage, sometimes referred to as material-based hydrogen storage, is the process of storing hydrogen on solid materials by chemisorption or physisorption [26]. By doing away with the requirement for high-pressure tanks, this technique allows for a safer and more regulated release of hydrogen at particular temperatures [26, 39]. Metal hydrides, complex hydrides, ammonia-based compounds, carbon materials, metal-organic frameworks, covalent organic frameworks, and porous bio-based materials are among the materials that have been investigated for this purpose. For instance, complex hydrides provide high storage capacities through multi-element chemical reactions, whereas metal hydrides store hydrogen through strong metal-hydrogen bonds [15, 39]. No single material satisfies all optimal requirements; instead, each material differs in terms of capacity, operating conditions, reversibility, cost, and environmental impact [26]. The goal of ongoing research is to optimize these materials for useful and effective hydrogen storage applications.

## 4.7 Novel Developments in Hydrogen Storage Using Cryogenic Hydrogen and LOHC

Particularly in the areas of cryogenic hydrogen storage and LOHCs, recent developments in hydrogen storage technologies have offered encouraging answers to some of the industry's most pressing problems.

### 4.7.1 Cryogenic Hydrogen Storage Improvements

Enhancing insulation and energy efficiency has been a key priority in an effort to lower heat transfer and hydrogen boil-off. Variable density multilayer insulation with vapor-cooled shield is one of the best insulation techniques [40]. In order to increase energy efficiency, NASA has also incorporated the internal integrated refrigeration system into storage tank designs. However, because of the stringent insulation requirements and extremely low temperatures involved, integrating such systems is still difficult [23]. Furthermore, thin-film technologies are becoming more significant for cryogenic tanks because they enhance surface qualities like reflectivity and corrosion resistance, which support insulation and tank longevity [30].

### 4.7.2 Advanced Materials for Tanks

The development of robust and lightweight materials, such as aerogels, cryogenic composites, and particularly titanium alloys, is still being researched. Because of their exceptional strength-to-weight ratio and ability to maintain high strength at low temperatures, titanium alloys hold great promise for making storage tanks lighter and more effective [29]. To get past financial and technological obstacles in liquid hydrogen storage, these materials must be optimized.

### 4.7.3 Cryogenic Storage in Practice

Hydrogen storage for automotive and aerospace applications now uses cryogenic cooling and super-insulated storage techniques, which were initially validated in space technology. Hydrogen engines have been used in military and supersonic aircraft, and liquid hydrogen powers rockets such as the Space Shuttle and Ariane. BMW has tested vehicles with upgraded storage vessels that lower daily evaporation losses to less than 1.5% and created automated liquid hydrogen filling stations [32].

### 4.7.4 Advances in LOHC Systems

LOHCs offer a denser, safer method of chemically storing hydrogen in liquid form at room temperature, making handling and transportation simpler. In order to increase the efficiency of hydrogen release, recent catalyst advancements concentrate on functioning at lower temperatures with greater stability. According to research, platinum-based catalysts need to be properly engineered to avoid adverse reactions and irreparable harm to the carrier molecules [27]. To better mimic real-world conditions and close the gap between lab and real-world systems, ionic liquid forms of LOHCs are also being investigated.

### 4.7.5 Environmental and Safety Considerations for LOHCs

Notwithstanding their benefits, certain LOHC compounds are concerning because of possible negative effects on the environment and human health. Certain LOHCs have the potential to bioaccumulate or be carcinogenic and toxic to reproduction (CMR), which are risks that are difficult to forecast using current models. To guarantee safer LOHC development, researchers thus place a strong emphasis on comprehensive ecotoxicological screening and CMR evaluations [18]. Selecting LOHCs that are less harmful to the environment will be essential to their function in a sustainable hydrogen economy, providing an eco-friendlier substitute for fossil fuels. Table 4.2 shows the recent advancements in hydrogen storage technologies.

**Table 4.2:** Recent advancements in hydrogen storage technologies [43].

| Area | Focus/advancements | Benefits | Challenges |
|---|---|---|---|
| Cryogenic hydrogen storage | Improved insulation (variable density multilayer insulation with vapor-cooled shield), Internal integrated refrigeration system, and thin-film technology | Reduced boil-off losses, enhanced tank durability, and efficiency | Complex integration due to low temperatures and insulation needs |
| Advanced tank materials | Use of aerogels, cryogenic composites, and titanium alloys | Lightweight tanks with a high strength-to-weight ratio | Economic and technical limitations in material application |
| Cryogenic storage applications | Used in rockets (Space Shuttle, Ariane), military aircraft, and BMW cars | Proven technology with reduced evaporation losses | Still faces evaporation and high infrastructure costs |
| LOHC systems | Catalyst optimization and ionic liquid LOHCs | Safe, ambient condition storage and easier transport | Side reactions, catalyst degradation |
| LOHC safety and environment | Toxicological testing and CMR risk assessments | Environmentally safer hydrogen storage alternatives | Potential for toxicity and bioaccumulation in some LOHCs |

## 4.8 Conclusion

In the global shift to clean energy, developing a hydrogen-based energy system is not only a scientific challenge but also a social necessity. LOHCs and cryogenic hydrogen storage stand out as two of the most promising technologies among the different approaches to hydrogen storage; each has unique benefits and drawbacks. Cryogenic hydrogen storage offers high volumetric energy density and is ideal for large-scale, centralized applications such as aerospace and international shipping. High energy demands, intense infrastructure, boil-off losses, and intricate safety requirements, especially in small-scale or decentralized settings, present challenges. On the other hand, LOHC systems are a strong choice for decentralized and mobile applications because they provide ambient condition storage that is safe and compatible with the current fuel infrastructure. They get rid of the flammability and boil-off hazards that are common in cryogenic systems. The energy-intensive nature of the hydrogenation and dehydrogenation processes, catalyst deterioration, possible environmental toxicity, and present scalability problems, however, limit LOHC technologies. Comparatively speaking, neither technology can meet every need of a hydrogen economy on its own. Rather, their complementary nature – LOHCs allow for safer, more flexible local transport and decentralized use, while cryogenic hydrogen supports high-capacity, long-distance storage – highlights the necessity of a diverse hydrogen infrastructure. Recent developments in materials science, system integration, catalyst design, and thermal insulation have greatly enhanced the feasibility and performance of both technologies. Nevertheless, further innovation and policy support are essential to address cost, efficiency, and safety hurdles.

In the future, a hybrid approach that is adapted to particular use cases and local infrastructure capabilities will probably be the best course of action for hydrogen storage. We can build a strong and resilient hydrogen value chain by utilizing both cryogenic and LOHC systems strategically and integrating them with cutting-edge storage techniques like cryo-compression and material-based technologies. In addition to expediting the use of hydrogen as a clean energy vector, this multi-pathway strategy guarantees a more sustainable, scalable, and equitable energy future.

# References

[1] Abe, J.O., Popoola, A.P.I., Ajenifuja, E. and Popoola, O.M., 2019. Hydrogen energy, economy and storage: Review and recommendation. International Journal of Hydrogen Energy, 44(29), pp.15072–15086.

[2] Al Ghafri, S.Z., Munro, S., Cardella, U., Funke, T., Notardonato, W., Trusler, J.M., Leachman, J., Span, R., Kamiya, S., Pearce, G. and Swanger, A., 2022. Hydrogen liquefaction: a review of the fundamental physics, engineering practice and future opportunities. Energy & Environmental Science, 15(7), pp.2690–2731.

[3] AlZohbi, G., Almoaikel, A. and AlShuhail, L., 2023. An overview on the technologies used to store hydrogen. Energy Reports, 9, pp.28–34.

[4] Aziz, M., 2021. Liquid hydrogen: A review on liquefaction, storage, transportation, and safety. Energies, 14(18), p.5917.

[5] Barthélémy, H., Weber, M. and Barbier, F., 2017. Hydrogen storage: Recent improvements and industrial perspectives. International Journal of Hydrogen Energy, 42(11), pp.7254–7262.

[6] Cai, Y., Zhou, D., Ye, B., Ding, M., Li, N., Cao, J., Yang, Y., Zhang, Y. and He, K., 2024. Safety considerations for cryogenic hydrogen circulation system in China spallation neutron source phase II project. International Journal of Hydrogen Energy, 84, pp.28–38.

[7] Calabrese, M., Portarapillo, M., Di Nardo, A., Venezia, V., Turco, M., Luciani, G. and Di Benedetto, A., 2024. Hydrogen safety challenges: a comprehensive review on production, storage, transport, utilization, and CFD-based consequence and risk assessment. Energies, 17(6), p.1350.

[8] Crabtree, R.H., 2008. Hydrogen storage in liquid organic heterocycles. Energy & Environmental Science, 1(1), pp.134–138.

[9] Di Profio, P., Arca, S., Rossi, F. and Filipponi, M., 2009. Comparison of hydrogen hydrates with existing hydrogen storage technologies: Energetic and economic evaluations. International Journal of Hydrogen Energy, 34(22), pp.9173–9180.

[10] Gemechu, D.N., Mohammed, A.M., Redi, M., Bessarabov, D., Mekonnen, Y.S. and Obodo, K.O., 2023. First principles-based approaches for catalytic activity on the dehydrogenation of liquid organic hydrogen carriers: a review. International Journal of Hydrogen Energy, 48(85), pp.33186–33206.

[11] Greene, D.L., Ogden, J.M. and Lin, Z., 2020. Challenges in the designing, planning and deployment of hydrogen refueling infrastructure for fuel cell electric vehicles. ETransportation, 6, p.100086.

[12] He, T., Pei, Q. and Chen, P., 2015. Liquid organic hydrogen carriers. Journal of Energy Chemistry, 24(5), pp.587–594.

[13] Hosseini, S.E., 2023. Fundamentals of Hydrogen Production and Utilization in Fuel Cell Systems. Elsevier: Amsterdam, The Netherlands.

[14] Huang, Z. and Autrey, T., 2012. Boron–nitrogen–hydrogen (BNH) compounds: Recent developments in hydrogen storage, applications in hydrogenation and catalysis, and new syntheses. Energy & Environmental Science, 5(11), pp.9257–9268.

[15] Lai, Q., Pratthana, C., Yang, Y., Rawal, A. and Aguey-Zinsou, K.F., 2021. Nanoconfinement of complex borohydrides for hydrogen storage. ACS Applied Nano Materials, 4(2), pp.973–978.

[16] Lavanya, M., Shrivastava, S., Lakshmi, T., Sandadi, E.R., Gour, S., Thomas, N.A., Priya, S.S. and Sudhakar, K., 2024. An overview of hydrogen storage technologies–Key challenges and opportunities. Materials Chemistry and Physics, 325, p.129710.

[17] Li, H., Zhang, X., Zhang, C., Ding, Z. and Jin, X., 2024. Application and analysis of liquid organic hydrogen carrier (LOHC) technology in practical projects. Energies, 17(8), p.1940.

[18] Markiewicz, M., Zhang, Y.Q., Bösmann, A., Brückner, N., Thöming, J., Wasserscheid, P. and Stolte, S., 2015. Environmental and health impact assessment of Liquid Organic Hydrogen Carrier (LOHC) systems–challenges and preliminary results. Energy & Environmental Science, 8(3), pp.1035–1045.

[19] Matveev, K.I. and Leachman, J.W., 2022. Thermoacoustic Flow-Through Cooler for Cryogenic Hydrogen. In Computer Aided Chemical Engineering, Vol. 49, pp.1891–1896. Elsevier: Amsterdam, The Netherlands.

[20] Mekonnin, A.S., Wacławiak, K., Humayun, M., Zhang, S. and Ullah, H., 2025. Hydrogen storage technology, and its challenges: a review. Catalysts, 15(3), p.260.

[21] Ministry of New and Renewable Energy (MNRE). (2023). National Green Hydrogen Mission: New Delhi, India.

[22] Müller, K., Völkl, J. and Arlt, W., 2013. Thermodynamic evaluation of potential organic hydrogen carriers. Energy Technology, 1(1), pp.20–24.

[23] Naquash, A., Agarwal, N. and Lee, M., 2024. A review on liquid hydrogen storage: current status, challenges and future directions. Sustainability, 16(18), p.8270.

[24] Navlani-García, M., Mori, K., Salinas-Torres, D., Kuwahara, Y. and Yamashita, H., 2019. New approaches toward the hydrogen production from formic acid dehydrogenation over Pd-based heterogeneous catalysts. Frontiers in Materials, 6, p.44.

[25] Niermann, M., Beckendorff, A., Kaltschmitt, M. and Bonhoff, K., 2019. Liquid Organic Hydrogen Carrier (LOHC)–Assessment based on chemical and economic properties. International Journal of Hydrogen Energy, 44(13), pp.6631–6654.

[26] Osman, A.I., Nasr, M., Eltaweil, A.S., Hosny, M., Farghali, M., Al-Fatesh, A.S., Rooney, D.W. and Abd El-Monaem, E.M., 2024. Advances in hydrogen storage materials: harnessing innovative technology, from machine learning to computational chemistry, for energy storage solutions. International Journal of Hydrogen Energy. 67, 1270–1294.

[27] Preuster, P., Papp, C. and Wasserscheid, P., 2017. Liquid organic hydrogen carriers (LOHCs): toward a hydrogen-free hydrogen economy. Accounts of Chemical Research, 50(1), pp.74–85.

[28] Pudukudy, M., Yaakob, Z., Mohammad, M., Narayanan, B. and Sopian, K., 2014. Renewable hydrogen economy in Asia–opportunities and challenges: an overview. Renewable and Sustainable Energy Reviews, 30, pp.743–757.

[29] Qiu, Y., Yang, H., Tong, L. and Wang, L., 2021. Research progress of cryogenic materials for storage and transportation of liquid hydrogen. Metals, 11(7), p.1101.

[30] Rampai, M.M., Mtshali, C.B., Seroka, N.S. and Khotseng, L., 2024. Hydrogen production, storage, and transportation: recent advances. RSC Advances, 14(10), pp.6699–6718.

[31] Rivard, E., Trudeau, M. and Zaghib, K., 2019. Hydrogen storage for mobility: a review. Materials, 12(12), p.1973.

[32] Schlapbach, L. and Züttel, A., 2001. Hydrogen-storage materials for mobile applications. Nature, 414(6861), pp.353–358.

[33] Sheffield, J.W., Martin, K.B. and Folkson, R., 2014. Electricity and hydrogen as energy vectors for transportation vehicles. In Alternative Fuels and Advanced Vehicle Technologies for Improved Environmental Performance (pp.117–137). Woodhead Publishing: Cambridge, UK.

[34] Simanullang, M., 2025. Liquid hydrogen storage and insulation materials for liquid hydrogen storage tanks: Trends and challenges. International Journal of Hydrogen Energy. 140, 881–888.

[35] Sotoodeh, K. and Gudmestad, O.T., 2023. Valve design considerations in liquid hydrogen systems to prevent failure. International Journal on Interactive Design and Manufacturing (IJIDeM), 17(3), pp.1429–1441.

[36] Teichmann, D., Arlt, W., Wasserscheid, P. and Freymann, R., 2011. A future energy supply based on Liquid Organic Hydrogen Carriers (LOHC). Energy & Environmental Science, 4(8), pp.2767–2773.

[37] Usman, M.R., 2022. Hydrogen storage methods: Review and current status. Renewable and Sustainable Energy Reviews, 167, p.112743.

[38] Valenti, G., 2016. Hydrogen liquefaction and liquid hydrogen storage. In Compendium of Hydrogen Energy (pp.27–51). Woodhead Publishing: Cambridge, UK.

[39] Von Colbe, J.B., Ares, J.R., Barale, J., Baricco, M., Buckley, C., Capurso, G., Gallandat, N., Grant, D.M., Guzik, M.N., Jacob, I. and Jensen, E.H., 2019. Application of hydrides in hydrogen storage and compression: Achievements, outlook and perspectives. International Journal of Hydrogen Energy, 44(15), pp.7780–7808.

[40] Wang, Z., Wang, Y., Afshan, S. and Hjalmarsson, J., 2021. A review of metallic tanks for $H_2$ storage with a view to application in future green shipping. International Journal of Hydrogen Energy, 46(9), pp.6151–6179.

[41] Yanxing, Z., Maoqiong, G., Yuan, Z., Xueqiang, D. and Jun, S., 2019. Thermodynamics analysis of hydrogen storage based on compressed gaseous hydrogen, liquid hydrogen and cryo-compressed hydrogen. International Journal of Hydrogen Energy, 44(31), pp.16833–16840.

[42] Zhou, Y., Li, R., Lv, Z., Liu, J., Zhou, H. and Xu, C., 2022. Green hydrogen: A promising way to the carbon-free society. Chinese Journal of Chemical Engineering, 43, pp.2–13.

[43] Preuster, P., Alekseev, A. and Wasserscheid, P., 2017. Hydrogen storage technologies for future energy systems. Annual review of chemical and biomolecular engineering, 8, pp.445–471.

Deoashish Panjiara, Neeraj Kumar Yadav, Abhay Pratap Singh, Abhay Kumar Choudhary

# 5 Emerging Catalyst Materials for Fuel Cells

**Abstract:** Fuel cells have emerged as a viable chemical energy to electrical conversion technology due to their minimal environmental impact and high energy efficiency, as well as their ability to use renewable energy sources. However, researchers still find challenges for the fully commercialized use of fuel cells due to the higher cost of catalysts, slow reaction kinetics, and durability. Catalysts improve fuel cell performance by boosting electrochemical oxidation and reduction reactions. Due to the high cost of platinum (Pt) noble catalysts, research is going on for innovative and low-cost catalyst materials, such as non-platinum-group metals. The emerging catalytic material consists of non-precious metal catalysts, such as transition metal nitrogen-carbon (M-N-C) catalysts, metal-organic framework (MOF)-derived materials, and single-atom catalysts, which are gaining popularity due to low cost and environmental benefits. N-doped catalysts (M-N-C) are commonly used in acids, replacing Pt with earth-abundant metals, such as Fe, Cu, Ni, Co, Mn, and Sn. Fe-N-C, which has iron (Fe) coupled to nitrogen (N) and carbon (C), has particular iron-nitrogen ($FeN_4$) sites, making it highly effective for the oxygen reduction reaction in acidic conditions. New techniques, such as alloying, core-shell designs, and defect engineering, have significantly improved catalytic performance and durability by lowering platinum content while retaining performance. Metal-free catalysts, such as graphene-based materials and carbon nanotubes, MOFs, perovskite oxides, and carbon structures doped with heteroatoms and hybrid nanomaterials, have boosted the variety of high-performance catalysts. This chapter provides details about the latest developments, challenges, and future directions in fuel cell catalyst research.

**Keywords:** Fuel cell, catalysts materials, non-precious metal, catalysts, metal-organic frameworks, nanomaterials, carbon nanotubes

## 5.1 Introduction

Fuel cell technology is considered one of the most viable next-generation energy conversion/storage devices because of its high efficiency and minimal emissions, facilitating the transition to a fossil fuel-free future [1, 2]. Fuel cells are electrochemical devices that transform the chemical energy of a fuel (typically hydrogen) and an oxidant (usually oxygen) directly into electricity, producing water and heat as byproducts. In fuel cells, electricity is produced from the energy contained in fuels in the form of chemical bonds via electrochemical reactions at the electrodes, which results in low emissions and minimal environmental harm [3, 4]. Fuel cells have multiple benefits, including high efficiency (up to 70–80% use of produced heat), high energy density,

 | https://doi.org/10.1515/9783112181072-005

reliability, quiet operation, modular and small design, fuel refilling for reuse, and use of nonconventional fuels including hydrogen, methanol, ethanol, and formic acid [5, 6]. Furthermore, this technology offers a cleaner and more efficient alternative to traditional heat engines that employ multi-step combustion processes, as well as energy security and environmental benefits [1]. They have now proven successful in a wide range of applications, including transportation systems (particularly fuel cell electric vehicles – FCEVs), household use, portable energy devices, military projects, the automobile market, and even wastewater treatment [1, 6].

## 5.1.1 Overview of Fuel Cell Technology

A fuel cell is made up of three main components: two electrodes (anode and cathode), and an electrolyte membrane. Figure 5.1 depicts a typical schematic representation of a $H_2$-fueled proton exchange membrane (PEM) fuel cell ($H_2$-fueled PEMFC). The anode and cathode of the fuel cell are made up of active catalyst materials, primarily platinum (Pt)-based, which catalyze electrochemical reactions. The active components, i.e., catalyst materials, are coated on an electrically conducting porous substrate or gas diffusion layer with a carbon microporous layer to manufacture anode and cathode electrodes using a suitable catalyst, depending upon the fuel cell or oxidant to be used in fuel cells. The PEMFC makes use of polymer electrolyte membranes, notably solid Nafion®, which is highly conducting (0.1 S $cm^{-1}$) and transports ions (protons, $H^+$) while preventing the direct mixing of reactants on the anode and cathode sides [7]. The ion-permeable membrane electrolyte medium is sandwiched between an anode and cathode, which is known as the membrane electrode assembly (MEA) (Figure 5.1). Pure hydrogen is delivered to the anode side as fuel (usually a reducing agent), while oxygen, an oxidizing agent, is supplied to the cathode side. When the acidic medium (Nafion®) is employed as an electrolyte, the electrooxidation of fuel at the anode produces hydrogen ions/protons ($H^+$) and electrons ($e^-$), as illustrated in Figure 5.1 by the following equation (5.1):

Anode:

$$H_2(g) \rightarrow 2H^+ + 2e^- \tag{5.1}$$

The electrons pass through the external circuit in the direction of the cathode and create an electrical current. The hydrogen ions ($H^+$) pass through the conductive membrane electrolyte from the anode side to the cathode side. The hydrogen ions ($H^+$) and electrons ($e^-$) combine at the cathode with oxidants such as oxygen supplied from an external source resulting in the formation of water, as shown in Figure 5.1 by the following equation (5.2):

Cathode:

$$\frac{1}{2}O_2(g) + 2H^+ + 2e^- \rightarrow H_2O \tag{5.2}$$

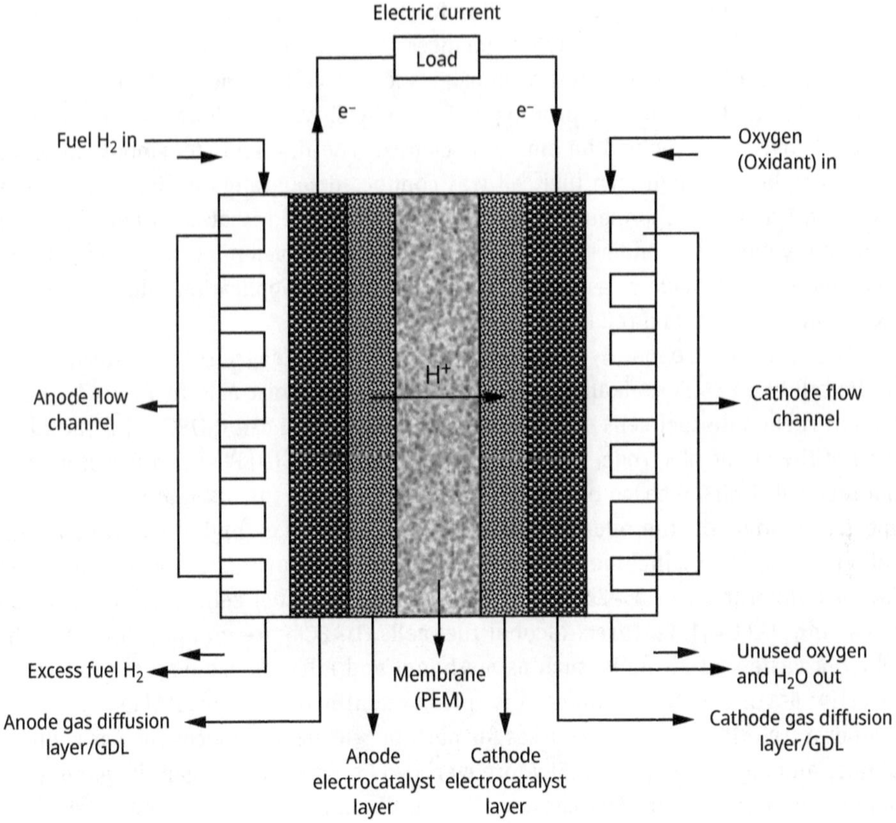


**Figure 5.1:** Typical scheme of a $H_2$-fueled PEM fuel cell ($H_2$-fueled PEMFC).

The following equation (5.3) summarizes the overall process in the $H_2$-fueled PEMFC that produces water, heat, and electrical work:

Overall:

$$H_2(g) + \frac{1}{2}O_2(g) \rightarrow H_2O + \text{electricity} + \text{heat} \tag{5.3}$$

Fuel cells operate by electrochemically splitting the fuel (mostly hydrogen or hydrogen-rich molecules such as methanol, ethanol, acetic acid, $NaBH_4$, etc.) at the electrode surface with high electrical power, resulting in lower energy losses than direct combustion of fuels in internal combustion (IC) engines. The architecture of the single-step conversion of chemical energy from supplied fuels to electrical energy in fuel cells, compared to the multi-step involved in conventional combustion heat engines, offers numerous unique advantages [2]. In short, fuel cells provide a cleaner, more efficient, and potentially more flexible conversion of chemical energy to electrical energy via electrochemical processes that are not limited by Carnot cycle effects. Thus, it has a high potential for replacing existing fossil fuel-based combustion engines operating on powering vehicles, potable appliances, and stationary power sources [1, 6, 7]. Although, fuel cells are the oldest technologies for converting energy, the exact origin is obscure. The first working prototype of a fuel cell was developed by Sir William Robert Grove, a lawyer and an amateur scientist. The first known demonstration of the principles of a hydrogen fuel cell was conducted separately by W.R. Grove and Christian Friedrich Schonbein in 1839 [2]. Since then, the researchers and scientific community have been endlessly working to design and develop a wide variety of fuel cells that could provide a feasible solution for practical applications along with low cost, high efficiency, and reliability.

The fuel cells are mainly classified according to (i) the electrolyte material used in the cell like PEMFCs, alkaline fuel cells (AFCs), phosphoric acid fuel cells (PAFCs), molten carbonate fuel cells (MCFCs) and solid oxide fuel cell (SOFCs), (ii) the fuels used at the anode electrode, i.e., direct ethanol fuel cells (DEFCs) and direct methanol fuel cell (DMFCs) based on the ethanol and methanol fuel used at the anode, respectively, and (iii) the operating temperature range, i.e., high-temperature fuel cells (600–1,000 °C) includes SOFCs and MCFCs, and low-temperature fuel cells (room temperature of 25–220 °C) includes remaining fuel cells such as PEMFCs, AFCs, and PAFCs [1, 2]. Direct alcohol fuel cells (DAFCs) are an appealing form in which a variety of alcohols, such as methanol and ethanol, are employed as fuels directly at the anode without any prior modification. The DAFCs are low-temperature (<100 °C) power sources for portable electronic devices such as cellular phones and laptop computers [5]. Moreover, DAFCs are currently gaining popularity among researchers and scientists due to the quick start up, high energy densities of alcohols (6.09 kW h $kg^{-1}$ for methanol and 8.01 kW h $kg^{-1}$ for ethanol) and the ease of storage and transportation [8, 9]. The DAFCs are further mainly classified based on the anode's inlet fuels, such as DMFCs that use methanol as anodic fuel and DEFCs that use ethanol as anodic fuel. On the contrary, high-temperature operation of fuel cells such as SOFCs and MCFCs increases electrical efficiency and reducing the demand for expensive catalysts [1]. These high-temperature fuel cells are especially attractive for usage in distributed generation, auxiliary power, electric utilities, and high-efficiency systems. However, the fabrication of the cell components and operation of these fuel cells are quite complicated due to high-temperature-induced thermal stresses and corrosion, mechanical and chemical compatibility issues with cells, and lengthy start-up and cooling periods [5]. In contrast, low-temperature fuel cells (PEMFCs, AFCs, and DAFCs) are relatively easy to construct and operate, and the selection of materials is not a factor [1, 2]. The fundamental working concept of all of these fuel cell types is very similar. Table 5.1 provides a detailed summary of the most prominent fuel cells and their operating aspects, as well as their various advantages and disadvantages.

**Table 5.1:** Comparison of most prominent fuel cells and their operational aspects [1, 2, 5].

| Fuel cell type | PEMFCs | AFCs | DAFCs | PAFCs | MCFCs | SOFCs |
|---|---|---|---|---|---|---|
| Typical electrolyte | Polymeric membrane or Nafion | KOH solution (30–50%) or alkaline polymer membrane | Polymeric membrane or Nafion | Phosphoric acid (100%) | Molten carbonate | Yttria-stabilizedzirconia (YSZ) or ceramics |
| Typical anode/ cathode catalyst | Anode: Pt/C<br>Cathode: Pt/C | Anode: Ni<br>Cathode: Ag/C | Anode: Pt-Ru/C<br>Cathode: Pt/C | Anode: Pt/C<br>Cathode: Pt/C | Anode: NiCr<br>Cathode: lithiated NiO | Anode: Ni-YSZ composite<br>Cathode: strontium-doped lanthanum manganite (LSM) |
| Fuel | $H_2$ | $H_2$ | Methanol-water solution, ethanol-water solution | $H_2$ | $H_2/CO/CH_4$ | $H_2/CO/CH_4$ |
| Charge carrier | $H^+$ | $OH^-$ | $H^+$ | $H^+$ | $CO_3^{2-}$ | $O^{2-}$ |
| Operating temperature (°C) | 30–100 | 50–200 | DMFC: Ambient-110<br>DEFC: Ambient-120 | 150–220 | 600–700 | 500–1,000 |
| Electrical efficiency (%) | 40–50 | 50–60 | DMFC: 35–60<br>DEFC: 20–40 | 40–50 | >60 | 50–60 |

(continued)

**Table 5.1** (continued)

| Fuel cell type | PEMFCs | AFCs | DAFCs | PAFCs | MCFCs | SOFCs |
|---|---|---|---|---|---|---|
| Advantages | Short start-up time, high power density, easy scale-up, vast power range, use of solid electrolyte lowers problems with electrolyte management and corrosion | Inexpensive catalyst and its flexibility, short start-up time, fast kinetics, and relatively low costs | Ease fuel storage and delivery, high fuel volumetric energy density, compact size, no $CO_2$ emission, and low start-up time | Mature technology, higher tolerance to fuel impurities, and suitable for combined heat and power (CHP) systems | Variety of fuel, high efficiency, good tolerance to impurities, inexpensive catalyst, supports internal reforming, usable with hybrid/gas turbine cycles | Variety of fuel, high efficiency, good tolerance to impurities, inexpensive catalyst, and supports internal reforming, usable with hybrid/gas turbine cycles |
| Disadvantages | Expensivecatalysts, sensitive to fuel impurities, and complex heat and water management | Highly pure $H_2$ and $O_2$ required, electrolyte management and corrosion issues, and intolerance to $CO_2$ | Expensive (using Ru and Pt), fuel crossover, low power density, low efficiency, and highly flammable and toxic fuel (methanol) | Long start-up time, expensive catalyst, low power density, low ionic conductivity, and relatively large and expensive auxiliary system | Long start-up time, low power density, handling liquid electrolyte issues, cathode dissolution, and hardware corrosion | Long start-up time, low power density, sealing and durability issues, high-temperature corrosion, and thermal stresses |
| Power range | 1 W to 500 kW | 10 W to 200 kW | 100 mW to 1 kW | 50 kW to 1 MW | <1 kW to 1 MW | 5 kW to 3 MW |
| Applications | Portable power, backup power, small distributed generation, and vehicles | Submarine, spacecraft, military, and backup power | Portable electronic gadgets (mobile and laptops), and vehicles | Distributed generation | Electric utility, auxiliary power, and large distributed generation | Electric utility, auxiliary power, and distributed generation |

## 5.1.2 Importance of Catalysts in Electrochemical Reactions

In fuel cells, the chemical energy contained in the chemical bonds of a fuel and an oxidant is converted into electrical energy through the redox reactions. Catalysts are essential components that drive electrochemical reactions at the anode and cathode electrodes [10]. The catalyst materials accelerate the electrochemical reactions occurring at electrodes, increasing efficiency and overall system performance. Electrochemical reactions in fuel cells have kinetic limitations, particularly the oxygen reduction reaction (ORR), which is a complex electrochemical process involving multiple steps [11]. Because of the ORR's slow reaction kinetics at the cathode electrode, it is considered a rate-controlling step for overall fuel cell performance [12]. A catalyst is a substance that lowers the activation energy needed for a chemical reaction thermodynamically, significantly accelerating chemical processes and is not consumed in the process. Catalysts ensure that processes proceed closer to their thermodynamic potentials [13]. Furthermore, catalysts enable fuel cells to operate efficiently at lower temperatures (60–80 °C for PEMFCs), which improves durability and integration into portable devices. Currently, platinum (Pt) and Pt alloys supported on carbon catalysts are the most often utilized materials, which are expensive and scarce. Non-precious metal catalysts (NPMCs) and platinum-group metal-free (PGM-free) alternatives are currently being researched.

## 5.1.3 Drivers for Developing Emerging Catalyst Materials

A variety of technological, economic, environmental, and scalability issues drive the development of novel fuel cell catalyst materials. These drives seek to overcome the limitations of traditional catalysts (particularly platinum-based catalysts) and encourage the commercialization of fuel cell technology. Key drivers are as follows:

**a. Cost Reduction**

Platinum (Pt) and other PGM nanoparticles deposited onto carbon materials are the most commonly used as traditional fuel cell catalysts, which are rare and expensive and account for a significant portion of the manufacturing costs of PEMFCs, i.e., 55% of the overall cost [14]. Furthermore, price fluctuations in PGM-based catalysts have an impact on the long-term economic viability of fuel cell technology. Thus, it is critical to develop PGM-free or low-PGM catalysts to lessen reliance on limited and expensive materials.

**b. Improved Catalyst Durability and Stability**

The performance of typical Pt-based supported catalysts during PEMFC operation is influenced by Pt dissolution, Pt migration, and Pt nanoparticle sintering (particle growth) via the Ostwald ripening process at the electrode surface [10]. Furthermore, the presence of CO, sulfur, or other contaminants in the fuel supply poisons the active sites of the Pt catalyst, which leads to anodic deactivation [14]. Core-shell architectures, durable nanostructured materials, and CO-tolerant catalysts (e.g., Pt-Ru alloys and ceria-supported metals) or non-Pt materials with alternative supports (e.g., titanium oxides and doped carbon) are urgently needed to display long life and robustness.

**c. Performance Enhancement**
Using alloy catalysts, nanoengineering, or metal-organic frameworks (MOFs), new and emerging catalysts can be designed with several desirable properties, including improved hydrogen oxidation reaction (HOR) rates under low PGM loading, increased electrochemically active surface area (ECSA), and enhanced ORR kinetics, all of which can increase intrinsic activity.

**d. Environmental Sustainability**
During the extraction and development of PGMs, catalysts require a lot of energy. Furthermore, catalyst deterioration in fuel cells can release nanoparticles into the environment. The most promising alternative is to develop eco-friendly synthesis pathways and sustainable catalyst materials that have low environmental impact [15].

**e. Fuel Cell Commercialization and Scalability**
The extensive use of fuel cells in vehicles, portable electronics, and fixed electricity requires materials that are both economical and scalable. It is necessary to create scalable catalyst production methods (such as spray pyrolysis and inkjet printing) that are suitable for mass production.

## 5.1.4 Challenges in Current Catalyst Technologies

Catalyst technologies are critical components of fuel cells, influencing fuel cell performance and accounting for a significant percentage of total costs. Despite tremendous progress in catalyst design and development, many challenges need to be addressed until viable fuel cell technologies can be widely commercialized. Several important challenges are listed below:

a. High cost of PGMs
b. Catalyst durability and stability
c. Slow kinetics of the ORR
d. Mass transport limitations
e. Poor performance of non-PGM catalysts
f. Operating conditions and compatibility
g. Scalability and manufacturing challenges

## 5.2 Limitations of Platinum-Based Catalysts

Platinum (Pt)-based catalysts are widely employed as catalytic materials for both anode and cathode electrodes due to their high catalytic activity, chemical stability, good corrosion resistance, and electrochemical performance in acidic environments [16]. Furthermore, the electrode reactions in the two most prevalent fuel cell types, DAFCs, and PAFCs, rely substantially on the usage of Pt or Pt-based catalysts, particularly for ORRs [5, 9].

Noble metal-based compounds, particularly Pt-based catalysts, are the most practicable catalysts though highly effective, have limitations in fuel cells, including high cost, slow kinetics of ORR, and potential poisoning by impurities like CO or methanol. Additionally, Pt catalysts can experience degradation over time, including oxidation, migration, loss of active surface area, and corrosion of the carbon support. The PGMs consist of six d-block elements: platinum (Pt), palladium (Pd), rhodium (Rh), ruthenium (Ru), iridium (Ir), and osmium (Os). The chemical constitution of these elements is similar, with 4d electron shells for Ru, Rh, and Pd and 5d electron shells for Os, Ir, and Pt [17]. The vast applications of PGMs are as catalysts, which include fuel cells, catalytic converters, and electrolyzers. However, the price of precious metals is notoriously volatile, notably in the case of Pt, and their limited availability presents significant barriers to large-scale adoption of clean energy technology [18]. South Africa, Russia, the United States, Zimbabwe, and Canada are the top five producers of PGM metals. South Africa is the producer of 92% of Pt. While two countries, Russia (40%) and South Africa (37%), market 77% of the entire Palladium. Furthermore, South Africa dominates the rhodium market by 80% [19]. The mining and refining techniques are highly energy-intensive and detrimental to the environment, which contributes to the high market price. Moreover, geopolitical factors and supply chain constraints further exacerbate price volatility and availability issues [17, 19].

## 5.3 Non-precious Metal Catalysts

The limitation of the platinum-based precious metal catalyst forced the researcher to focus on the development of NPMCs. These alternatives typically utilize abundant and low-cost transition metals such as iron, cobalt, manganese, or nickel, which are often incorporated into nitrogen-doped carbon structures, metal oxides, carbides, or MOFs [20]. The use of NPMCs can effectively reduce the cost, enhance long-term stability, and assist the scalability of fuel cell technology for larger commercial applications. The most promising NPMC is the carbon frameworks with transition metal/nitrogen sites, often known as M-N-C materials [21]. Catalytically active centers are commonly defined as M-$N_x$/C, where M is a fourth-period d-transition metal (typically Fe, Co, or Mn) that coordinates with $x$ nitrogen atoms and is anchored to a carbon matrix. M-N-C systems have sparked widespread interest because they combine reputable oxidation-reduction (ORR) activity with abundant and affordable precursors. The NPMC, metal-lated macrocyclic compounds like porphyrins and phthalocyanines, for ORR, is first discovered in 1964 [22]. However, these compounds were found to decompose rapidly in acidic conditions. It was discovered that pyrolyzing macrocycles at high temperatures (about 700–1,100 °C) in an inert environment significantly improves catalytic activity and acid stability. During this heat treatment, the ligand scaffold carbonizes, and nitrogen and metal atoms become entrenched in the developing graphitic network, resulting in more durable and long-lasting M-$N_x$/C active motifs [23, 24]. To lessen reliance on costly macrocyclic molecules, a variety of nitrogen- and carbon-rich precursors, including polymers, have been coupled with inorganic metal salts to create effective catalysts. These materials, known as M-N-C catalysts, are normally made up of three fundamental elements: metal, nitrogen, and carbon. The metal (M) is usually iron, cobalt, or manganese. These ORRs of catalysts performance are thought to be due to the metal center's electrical properties and interaction with the surrounding carbon structure [25]. The activity and stability are the two important characteristics of the catalyst that must be improved for the successful and long-term implementation in the fuel cell.

Among the transition metals, iron has the most promising oxygen reduction activity in acidic conditions. As a result, several Fe-N-C catalysts have been synthesized utilizing various precursors and support materials. These catalysts are typically manufactured using high-temperature treatment techniques and are being actively researched as prospective replacements for platinum-based fuel cell systems [26–29]. The stability issue of the M-N-C catalyst has been reported in various researches [30–33]. The various possible reported mechanisms were proposed, including leaching of non-precious metal species from the catalyst structure [30–32], oxidative attack by hydrogen peroxide or other radical species [33], and protonation of nitrogen functionality or active sites, followed by anion adsorption [34]. Experimental evidence confirms the involvement of each of these processes, and, likely, many degrading paths occur concurrently, resulting in a decrease in catalyst durability. Furthermore, dissolved $Fe^{2+}$ ions can interact with $H_2O_2$ through Fenton-type reactions, releasing highly reactive radicals that can damage the catalyst, ionomer, and polymer membrane in the fuel cell [20]. To address these difficulties, researchers have begun to investigate other transition metals, such as cobalt and manganese, as alternatives to iron in NPMCs, to improve long-term stability and lower the danger of Fenton-induced deterioration. The different methods were investigated to increase the activity of the M-N-C catalyst. The active sites that are responsible for the ORR in M-N-C catalysts are precise in nature and a topic of active investigation. This is due to the structural complexity and the presence of various potential contributors such as M-$N_x$ sites, metal carbides, nitrides, and $CN_x$ species [35, 36]. Among these, M-$N_4$ coordination structures are widely considered as the primary contributors to ORR activity. The various studies are done on these catalysts and these studies suggest that the performance of M-$N_4$ sites can be influenced by their location on the carbon matrix, surrounding coordination environment, and the presence of nearby heteroatoms.

These characteristics affect the cell performance and suggest that activity can be modified for better fuel cell performance [37, 38]. For example, $NH_4Cl$ was utilized during synthesis to introduce a high density of pores and nitrogen-doped edge sites, which serve as favorable locations for anchoring Fe-$N_4$ active centers [39]. It is also found through various investigations that microporous structures in carbon supports could serve as active sites and that the micropore surface area influences the ORR performance of M-N-C catalysts [40]. Interestingly, it is seen that, although Fe-N-C catalysts synthesized using extremely porous carbon black with increased BET surface area did not have higher ORR activity than those prepared with ordinary carbon supports. This study suggests that ammonia ($NH_3$) alone may not successfully introduce nitrogen into the carbon surface to produce Fe-$N_4$ sites. This may be to lack of disordered carbon areas needed for $NH_3$ reaction. To solve this issue, the researchers used planetary ball milling to embed a mixture of iron precursors and pore-filling agents into the micropores. Further pyrolysis in an ammonia atmosphere resulted in a carbon structure with an increased density of accessible Fe-$N_4$ catalytic sites and a more favorable porous architecture [40].

## 5.4 Advanced and Emerging Catalyst Materials

To decrease the dependency on the precious metal catalyst, the advance non-precious materials were investigated for the catalyst development. The materials like MOFs and single-atom catalysts (SACs) show improved performance in the ORR. It is seen that these catalysts have a unique structure and electronic properties compared with other non-noble catalyst and NPMCs. The MOFs are adaptable substrates for developing active ORR sites due to their large surface area, variable porosity, and well-defined metal coordination environments [41]. The catalyst design improved using SACs by fixing single metal atoms to supports, boosting atomic efficiency and enabling precise control of the catalytic environment [42]. These novel materials have the potential to significantly improve ORR catalytic activity, selectivity, and durability. These characteristics are critical for making fuel cells of low cost and high performance.

MOFs are crystalline materials synthesized by using metal centers and organic linkers that form well-organized and porous structures at a molecular level. The structure of these MOFs often includes negatively charged counter ions within their pores to ensure charge neutrality, as well as solvent molecules that may remain trapped after production [43]. The structural versatility of MOFs is one of its most distinguishing features. The advantage of this feature is that researchers may accurately control the composition and properties of MOFs by carefully selecting the type of metal ion and organic ligand. This makes them very versatile for developing materials that meet a variety of functional needs [43]. The synthesis of M-N-C catalyst uses MOFs as one its effective precursor. This is due to straightforward fabrication and favorable structural characteristics MOFs. These frameworks are typically composed of metal ions coordinated with nitrogen-rich organic ligands, resulting in highly ordered architectures with abundant coordination sites. It is to be noted that, during pyrolysis process, parts of this ordered structure can be preserved. This inspired the concept of utilizing MOFs to generate a high density of atomically dispersed M-$N_x$/C active sites. This method helps to minimize metal particle aggregation, which otherwise leads to the formation of catalytically inactive clusters, thereby enhancing the overall activity and efficiency of the resulting catalyst [44, 45].

In search of synthesizing efficient M-N-C catalysts for the oxygen reduction process (ORR), a new type of MOF, i.e., zeolitic imidazolate framework (ZIF), is one of the most extensively researched materials. These frameworks are made up of transition metal ions coordinated with imidazole-based ligands in a tetrahedral configuration. 2-methylimidazole (MeIm) is often used ligand in this application. ZIF when combined with $Zn^{2+}$ ions and $Co^{2+}$, it easily produces, ZIF-8 and ZIF-67, respectively. These ZIF structures have exceptional thermal stability and nitrogen-rich environments. This makes them ideal precursors for producing high-performance catalysts [46]. Ding et al. [47] proposed a novel two-step approach for increasing the density of catalytic active sites. In this approach, manganese was first integrated into a ZIF-8 framework to form a precursor material. After pyrolysis and acid washing, an Mn-doped porous carbon structure was formed. In the following phase, an additional manganese source and cyanamide were injected into the porous carbon matrix. After heat treatment, the synthesis was completed and then used in the fuel cell. It was observed that the catalyst performed well in fuel cells, with a peak power density of 0.46 W $cm^{-2}$. The aggregation of Mn in this synthesis method was efficiently reduced, which are prevalent due to wide oxidation range (0 to + 7) of manganese [47]. Fe-induced Fenton reactions produce damaging radicals and hence; to reduce this, researchers are actively investigating Co- and Mn-based replacements. It has been discovered that the inclusion of zinc in the ZIF-8 framework aids in spatially isolating metal atoms such as cobalt, hence preventing cobalt particle agglomeration during high-temperature treatment. In addition, the volatilization of Zn during pyrolysis increases the surface area of the final catalyst. As a result, pyrolyzing Zn/Co bimetallic ZIF precursors produces porous, self-supporting M-N-C catalysts with increased porosity and active sites. Furthermore, changing the Zn to Co ratio allows you to control the composition of the catalyst, structure, and overall oxygen reduction performance.

The current result for a single-cell PEMFC using MOF-derived transition-metal-nitrogen-carbon (M-N-C) catalysts stands at 1.21 W $cm^{-2}$ under $H_2/O_2$ and 0.72 W $cm^{-2}$ under $H_2$/air. The results were obtained by carefully inserting a small quantity of zirconium into a porous Fe/Zr-N-C matrix. The Fe-$N_2$-Zr-$N_2$ coordination ensemble shifts the d-band center southward, decreasing the activation energy needed for $^*O_2$ hydrogenation [21]. Because this structure has a higher formation energy, it is more thermodynamically stable and delays Fe dissolution [48]. Separately, moving away from

MOF-derived systems, a core-shell Fe-N-C catalyst tailored to contain mostly low-spin Fe(II)-N-4 sites and virtually no Fe-O species has given a consistent power density of 1.26 W $cm^{-2}$ across a 500 h durability test [49]. Although M-N-C catalysts have oxygen reduction activity comparable to Pt/C, their long-term viability remains difficult. Demetallation [50], carbon support degradation [51], and pore blockage due to water accumulation are the most common causes of performance deterioration [50]. This degradation is largely attributed to the loss of active iron sites, which is frequently caused by structural changes and oxidation. While carbon corrosion is usually restricted in regular fuel cell operation, it becomes more noticeable during frequent start-stop cycles, particularly at high voltages. Durability tests of Fe-N-C materials showed a considerable $E_{1/2}$ potential drop after 5,000 accelerated cycles, indicating their fragility to protracted usage [51]. The addition of carbon nanotubes and graphene to M-N-C systems, either as supports or conductive bridges, can improve catalyst performance by improving particle dispersion, increasing porosity, and allowing for better exposure and transport to active sites. These conductive carbon frameworks also help to improve electron transport and structural stability [21].

Another emerging material for the ORRs is SACs have gained significant attention. This is due to their resistance to Fenton reactions and atomic sizes similar to transition metals. These characteristics make SACs (M–N–C) highly suitable for delivering improved catalytic efficiency and long-term stability in electrochemical environments [42]. The development of SACs has substantially increased electrocatalysis, particularly in the promotion of the oxygen reduction process (ORR) in fuel cells. SACs increase catalytic activity and selectivity while maximizing atomic usage by binding individual metal atoms to support materials. This method provides an efficient alternative to conventional bulk metal catalysts. The introduction of nitrogen-doped carbon-based M-N-C catalysts in this field achieves a significant attention, in which isolated metal atoms are uniformly distributed on customized supports. These materials showed good result for energy conversion devices, such as fuel cells and metal-air batteries, by efficiently driving the ORR [52]. Early research on transition-metal catalysts found that dispersing single metal atoms could significantly increase catalytic efficiency. When carbon black containing metal-phthalocyanines was initially used as an electrode, the macrocyclic ligands deteriorated in acid and the embedded metals leached out, indicating low durability [53]. Alt et al. [53] used molecular orbital principles to explain how $O_2$ is activated and why catalytic activity follows the Co > Fe > Ni sequence. Their investigation also indicated that both ligand substituents and the characteristics of the carbon support regulate the robustness and reactivity of the resultant chelate complexes. Various types of research have been performed to develop different SACs which include non-noble metals like Cr, Mo, Cu, Mn, Ni, etc., due to the numerous advantages of SACs over conventional catalysts [54, 55]. The advantages include high activity, selectivity, and well-defined structure [42].

## 5.5 Catalyst Design Techniques for Improvement of Performance

The advancement of fuel cells including PEMFCs and anion exchange membrane fuel cells (AEMFCs) depends on the development of high-performance catalysts. The recent design methods, such as alloying, core-shell nanostructure and defect engineering, enhance the catalyst durability, activity, and cost economy [56, 57]. These changes improve the reaction kinetics due to bifunctional mechanism; it also decreases the dependency on precious metal catalysts [58]. The core-shell nanostructures offer accurate surface control and increase the active-site exposure for the electrochemical reactions by ensuring structural stability of the catalyst. The process of defect engineering helps to creates active sites like vacancies and dislocations; it increases charge transfer and stability particularly when it is combined with hybrid nanomaterials like metal oxides and carbon supports [56]. The application of multidimensional designs like nanowire, nanorods, nanoframes, and nanocages helps to enhance the ECSA, performance and mass specific activity [59]. These collective techniques are very important for the synthesis of next generation of fuel cells system.

SACs featuring porous structural designs generate considerably easier usage atoms, consequently less precious metals are required. The SACs have very strong electrical and adsorption properties, increasing their activity and selectivity as a catalyst [60]. Moreover, other technique for catalyst design that can be used is alloying and composition tuning. Alloying of platinum with transition metals like nickel, cobalt, or iron can significantly modify the way its active-site function by fine-tuning their electronic structure. The alloying technique also improves the ORR kinetics and tolerance to CO poisoning [61]. By selecting the effective active intermediates and reaction pathways is also a critical aspect for the preparation of catalyst. Using density functional theory (DFT) calculations, electrochemical studies, high-resolution techniques, and in situ spectroscopy, the appropriate catalyst can be identified. The CO poisoning at the anode side is the main problem associated with low temperature fuel cells. By the application of bifunctional mechanism including electronic interactions that facilitate CO oxidation, has enhanced anode performance and reduced the poisoning [62]. The SACs continue to gain a lot of attention due to their exceptional catalytic activity, which is derived from isolated active sites and distinctive electronic structures. Methods for enhancing metal loading that can boost the density of single-atom sites without causing agglomeration are critical for reducing the consumption of noble metals while retaining excellent performance [60]. To achieve market-viable fuel cell systems, non-noble metals must be used as competitively cost replacements. Fe-N-C and Co-N-C have shown encouraging activity and stability, providing a practical route to lessen reliance on costly noble metals [63].

The hybrid catalysts work better, for example, the combination of MXenes as appropriate support (like $Ti_3C_2T_x$) with precious metals such as platinum (Pt) or rhodium

(Rh) creates desirable anode catalyst (e.g., Pt clusters/$Ti_3C_2T_x$). The strong connection between Pt particles and MXene sheets provides increased surface area and lets electrons travel quickly, making the catalyst more effective for methanol oxidation relatively at low overpotential and improving the resistance to CO poisoning [64]. Xiao et al. [65] synthesized Pd nanoworm (NWs) grown on poly (diallyldimethylammonium chloride) (PDDA) modified $Ti_3C_2T_X$ (positively charged MXene) support material via direct electrostatic attraction. The synthesized anode catalyst Pd NWs/PDDA-MX shown improved catalytic activity with a high ECSA (105.3 $m^2\ g^{-1}$) for methanol oxidation in alkaline medium when compared to typical Pd NP catalysts supported on carbon. The making of 3D hybrid structures, such as Pt on reduced graphene oxide (rGO) combined with $Ti_3C_2T_X$, gives more surface area and allows materials to move more easily during reactions, which also boosts the catalyst's performance [66].

The design criterion of great importance still is optimizing noble metal content. Rh nanosheet/ MXene catalysts with carefully regulated, for example, Rh content has shown high mass activity and durability over commercial Pt/C catalysts, so highlighting the balance between cost and performance [67]. Strong metal-support electronic interactions help reduce overpotentials and improve reaction kinetics; hence, charge transfer mechanisms are also important in MXene-supported systems [68].

Another necessary approach is addressing catalyst poisoning. $Ti_3C_2T_X$ nanosheets help to oxidize CO intermediates, so improving tolerance to poisoning and extending catalyst lifetime [69]. One of the promising areas in catalyst development is the use of MOFs. This technique includes the direct addition of metal atoms to MOFs to enhance their dispersion and interaction with the environment, thereby boosting performance [70]. A stable carbon-based material with active sites improving its efficiency and durability can be obtained by heating. These methods are very useful because they have the ability to simply modify their structure by metal loading and site placement. This retains things from clumping together and makes it likely to make SACs that are very active and selective [70]. It also helps to stabilize the SACs through the high-temperature processes and keeps the catalyst active for very long duration [71]. The structure performance relationship of the synthesized catalyst can be examined using characterization techniques like X-ray photoelectron spectroscopy (XPS) and X-ray absorption spectroscopy (XAS).

### 5.5.1 Metals and Bimetallic Catalysts

The advancement of high-temperature PEMFCs (HT-PEMFCs) is significantly dependent on the development of catalysts that are more active, durable, and poison-resistant. An appropriate catalyst can be synthesized by alloying Pt with non-noble metals like Ni, Co, Fe, Cu, Bi, or Zn. The alloying technique helps to form core-shell and Pt-skin nanostructures, hence increasing stability and Pt usage. The advanced structures such as nanoframes and nano dumbbells create greater active sites, whereas the bimetallic catalysts such as PtNi and PtCo provide more improved ORR activity and greater resistance to CO poisoning in comparison with pure platinum [72]. The use of MOF-derived porous carbon to bimetallic catalysts can further boost performance by attaching Pt atoms and enhancing electron transport [73]. Because of their high conductivity and surface area, MXenes (e.g., $Ti_3C_2T_x$) provide suitable support. Rh-active metal supported on MXene, for instance, shows better long-term stability than commercial Pt [67]. Non-noble bimetallic systems, such as FeNi-$N_6$ on nitrogen-doped carbon, have been developed to cut costs while maintaining performance, with improved activity linked to well-coordinated metal-nitrogen structures [74]. However, challenges like catalyst stability and preventing metal clumping remain. Ongoing research focuses on maximizing active site usage and improving structural coordination for better energy applications.

### 5.5.2 Core-Shell Nanostructures

The core-shell nanostructure is an emerging material for advancing catalyst technologies in HT-PEMFCs. In general, those that contain platinum (Pt) with transition metals, these designs have major benefits in catalytic activity, structural stability, and resistance to poisoning. PtFe@PtRuFe with intermetallic cores and Pt shells show improved CO tolerance and higher mass activity than conventional Pt catalysts [75]. The core-shell catalysts provide precise control over surface composition and electronic properties; it leads to an improved catalytic performance by tuning the d-band center and enhancing ORR kinetics [76]. Their nanostructure also increases active-site density and mechanical stability at high temperatures. A major advantage is reduced noble metal use typically a Pt-based shell over a cheaper metal core achieving high efficiency at lower cost [77]. Common synthesis methods include chemical reduction, electrochemical deposition, hydrothermal methods, and galvanic replacement, with thermal annealing and controlled reduction aiding shell formation and stability.

For both anodic and cathodic reactions in fuel cells, these design techniques have produced very active and robust noble metal-based catalysts. For ORR, M@Pt core-shell structures, for example, offer up to a tenfold increase in activity and up to 90 times greater performance than commercial Pt/C due to optimal facet exposure, so greatly outperforming conventional Pt catalysts [76]. Nonetheless, difficulties such as lower long-term stability in low Pt-loading alloys necessitate additional research before commercialization. Another key approach to maximize oxygen diffusion and active-site accessibility is hierarchical pore structure modulation. Connected porosity in these core-shell systems is introduced using techniques including hard templating and gas-phase etching. For instance, high specific surface areas (SSAs) shown by MOF-5-derived structures correspond with enhanced electrocatalytic performance [78].

One-pot wet-chemical techniques allow RhMo alloy nanosheets to be produced with a clear example of core-shell synthesis. Adsorbed CO molecules inhibit lateral growth in this process, so promoting nanosheet formation with improved hydrogen binding energies at the hexagonal close-packed and face-centered cubic interfaces. Accelerating the adsorption-desorption cycles of intermediate *H species [79], these structural characteristics help more effective HOR and hydrogen evolution reaction (HER). Notwithstanding their performance, problems including morphological control and environmental issues from chemical byproducts call for the development of greener synthesis methods. In addition to noble metals (Pt or Pd), the P-block metal-based catalysts have been investigated for ORR in fuel cells. Core-shell structures composed of antimony (Sb), bismuth (Bi), and aluminum (Al) with nitrogen-doped carbon shells are emerging as cost-effective alternatives to PGMs.

The pyrolysis-processed SACs have a stable and evenly distributed active site. The advanced imaging methods like transmission electron microscopy (TEM) and high-angle annular dark-field scanning transmission electron microscopy (HAADF-STEM). HAADF-STEM delivers clear visions into the atomic arrangement and structural stability of these catalysts.

Furthermore, metal sulfide core-shell structures, such as those with Ni-Co sulfide, have confirmed improved performance in electrochemical applications. The orderly nanoarrays with controlled shapes suggest excellent conductivity and a large number of redox-active sites. These materials with long-term stability and high energy efficiency appealing as bifunctional catalysts for metal-air batteries and water-splitting reactions [80]. According to Bai et al. [81], hybrid materials such as $Co_9S_8$@N-C have more catalytic activity than conventional Pt/C in ORR and oxygen evolution reaction (OER) in fuel cells and metal-air batteries because of their synergistic structure-property correlations.

Lastly, the core-shell nanostructures indicate to a highly adaptable and effective technique for developing catalysts for next-generation fuel cells. The simple modification in composition, structure, and morphology in the synthesized catalysts can offer remarkable activity, durability, and economy. The future research should mainly focus on scalable green synthesis, deeper mechanistic insights, and integration into full-cell systems to unlock their real-world energy potential.

### 5.5.3 Hybrids Nanomaterials and Defect Engineering

The defect engineering and hybrid nanomaterials enhance the performance, durability, and cost-effectiveness of fuel cell catalysts. These techniques are particularly applicable to high-temperature fuel cell systems like SOFCs and low-temperature systems like DMFCs and PEMFCs. Furthermore, during the defect engineering process, we produce tiny flaws or imperfections in the catalyst material, such as missing atoms, fissures, or boundaries, which serves to optimize the electronic behavior of the catalyst while in use. In addition, active sites boost catalytic efficiency, whereas defects speed charge transfer. The exact etching procedure employed for the manufacture of such flaws, boosting the adsorption and catalytic activity, aiding in the fabrication of ultrathin materials with high coordination active sites [82]. Recently, there has been a lot of interest in MXenes 2D transition metal carbides and nitrides in fuel cell applications. Due to their high conductivity and modifiable surfaces, defective MXene-based catalysts use selective etching to add vacancies and functional groups, which significantly improve electrocatalytic activity. The generated defects enhance activity and stability by serving as active sites for catalytic processes [83]. The recently synthesized 2D $Ti_3C_2T_x$ nanosheets containing hydroxyl groups (–OH) can improve the CO tolerance of catalysts used in DMFCs by facilitating in the breakdown of carbon monoxide (CO). Furthermore, the methanol oxidation processes are enhanced by the hybrid produced by Rh nanoparticles anchored on defect-rich MXenes, which exhibits considerably greater mass current density [84].

Hybrid nanomaterials, which combine various nanostructures or elements, have higher catalytic activity, corrosion resistance, and electrical conductivity. The use of these materials, which are created at the atomic or nanoscale, enables the combination effects that are not achievable with single-phase counterparts. When metal carbides or oxides are added to carbon supports in acidic fuel cell applications, they improve stability and resistance to corrosion, which makes them an appropriate catalyst for PEMFCs [85]. Moreover, the micro-mesoporous carbon structure enhances the ionic and electronic conductivity by facilitating proton and oxygen transfer.

The noble metals like Pt or Rh can be combined with supports like rGO or MWCNTs to enhance dispersion, avoid agglomeration, and boost active-site availability. A high-surface-area, conductive hybrid support with enhanced mass transport and expanded triple-phase boundaries is produced by the combination of rGO and MWCNTs. One of the examples is Pt/rGO/MWCNT hybrid, obtained a power density of 675 m W $cm^{-2}$, about double that of Pt/rGO alone [86]. The catalyst-support interactions are further regulated by the surface functionalization. The functional groups on rGO and MXenes improve metal anchoring and nanoparticle stability, enhancing durability and electrochemical performance [83]. Furthermore, complex morphologies with high aspect ratios, more surface-active sites, and effective charge and mass transport can be produced by combining 1D nanowires with 2D nanosheets [87].

Despite these advantages, there are still practical challenges, such as achieving consistent, scalable synthesis, ensuring long-term electrochemical stability, and revealing defect function relationships. Future research should focus on (i) developing cost-effective and environmentally benign synthesis techniques, (ii) employing in situ and operando characterization to probe structure-activity relationships, (iii) tailoring defect density and type for target reactions such as ORR and CO oxidation, and (iv) integrating computational modeling with experimental approaches to facilitate rational catalyst design. The continued advancement of hybrid and defect-engineered nanomaterials is expected to play a crucial role in the development of next-generation, efficient, and durable fuel cell technologies.

## 5.6 Future Prospects and Difficulties

The advancement of catalyst materials has a significant impact on the future of fuel cell technology. Especially for systems including AEMFCs, PEMFCs, DEFCs, and DMFCs, the development of highly active, stable, and reasonably priced catalysts is fundamental to overcoming current limitations. While the design of creative catalysts has advanced significantly, some difficulties remain that limit widespread commercialization and practical application. A major research focus is on reducing the dependence on PGMs through the use of NPMCs. Nitrogen-doped carbon frameworks, as well as cobalt and transition metal integration, have contributed to improve catalyst durability and activity [88]. Catalytic performance can be improved by employing creative design strategies such as defect engineering, interface tuning, strain modulation, and facet control. SACs have gained popularity as a solution to this problem. Unique active sites and high metal usage efficiency provided by SACs could help to increase general catalytic behavior [82] and CO tolerance. Still, there are difficulties adjusting their structure, stabilizing single atoms, and deciphering the fundamental reaction routes. Non-noble metal-based catalysts, such as Fe-N-C or Co-N-C, are being investigated as potential replacements for expensive PGMs. Although these materials exhibit potential, their catalytic activities lag behind those of their PGM counterparts. Hybrid nanostructures incorporating metal oxides, carbon supports, and conductive materials are being developed to improve catalytic activity as well as electronic conductivity.

The scalability of highly performing catalysts is one of the main obstacles to commercialization. Although many laboratory-scale synthesis techniques produce interesting materials, industrial-scale reproduction of these features remains difficult. Market viability depends on ensuring that affordable, repeatable, large-scale catalysts without sacrificing structural or electrochemical integrity are produced. Furthermore, designing modern catalysts necessitates a fundamental understanding of reaction mechanisms as well as the identification of actual active sites. Many times, empirical, existing methods yield ineffective material optimization. Essential tools for clearly understanding catalyst structure-activity relationships at the atomic level are synchrotron radiation-based XAS, in situ XPS, and electron microscopy. Even with high initial activity, many novel catalysts degrade when used in fuel cells. Long-term running can cause problems, including structural breakdown, mass and heat transfer restrictions, and leaching of active metals. Reliable and consistent performance depends on increasing durability and turning frequency. Dessie et al. [89] noticed that using a composite $MnO_2/Fe_3O_4$ (1:1) at 8% activated carbon as anode catalyst in a DEFC resulted in a maximum power density of 44.41 m W $cm^{-2}$ for electro-oxidation of ethanol in a 1 M KOH electrolyte. Additionally, advancements in catalyst supports are vital. Biomass-derived carbon supports offer a sustainable way ahead with hierarchical porosity that facilitates improved charge-transfer kinetics and faster ion transfer when paired with transition metal oxides such as $MnO_2$ and $Fe_3O_4$. Similarly, the creation of highly porous carbon substrates for Pt nanoparticle dispersion can enhance mass transfer and catalytic usage.

Additionally, the performance of fuel cells is constrained by the materials utilized in other components. For example, despite their widespread use, Nafion® membranes are expensive and have limited use at high temperatures. Creating substitute PEMs with enhanced mechanical and thermal stability is a major area of study need [90]. Long-term operation of fuel cells is dependent on similarly corrosion-resistant bipolar plates. MXenes have recently gained popularity as DMFCs components due to their high methanol tolerance and catalytic activity. However, their usefulness is restricted by issues such as poor dispersion, sheet restacking, and decreased stability in acidic environments. Surface modification, the use of interlayer spacers, and hybridization with conductive materials are being investigated as ways to overcome these limitations. Finally, the use of sustainable materials and ecologically friendly synthesis methods becomes increasingly important as the fuel cell industry pursues greener solutions. Not only do catalysts derived from renewable sources and abundant elements have a lesser environmental impact, but they also align with the global effort to achieve carbon neutrality. To ensure their overall sustainability, future studies on fuel cell materials must combine eco-design concepts with life-cycle assessment.

## 5.7 Conclusions

The development of fuel cells on a commercial scale is going through a crucial phase. For this purpose, the development of an advanced and reliable electrocatalyst for the fuel cell is extremely crucial. The limitations of platinum-based electrocatalysts, particularly their high cost, limited availability, and susceptibility to poisoning are the major hindrances. These limitations have prompted a search for non-noble metal alternatives, with a particular emphasis on transition M-N-C catalysts. Fe-N-C catalysts with active $FeN_4$ sites show great promise in oxygen reduction processes (ORR), providing a cost-effective and scalable path to Pt-free systems. The MOFs and SACs are emerging catalysts that represent the forefront of precision catalyst design. These emerging materials enable atomic-level control of overactive sites, which improves both performance and understanding of reaction mechanisms. To improve catalytic efficiency even further, numerous design strategies were investigated, including alloying and bimetallic catalysts, core-shell nanostructures, defect engineering, and hybrid nanomaterials. These methods offer effective tools for adjusting the electronic structure, increasing durability, and boosting mass and charge transport inside the catalyst layer. Overall, the development of non-precious and structurally enhanced electrocatalysts represents a significant step toward economically viable and high-performance fuel cell devices. Continued interdisciplinary research combining materials science, electrochemistry, and engineering is required to address the remaining issues of long-term stability, synthesis scalability, and integration into complete fuel cell systems. With these advancements, next-generation catalysts have the potential to speed up the general adoption of clean and sustainable energy systems.

## References

[1] Sebbani I, Ettouhami MK, Boulakhbar M. Fuel cells: A technical, environmental, and economic outlook. Cleaner Energy Systems. 2025 Jan 1:100168.

[2] Sharaf OZ, Orhan MF. An overview of fuel cell technology: Fundamentals and applications. Renewable and sustainable energy reviews. 2014 Apr 1;32:810–53.

[3] Othman R, Dicks AL, Zhu Z. Non precious metal catalysts for the PEM fuel cell cathode. International Journal of Hydrogen Energy. 2012 Jan 1;37(1):357–72.

[4] Choudhary AK, Pramanik H. Synthesis of low-cost $HNO_3$-functionalized acetylene black carbon supported Pt-Ru/C AB nano electrocatalysts for the application in direct ethanol fuel cell (DEFC). Korean Journal of Chemical Engineering. 2019 Oct;36(10):1688–707.

[5] Abdelkareem MA, Elsaid K, Wilberforce T, Kamil M, Sayed ET, Olabi A. Environmental aspects of fuel cells: A review. Science of The Total Environment. 2021 Jan 15;752:141803.

[6] Aminudin MA, Kamarudin SK, Lim BH, Majilan EH, Masdar MS, Shaari N. An overview: Current progress on hydrogen fuel cell vehicles. International Journal of Hydrogen Energy. 2023 Feb 5;48 (11):4371–88.

[7] Wang Y, Seo B, Wang B, Zamel N, Jiao K, Adroher XC. Fundamentals, materials, and machine learning of polymer electrolyte membrane fuel cell technology. Energy and AI. 2020 Aug 1;1:100014.

[8] Choudhary AK, Pramanik H. Addition of rhenium (Re) to Pt-Ru/f-MWCNT anode electrocatalysts for enhancement of ethanol electrooxidation in half cell and single direct ethanol fuel cell. International Journal of Hydrogen Energy. 2020 May 5;45(24):13300–21.

[9] Choudhary AK, Pramanik H. Enhancement of ethanol electrooxidation in half cell and single direct ethanol fuel cell (DEFC) using post-treated polyol synthesized Pt-Ru nano electrocatalysts supported on $HNO_3$-functionalized acetylene black carbon. International Journal of Hydrogen Energy. 2020 Jan 1;45(1):574–94.

[10] Sinniah JD, Wong WY, Loh KS, Yunus RM, Timmiati SN. Perspectives on carbon-alternative materials as Pt catalyst supports for a durable oxygen reduction reaction in proton exchange membrane fuel cells. Journal of Power Sources. 2022 Jun 30;534:231422.

[11] Bouleau L, Pérez-Rodríguez S, Quilez-Bermejo J, Izquierdo MT, Xu F, Fierro V, Celzard A. Best practices for ORR performance evaluation of metal-free porous carbon electrocatalysts. Carbon. 2022 Apr 15;189:349–61.

[12] An F, Bao XQ, Deng XY, Ma ZZ, Wang XG. Carbon-based metal-free oxygen reduction reaction electrocatalysts: past, present and future. New Carbon Materials. 2022 Apr 1;37(2):338–54.

[13] Banoth P, Kandula C, Kollu P. Introduction to electrocatalysts. In Noble Metal-Free Electrocatalysts: New Trends in Electrocatalysts for Energy Applications. Volume 2 2022 (pp. 1–37). American Chemical Society: Washington, DC, USA.

[14] Sudarsono W, Tan SY, Wong WY, Omar FS, Ramya K, Mehmood S, Numan A, Walvekar R, Khalid M. From catalyst structure design to electrode fabrication of platinum-free electrocatalysts in proton exchange membrane fuel cells: A review. Journal of Industrial and Engineering Chemistry. 2023 Jun 25;122:1–26.

[15] Wu G, More KL, Johnston CM, Zelenay P. High-performance electrocatalysts for oxygen reduction derived from polyaniline, iron, and cobalt. science. 2011 Apr 22;332(6028):443–7.

[16] Debe MK. Electrocatalyst approaches and challenges for automotive fuel cells. Nature. 2012 Jun 7;486(7401):43–51.

[17] Hughes AE, Haque N, Northey SA, Giddey S. Platinum group metals: A review of resources, production and usage with a focus on catalysts. Resources. 2021 Sep 20;10(9):93.

[18] Sealy C. The problem with platinum. Materials Today. 2008 Dec 1;11(12):65–8.

[19] Sverdrup HU, Ragnarsdottir KV. A system dynamics model for platinum group metal supply, market price, depletion of extractable amounts, ore grade, recycling and stocks-in-use. Resources, Conservation and Recycling. 2016 Nov 1;114:130–52.

[20] He Y, Tan Q, Lu L, Sokolowski J, Wu G. Metal-nitrogen-carbon catalysts for oxygen reduction in PEM fuel cells: self-template synthesis approach to enhancing catalytic activity and stability. Electrochemical Energy Reviews. 2019 Jun 1;2(2):231–51.

[21] Wong WY, Rani MA, Loh KS, Lim KL, Minggu LJ. Current progress on rational design of porous MOF-derived transition metal-nitrogen-carbon as oxygen reduction reaction catalysts for proton exchange membrane fuel cells. Current Opinion in Green and Sustainable Chemistry. 2025 Jan 25;45:101001.

[22] Jasinski R. A new fuel cell cathode catalyst. Nature. 1964 Mar 21;201(4925):1212–3.

[23] Gupta S, Tryk D, Bae I, Aldred W, Yeager E. Heat-treated polyacrylonitrile-based catalysts for oxygen electroreduction. Journal of applied electrochemistry. 1989 Jan;19(1):19–27.

[24] Zagal JH, Páez MA, Silva JF. Fundamental aspects on the catalytic activity of metallomacrocyclics for the electrochemical reduction of $O_2$. InN$_4$-macrocyclic metal complexes 2006 (pp. 41–82). New York, NY: Springer New York.

[25] Peng H, Liu F, Liu X, Liao S, You C, Tian X, Nan H, Luo F, Song H, Fu Z, Huang P. Effect of transition metals on the structure and performance of the doped carbon catalysts derived from polyaniline and melamine for ORR application. Acs Catalysis. 2014 Oct 3;4(10):3797–805.

[26] Gupta S, Qiao L, Zhao S, Xu H, Lin Y, Devaguptapu SV, Wang X, Swihart MT, Wu G. Highly active and stable graphene tubes decorated with FeCoNi alloy nanoparticles via a template-free graphitization for bifunctional oxygen reduction and evolution. Advanced Energy Materials. 2016 Nov;6 (22):1601198.

[27] Li J, Ghoshal S, Liang W, Sougrati MT, Jaouen F, Halevi B, McKinney S, McCool G, Ma C, Yuan X, Ma ZF. Structural and mechanistic basis for the high activity of Fe–N–C catalysts toward oxygen reduction. Energy & Environmental Science. 2016;9(7):2418–32.

[28] Wang X, Zhang H, Lin H, Gupta S, Wang C, Tao Z, Fu H, Wang T, Zheng J, Wu G, Li X. Directly converting Fe-doped metal–organic frameworks into highly active and stable Fe-NC catalysts for oxygen reduction in acid. Nano Energy. 2016 Jul 1;25:110–9.

[29] Li J, Chen S, Li W, Wu R, Ibraheem S, Li J, Ding W, Li L, Wei Z. A eutectic salt-assisted semi-closed pyrolysis route to fabricate high-density active-site hierarchically porous Fe/N/C catalysts for the oxygen reduction reaction. Journal of Materials Chemistry A. 2018;6(32):15504–9.

[30] Choi CH, Baldizzone C, Polymeros G, Pizzutilo E, Kasian O, Schuppert AK, Ranjbar Sahraie N, Sougrati MT, Mayrhofer KJ, Jaouen F. Minimizing operando demetallation of Fe-NC electrocatalysts in acidic medium. ACS catalysis. 2016 May 6;6(5):3136–46.

[31] Varnell JA, Tse EC, Schulz CE, Fister TT, Haasch RT, Timoshenko J, Frenkel AI, Gewirth AA. Identification of carbon-encapsulated iron nanoparticles as active species in non-precious metal oxygen reduction catalysts. Nature communications. 2016 Aug 19;7(1):12582.

[32] Choi CH, Baldizzone C, Grote JP, Schuppert AK, Jaouen F, Mayrhofer KJ. Stability of Fe-N-C catalysts in acidic medium studied by operando spectroscopy. Angewandte Chemie International Edition. 2015 Oct 19;54(43):12753–7.

[33] Goellner V, Armel V, Zitolo A, Fonda E, Jaouen F. Degradation by hydrogen peroxide of metal-nitrogen-carbon catalysts for oxygen reduction. Journal of the electrochemical society. 2015 Mar 23;162(6):H403.

[34] Herranz J, Jaouen F, Lefèvre M, Kramm UI, Proietti E, Dodelet JP, Bogdanoff P, Fiechter S, Abs-Wurmbach I, Bertrand P, Arruda TM. Unveiling N-protonation and anion-binding effects on Fe/N/C catalysts for $O_2$ reduction in proton-exchange-membrane fuel cells. The Journal of Physical Chemistry C. 2011 Aug 18;115(32):16087–97.
[35] Chung HT, Cullen DA, Higgins D, Sneed BT, Holby EF, More KL, Zelenay P. Direct atomic-level insight into the active sites of a high-performance PGM-free ORR catalyst. Science. 2017 Aug 4;357 (6350):479–84.
[36] Li Q, Wang T, Havas D, Zhang H, Xu P, Han J, Cho J, Wu G. High-performance direct methanol fuel cells with precious-metal-free cathode. Advanced Science. 2016 Nov;3(11):1600140.
[37] Workman MJ, Serov A, Tsui LK, Atanassov P, Artyushkova K. Fe–N–C catalyst graphitic layer structure and fuel cell performance. ACS Energy Letters. 2017 Jul 14;2(7):1489–93.
[38] Li Z, Wei L, Jiang WJ, Hu Z, Luo H, Zhao W, Xu T, Wu W, Wu M, Hu JS. Chemical state of surrounding iron species affects the activity of Fe-Nx for electrocatalytic oxygen reduction. Applied Catalysis B: Environmental. 2019 Aug 15;251:240–6.
[39] Fu X, Li N, Ren B, Jiang G, Liu Y, Hassan FM, Su D, Zhu J, Yang L, Bai Z, Cano ZP. Tailoring $FeN_4$ sites with edge enrichment for boosted oxygen reduction performance in proton exchange membrane fuel cell. Advanced Energy Materials. 2019 Mar;9(11):1803737.
[40] Jaouen F, Lefèvre M, Dodelet JP, Cai M. Heat-treated Fe/N/C catalysts for $O_2$ electroreduction: are active sites hosted in micropores?. The Journal of Physical Chemistry B. 2006 Mar 23;110(11):5553–8.
[41] Zhao L, Zhang Y, Huang LB, Liu XZ, Zhang QH, He C, Wu ZY, Zhang LJ, Wu J, Yang W, Gu L. Cascade anchoring strategy for general mass production of high-loading single-atomic metal-nitrogen catalysts. Nature communications. 2019 Mar 20;10(1):1278.
[42] Chachuli SH, Timmiati SN, Loh KS, Wong WY, Lim KL, Yunus RM. P-block cathode electrocatalysts: A critical review of their role and impact on oxygen reduction reaction in fuel cells applications. Journal of Industrial and Engineering Chemistry. 2025 Sept 25;149. 176–208.
[43] Osmieri L. Transition metal–nitrogen–carbon (M–N–C) catalysts for oxygen reduction reaction. Insights on synthesis and performance in polymer electrolyte fuel cells. ChemEngineering. 2019 Feb 11;3(1):16.
[44] Wu G. Current challenge and perspective of PGM-free cathode catalysts for PEM fuel cells. Frontiers in Energy. 2017 Sep;11(3):286–98.
[45] Zhang H, Hwang S, Wang M, Feng Z, Karakalos S, Luo L, Qiao Z, Xie X, Wang C, Su D, Shao Y. Single atomic iron catalysts for oxygen reduction in acidic media: particle size control and thermal activation. Journal of the American Chemical Society. 2017 Oct 11;139(40):14143–9.
[46] Xia W, Zhu J, Guo W, An L, Xia D, Zou R. Well-defined carbon polyhedrons prepared from nano metal–organic frameworks for oxygen reduction. Journal of Materials Chemistry A. 2014;2 (30):11606–13.
[47] Ding L, Tang T, Hu JS. Recent progress in proton-exchange membrane fuel cells based on metal-nitrogen-carbon catalysts. Acta Phys. Chim. Sin. 2020;37(2010048):10–3866.
[48] Chi B, Zhang L, Yang X, Zeng Y, Deng Y, Liu M, Huo J, Li C, Zhang X, Shi X, Shao Y. Promoting ZIF-8-derived Fe–N–C oxygen reduction catalysts via Zr doping in proton exchange membrane fuel cells: durability and activity enhancements. ACS catalysis. 2023 Mar 13;13(7):4221–30.
[49] Wan Y, Yu L, Yang B, Li C, Fang C, Guo W, Xiao FX, Lin Y. Fe-NC core–shell catalysts with single low-spin Fe (II)-N4 species for oxygen reduction reaction and high-performance proton exchange membrane fuel cells. Journal of Energy Chemistry. 2024 Jun 1;93:538–46.
[50] Xu X, Zhang X, Kuang Z, Xia Z, Rykov AI, Yu S, Wang J, Wang S, Sun G. Investigation on the demetallation of Fe-NC for oxygen reduction reaction: The influence of structure and structural evolution of active site. Applied Catalysis B: Environmental. 2022 Jul 15;309:121290.

[51] Junaidi NH, Wong WY, Loh KS, Rahman S, Choo TF, Wu B. Enhanced oxygen reduction reaction catalyst stability and durability of MXene-supported Fe-NC catalyst for proton exchange membrane fuel cell application. Journal of Alloys and Compounds. 2023 Dec 15;968:171898.
[52] Lucchetti LE, de Almeida JM, Siahrostami S. Revolutionizing ORR catalyst design through computational methodologies and materials informatics. EES Catalysis. 2024;2(5):1037–58.
[53] Alt H, Binder H, Sandstede G. Mechanism of the electrocatalytic reduction of oxygen on metal chelates. Journal of Catalysis. 1973 Jan 1;28(1):8–19.
[54] Luo E, Zhang H, Wang X, Gao L, Gong L, Zhao T, Jin Z, Ge J, Jiang Z, Liu C, Xing W. Single-atom $Cr-N_4$ sites designed for durable oxygen reduction catalysis in acid media. Angewandte Chemie. 2019 Sep 2;131(36):12599–605.
[55] Li Z, Chen Y, Ji S, Tang Y, Chen W, Li A, Zhao J, Xiong Y, Wu Y, Gong Y, Yao T. Iridium single-atom catalyst on nitrogen-doped carbon for formic acid oxidation synthesized using a general host–guest strategy. Nature chemistry. 2020 Aug;12(8):764–72.
[56] Kong Z, Wu J, Liu Z, Yan D, Wu ZP, Zhong CJ. Advanced electrocatalysts for fuel cells: Evolution of active sites and synergistic properties of catalysts and carrier materials. InExploration 2025 Feb (Vol. 5, No. 1, p. 20230052).
[57] Goswami N, Chen K, Wang X, Spendelow JS, Borup RL, Mukherjee PP. Mechanistic understanding of thermodynamic metastability of core-shell catalysts in the polymer electrolyte fuel cell catalyst layer durability. Chemical Engineering Journal. 2024 Mar 15;484:149672.
[58] Fu H, Chen Z, Chen X, Jing F, Yu H, Chen D, Yu B, Hu YH, Jin Y. Modification Strategies for Development of 2D Material-Based Electrocatalysts for Alcohol Oxidation Reaction. Advanced Science. 2024 Oct;11(37):2306132.
[59] Wang YZ, Yang M, Ding YM, Li NW, Yu L. Recent advances in complex hollow electrocatalysts for water splitting. Advanced Functional Materials. 2022 Feb;32(6):2108681.
[60] Zhang S, Wang R, Zhang X, Zhao H. Recent advances in single-atom alloys: preparation methods and applications in heterogeneous catalysis. RSC advances. 2024;14(6):3936–51.
[61] Shao M, Chang Q, Dodelet JP, Chenitz R. Recent advances in electrocatalysts for oxygen reduction reaction. Chemical reviews. 2016 Mar 23;116(6):3594–657.
[62] Cai B, Chen X, Wang L, Fu H. Advanced progress for promoting anodic hydrogen oxidation activity and anti-CO poisoning in fuel cells. ACS Catalysis. 2024 Aug 28;14(18):13602–29.
[63] Xu HM, Zhu HR, Huang CJ, Zhang ZJ, Shuai TY, Zhan QN, Li GR. Recent advances in Fe-NC-and Co-NC-based materials as bifunctional electrocatalysts for oxygen reduction and oxygen evolution. Science China Chemistry. 2024 Apr;67(4):1137–60.
[64] Zhu J, Xia L, Yu R, Lu R, Li J, He R, Wu Y, Zhang W, Hong X, Chen W, Zhao Y. Ultrahigh stable methanol oxidation enabled by a high hydroxyl concentration on Pt clusters/MXene interfaces. Journal of the American Chemical Society. 2022 Aug 9;144(34):15529–38.
[65] Xiao D, Jiang Q, Xu C, Yang C, Yang L, He H, Huang H. Interfacial engineering of worm-shaped palladium nanocrystals anchored on polyelectrolyte-modified MXene nanosheets for highly efficient methanol oxidation. Journal of Colloid and Interface Science. 2022 Jun 15;616:781–90.
[66] Yang C, Jiang Q, Li W, He H, Yang L, Lu Z, Huang H. Ultrafine Pt nanoparticle-decorated 3D hybrid architectures built from reduced graphene oxide and MXene nanosheets for methanol oxidation. Chemistry of Materials. 2019 Oct 23;31(22):9277–87.
[67] Xiong J, Zhang Q, He H, Huang H. Ultrasmall Rh-decorated porous heterostructures stereo assembled from MXene and graphene for efficient methanol electrooxidation. International Journal of Hydrogen Energy. 2024 Nov 26;92:535–43.
[68] Zhou S, Yang X, Pei W, Liu N, Zhao J. Heterostructures of MXenes and N-doped graphene as highly active bifunctional electrocatalysts. Nanoscale. 2018;10(23):10876–83.

[69] Yang C, He H, Jiang Q, Liu X, Shah SP, Huang H, Li W. Pd nanocrystals grown on MXene and reduced graphene oxide co-constructed three-dimensional nanoarchitectures for efficient formic acid oxidation reaction. International Journal of Hydrogen Energy. 2021 Jan 1;46(1):589–98.

[70] Cui M, Xu B, Shi X, Zhai Q, Dou Y, Li G, Bai Z, Ding Y, Sun W, Liu H, Dou S. Metal–organic framework-derived single-atom catalysts for electrocatalytic energy conversion applications. Journal of Materials Chemistry A. 2024;12(30):18921–47.

[71] Zou L, Wei YS, Hou CC, Li C, Xu Q. Single-atom catalysts derived from metal–organic frameworks for electrochemical applications. Small. 2021 Apr;17(16):2004809.

[72] Brandiele R, Guadagnini A, Parnigotto M, Pini F, Coviello V, Badocco D, Pastore P, Rizzi GA, Vittadini A, Forrer D, Amendola V. Laser-optimized Pt-Y alloy nanoparticles embedded in Pt-Y oxide matrix for high stability and ORR electrocatalytic activity. Journal of Energy Chemistry. 2024 May 1;92:508–20.

[73] Chai L, Song J, Kumar A, Miao R, Sun Y, Liu X, Yasin G, Li X, Pan J. Bimetallic-MOF Derived Carbon with Single Pt Anchored C4 Atomic Group Constructing Super Fuel Cell with Ultrahigh Power Density And Self-Change Ability. Advanced Materials. 2024 Jan;36(1):2308989.

[74] Lun S, Wang H, Deng Y, Cui J, Liang P, Wang K, Lv L, Wan H, Wang H. FeNi decorated nitrogen-doped hollow carbon spheres as ultra-stable bifunctional oxygen electrocatalyst for rechargeable zinc–air battery with 2.7% decay after 300 h cycling. RSC advances. 2024;14(6):3857–66.

[75] Wang Q, Chen S, Li P, Ibraheem S, Li J, Deng J, Wei Z. Surface Ru enriched structurally ordered intermetallic PtFe@ PtRuFe core-shell nanostructure boosts methanol oxidation reaction catalysis. Applied Catalysis B: Environmental. 2019 Sep 5;252:120–7.

[76] Zhao X, Sasaki K. Advanced Pt-based core–shell electrocatalysts for fuel cell cathodes. Accounts of Chemical Research. 2022 Apr 22;55(9):1226–36.

[77] Wang X, Du P, Cheng K, Hua X, Xie M, Li Y, Zheng Y, Wang Y, Pi C, Zhang S. Structural Regulation of Advanced Platinum-Based Core-Shell Catalysts for Fuel Cell Electrocatalysis. Minerals. 2025 Feb 26;15(3):235.

[78] Khan IA, Qian Y, Badshah A, Nadeem MA, Zhao D. Highly porous carbon derived from MOF-5 as a support of ORR electrocatalysts for fuel cells. ACS applied materials & interfaces. 2016 Jul 13;8 (27):17268–75.

[79] Zhang J, Liu X, Ji Y, Liu X, Su D, Zhuang Z, Chang YC, Pao CW, Shao Q, Hu Z, Huang X. Atomic-thick metastable phase RhMo nanosheets for hydrogen oxidation catalysis. Nature Communications. 2023 Mar 30;14(1):1761.

[80] Chen TW, Anushya G, Chen SM, Kalimuthu P, Mariyappan V, Gajendran P, Ramachandran R. Recent advances in nanoscale based electrocatalysts for metal-air battery, fuel cell and water-splitting applications: An overview. Materials. 2022 Jan 8;15(2):458.

[81] Bai F, Qu X, Wang J, Chen X, Yang W. Confinement catalyst of $Co_9S_8$@ N-doped carbon derived from intercalated $Co(OH)_2$ precursor and enhanced electrocatalytic oxygen reduction performance. ACS Applied Materials & Interfaces. 2020 Jul 7;12(30):33740–50.

[82] Ni W, Chen H, Zeng J, Zhang Y, Younus HA, Zeng Z, Dai M, Zhang W, Zhang S. Regulating nonmetallic species beyond the first coordination shell of single-atom catalysts for high-performance electrocatalysis. Energy & Environmental Science. 2023;16(9):3679–710.

[83] Naguib M, Mochalin VN, Barsoum MW, Gogotsi Y. 25th anniversary article: MXenes: a new family of two-dimensional materials. Advanced materials. 2014 Feb;26(7):992–1005.

[84] Wang Y, Chen Z, Zhang C, Yang L, Jiang Q, Zhang J, He H, Huang H. Mesoporous hollow carbon sphere-embedded MXene architectures decorated with ultrafine Rh nanocrystals toward methanol electrooxidation. Inorganic Chemistry. 2024 Aug 28;63(36):16888–96.

[85] Liu H, Zhao J, Li X. Controlled synthesis of carbon-supported Pt-based electrocatalysts for proton exchange membrane fuel cells. Electrochemical Energy Reviews. 2022 Dec;5(4):13.

[86] Suter TA, Smith K, Hack J, Rasha L, Rana Z, Angel GM, Shearing PR, Miller TS, Brett DJ. Engineering catalyst layers for next-generation polymer electrolyte fuel cells: A review of design, materials, and methods. Advanced Energy Materials. 2021 Oct;11(37):2101025.

[87] Li X, Wang J. One-dimensional and two-dimensional synergized nanostructures for high-performing energy storage and conversion. InfoMat. 2020 Jan;2(1):3–2.

[88] Li B, Zhang L, Zhang J, Su Y. Recent insight in transition metal anchored on nitrogen-doped carbon catalysts: preparation and catalysis application. Electrochem. 2022 Sep 1;3(3):520–37.

[89] Dessie Y, Bekele ET, Gonfa BA, Ravikumar CR, Khasim S, Abraham G. Electrochemical performances of $MnO_2/Fe_3O_4$/activated carbon ternary composites for supercapacitor and direct ethanol fuel cell catalyst application. RSC advances. 2025;15(21):16493–509.

[90] Haider R, Wen Y, Ma ZF, Wilkinson DP, Zhang L, Yuan X, Song S, Zhang J. High temperature proton exchange membrane fuel cells: progress in advanced materials and key technologies. Chemical Society Reviews. 2021;50(2):1138–87.

Avanish kumar, Amit Kumar Rathoure, Dan Bahadur Pal

# 6 Portable and Wearable Fuel Cell Technologies: Innovations and Applications

**Abstract:** The increasing demand for energy in portable and wearable electronics or fuel cell has motivated the consideration of alternative power sources beyond conventional batteries because it has limitations of short lifespan, lengthy recharging times and environmental impact. Fuel cell technology appears encouraging solution due to its high efficacy, environmentally friendly, and potential for continuous power generation. This chapter offers inclusive overview of different type of fuel cell technology and highlights its fundamental working principles and their key components like anode, cathode, electrolyte, catalyst, and fuel. The types of fuel cells are available, which can be used as portable and wearable applications, such as proton exchange membrane fuel cells (PEMFCs), direct methanol fuel cells (DMFCs), solid oxide fuel cells (SOFCs), and microbial fuel cells (MFCs). The key design considerations for such type of fuel require as reduction in size, fuel storage and delivery, durability, and safety are examined to address the unique challenges modeled by these applications. The chapter further examined the diverse applications of portable and wearable fuel cells in consumer electronics, healthcare devices, military equipment, and remote power solutions. The performance analysis and efficiency optimization strategies are reviewed properly along with its recent advancements in materials and catalysts. The chapter content also includes the insights into market trends, future innovations, and the sustainability of fuel cell technology, which emphasize on potential integration with renewable energy sources for a greener future.

**Keywords:** Portable, wearable, environmentally, power generation, proton exchange membrane fuel cells, methanol fuel, microbial fuel cells, energy storage

## 6.1 Introduction

Currently, the biofuel cell is a developing and sustainable energy technology, which generates electricity through the biological oxidation of organic materials such as glucose or other bio-based fuels [1]. It uses enzymes or microorganisms as catalysts for increasing the efficiency of fuel cell [2]. Historically, the concept of biofuel cells came in to existence the early twentieth century when researchers discovered the concept of microbial fuel cells for power generation. The recent advancements in nanotechnology and biotechnology have revived interest in biofuel cells as a clean and efficient energy source [3]. With the increasing demand for portable and wearable electronics, such as smart watches, fitness trackers, medical monitoring devices as bio sensor,

 | https://doi.org/10.1515/9783112181072-006

which are required for compact, lightweight, and long-lasting power solutions. The conventional lithium-ion and nickel-metal hydride batteries that are currently lead the market, although it faces several limitations as short lifespan due to repeated charging cycles. It has also long recharging times and significant environmental impacts accompanying with the extraction of raw materials like lithium and cobalt [4]. Additionally, these all batteries contribute to electronic waste and pose recycling challenges. Moreover, their bulky design limits the flexibility and functionality of wearable devices. In contrast, biofuel cells offer a capable alternative by directly converting chemical energy of organic fuels into electricity through an eco-friendly and efficient process [5]. They provide higher energy density, longer operational life, and continuous power generation. Presently, the global fuel cell market size was valued at USD 4.78 billion in 2023 and is expected to be worth around USD 62.17 billion by 2034 [6], at a noteworthy compound annual growth rate (CAGR) of 27% from 2024 to 2034. Unlike conventional batteries, biofuel cells produce minimal pollution and operate at lower temperatures, which make them ideal for powering biomedical implants, biosensor, and other reduced electronic devices [6].

Furthermore, biofuel cells can use glucose from bodily fluids or organic compounds from the environment [7]. Despite challenges such as limited power output and the need for advanced electrode materials, there were research is going in development enzyme-based and microbial biofuel cells. Presently, the integration of nanomaterials is enhancing their performance and scalability. Even the demand for flexible, efficient, and eco-friendly energy solutions continues in demand, biofuel cells are developing as a revolutionary technology that can address the limitations of conventional batteries while supporting the development of next-generation portable and wearable electronics [8]. In the fuel cell with the reaction between hydrogen and oxygen, the cell generates water and heat as by-products. This process not only eliminates harmful emissions but also provides a continuous and reliable power source so long as the fuel is supplied [9]. In the context of portable and wearable devices, such as smart watches, fitness trackers, medical sensors, and other technologies like fuel cells, offer significant advantages. They have a higher energy density, which allows devices to operate for longer durations without frequent recharging. Additionally, fuel cells can be refueled quickly, unlike conventional batteries that require lengthy charging times. This makes them particularly useful in critical applications like healthcare monitoring devices and military communication systems, where uninterrupted power supply is essential [10].

Another key benefit of biofuel cells is their lightweight and flexible design, which is ideal for compact and wearable devices. Microbial fuel cells and biofuel cells, which use organic compounds like glucose or ethanol as fuel, are being developed to power small sensors and biomedical implants [11]. Moreover, fuel cells have a minimal environmental impact, as they do not involve toxic materials like lithium or cobalt and produce only water as a by-product. Even challenges like hydrogen storage and cost-effective manufacturing remain, ongoing research in nanomaterials and advanced catalysts is improving the efficiency and scalability of fuel cell technology. With these advancements, fuel cells offering sustainable and efficient power solutions for the growing demand in portable and wearable electronics [12].

## 6.2 Fuel Cells: Basic Working Principles of Fuel Cells

The fuel cell is an electrochemical device that changes chemical energy from a fuel (such as hydrogen or biofuels) into electrical energy with the help of redox reaction [13]. It operates similarly to a battery but does not require recharging as long as the fuel is continuously supplied. The process involves the movement of electrons from the anode to the cathode and generating electricity, while ions move through an electrolyte to complete the reaction [14]. The by-products of fuel cell operation are water and heat, which are environmentally friendly. The working of fuel cell is shown in Figure 6.1.

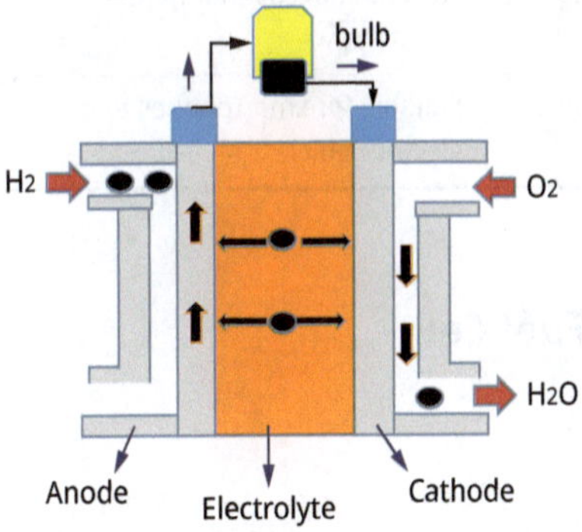


**Figure 6.1:** Schematic diagram for working of fuel cells.

Generally, all fuel cells consist of several key components that work together to generate electricity efficiently as shown in Figure 6.1 Here on the anode is where the fuel, such as hydrogen, undergoes oxidation and electrons and protons. These electrons march through an external circuit to run the electrical power [15]. At the cathode, the electrons combine with oxygen and protons to form water, which completes the reaction, and the reaction of fuel cell is shown in Figure 6.1. The electrolyte plays a vital role by allowing the transfer of ions, such as protons or hydroxide ions between the anode and cathode [16]. To enhance the efficiency of the reaction, a catalyst as platinum or other nanomaterials is used to accelerate the redox process at both electrodes. Finally, the fuel source, which can be hydrogen, methanol, or biofuels, serves as the energy carrier, ensuring continuous power generation [17]. Together, these components enable fuel cells to produce clean and efficient energy. The feature as comparison form is given in Table 6.1.

**Table 6.1:** Comparison of fuel cells with conventional batteries.

| Feature | Fuel cell | Conventional battery |
|---|---|---|
| Energy source | Continuous fuel supply (e.g., hydrogen) | Stored chemical energy |
| Operational life | Longer as long as fuel is supplied | Limited due to charge cycles |
| by-products | Water and heat (eco-friendly) | Conventional battery contains toxic waste and heavy metals |
| Recharging time | Instant refueling | Hours of recharging |
| energy density | Fuel cell has higher energy density [18] | Lower energy density |
| environmental impact | Minimal emissions | High environmental impact due to mining |
| Size and flexibility | Compact and flexible design | Bulky and rigid |
| Power output | Consistent and continuous | Decreases over time |
| Cost efficiency | High initial cost but cost-effective in the long run [19] | Lower initial cost but higher maintenance |
| Application | Ideal for portable and wearable devices, and biomedical implants | Suitable for smartphones and electric vehicles |

## 6.2.1 Types of Portable and Wearable Fuel Cell

### 6.2.1.1 PEMFC Working Principle

These cells are portable applications and low operating temperature (around 60–80 °C) and also having high efficiency [20]. At the anode, hydrogen molecules fragmented into protons and move to the proton-conducting electrolyte membrane, while electrons travel through an external circuit producing electricity [21]. The core part of the proton exchange membrane (PEM) allows only protons ($H^+$) to pass through while hindering electrons as shown in Figure 6.2, where the process begins at the anode, where hydrogen gas ($H_2$) is introduced and split into protons and electrons with the help of a platinum-based catalyst the reaction on anode and is shown by eqs. (6.1) and (6.2) as follows:

At the anode, the reaction is:

$$H_2 \rightarrow 2H + 2e^- \tag{6.1}$$

The protons move through the PEM to the cathode while the electrons are forced to move through an external circuit and produce electricity.

At the cathode, oxygen gas ($O_2$) reacts with the incoming protons and electrons produce water H2O as a by-product, represented by the following equation:

$$\frac{1}{2}O_2 + 2H^+ + 2e^- \rightarrow H_2O \tag{6.2}$$

While the overall reaction is written as the following equation:

$$H_2 + \frac{1}{2}O_2 \rightarrow H_2O + \text{electricity} \tag{6.3}$$

The proton exchange membrane fuel cells (PEMFCs) work in temperature about 50–80 °C. The working of PEMFCs is shown in Figure 6.2. PEMFCs develop huge energy with zero greenhouse gas emissions, lightweight design, and scalability for various applications [22]. However, the limitations include high production costs due to the application of expensive platinum catalysts, membrane degradation over time, and challenges in hydrogen storage and infrastructure [23]. Despite these challenges, PEMFCs are widely used in hydrogen-powered vehicles like cars, buses, and trucks, portable power generation for military and remote applications, stationary power plants, and even in aerospace applications such as NASA's space missions [24].

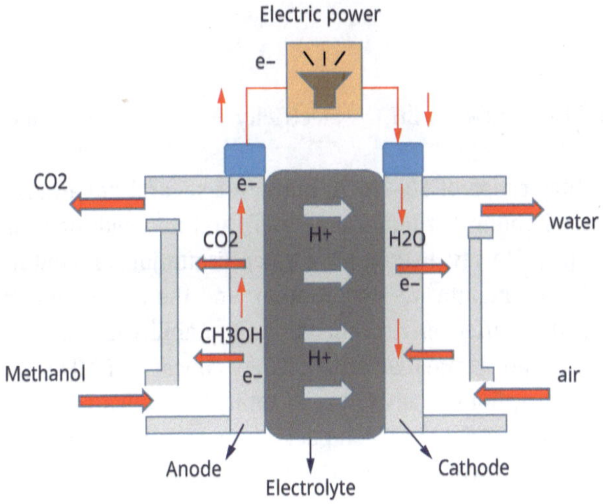


**Figure 6.2:** Schematic diagram of proton exchange membrane fuel cells.

The growing interest in clean energy solutions is driving advancements in PEMFC technology for a sustainable future.

#### 6.2.1.2 DMFCs

As shown in figure 6.3, direct methanol fuel cells (DMFCs) work same as PMFC but here liquid methanol is used in the place of hydrogen, which generates protons and electrons. In the fuel cell, the protons march through the membrane and the electrons flow through the external circuit, which generates electricity [25]. DMFCs are also a type of fuel cell that directly converts the chemical energy of liquid methanol into electrical energy through an electrochemical process [26]. Unlike other fuel cells that require hydrogen gas, DMFCs use methanol as a fuel source, which is used directly to the anode. The principle involves the oxidation of methanol at the anode, where it reacts with water to produce carbon dioxide and develop protons ($H^+$), and electrons ($e^-$), which is shown in the following equations:

The anode reaction is:

$$CH_3OH + H_2O \rightarrow CO_2 + 6H^+ + 6e^- \tag{6.4}$$

Protons move through membrane, while the electrons pass over an external circuit, producing electricity. At the cathode, oxygen reacts with the protons and electrons to produce water, which is represented by the following equations:

$$\frac{3}{2}O_2 + 6H^+ + 6e^- \rightarrow 3H_2O \tag{6.5}$$

The overall reaction is:

$$CH_3OH + \frac{3}{2}O_2 \rightarrow CO_2 + + 2H_2O + \text{electricity} \tag{6.6}$$

DMFCs operate in the temperature range of 50–120 °C and offer several advantages such as high energy density, easy handling of liquid methanol, and no need for complex hydrogen storage systems [27]. However, they have some limitations, including the slow oxidation rate of methanol, membrane degradation, and the production of carbon dioxide as a by-product, which affect environmental sustainability [28].

Additionally, methanol is toxic and flammable, posing safety concerns. DMFCs are primarily used in laptops and smartphones, as well as in military applications and backup power systems for remote areas [29]. Their compact design and lightweight nature make them suitable for applications where traditional batteries are less effective.

#### 6.2.1.3 SOFCs: Solid Oxide Fuel Cells

Solid oxide fuel cells (SOFCs) convert the chemical energy of hydrogen, natural gas, or biogas directly into electricity through an electrochemical process [30]. The core component is a solid ceramic electrolyte, typically made of yttrium-stabilized zirconia (YSZ), which allows only oxygen ions ($O^{2-}$) to pass through while blocking electrons.

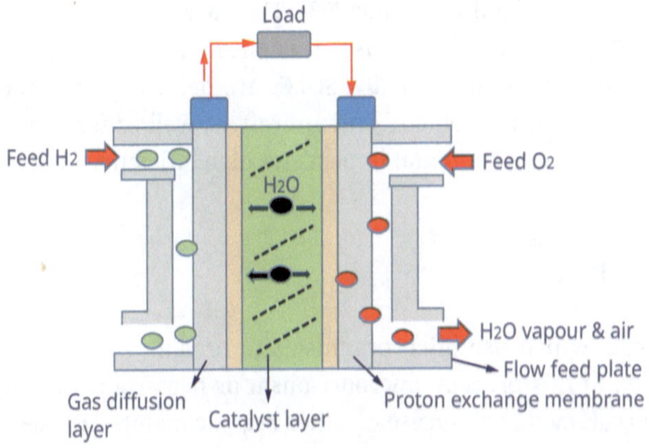


**Figure 6.3:** Schematic diagram of direct methanol fuel cells (DMFCs) [29].

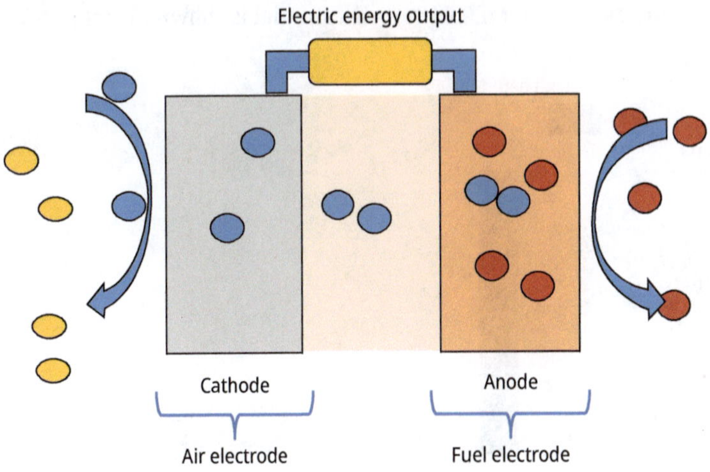


**Figure 6.4:** Schematic diagram of SOFCs.

The process begins at the cathode, where oxygen is reduced to form oxygen ions. These oxygen ions pass through the solid electrolyte to the anode, where they react with hydrogen fuel to produce water, electrons, and heat as the electrons produced at the anode flow through an external circuit, producing electricity [31]. The reaction on cathode and anode along with overall reaction is shown in SOFC reaction in Figure 6.4. SOFCs operate temperature between 700 and 1,000 °C, which permit them to achieve high efficiency (up to 60%) and enable internal reforming of hydrocarbon fuels like methane [32]. However, this high-temperature operation also leads to certain limitations, such as material degradation, long start-up times, and the need for expensive, heat-resistant materials. Despite these challenges, SOFCs offer advantages like fuel flexibility, long-term stability, and minimal emissions [33]. They are widely used in stationary power generation for residential and industrial applications, as well as in combined heat and power (CHP) systems. Additionally, SOFCs are being explored for the use in auxiliary power units (APUs) for vehicles and aircraft, as well as for large-scale power plants due to their little environmental impact and high efficiency [34].

#### 6.2.1.4 MFCs: Microbial Fuel Cells

MFCs are different types of cell, which use microorganisms to breakdown organic matter and generate electricity. In this process, microorganism or bacteria transfers electrons to the anode while breaking down glucose or other organic materials. These electrons flow through an external circuit to the cathode, where they react with oxygen to produce water. MFCs are particularly promising for biomedical applications, such as implantable biosensors due to their ability to generate power from biological waste [35]. The schematic diagram of MFC with their essential is shown in Figure 6.5.

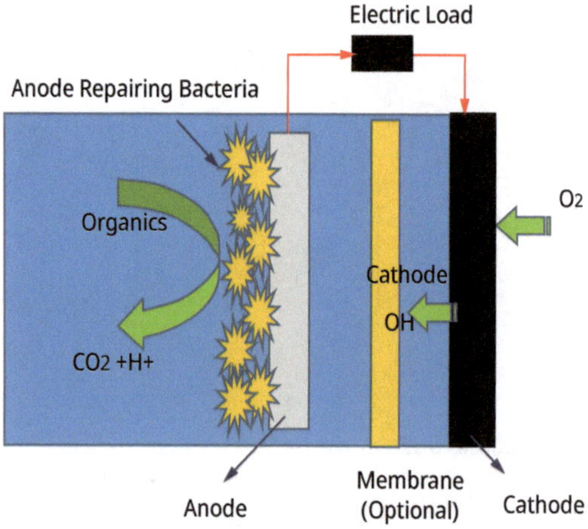


**Figure 6.5:** Schematic diagram of MFCs.

The MFCs use the metabolic activity of microorganisms to transform organic matter directly into electrical energy [36]. The principle is based on the ability of certain bacteria, known as exoelectrogens that transfer electrons to an electrode during the breakdown of organic substrates where bacteria oxidize organic compounds, such as glucose or wastewater contaminants through their metabolic processes [37]. During this oxidation, electrons and protons are released and the electrons are transferred to the anode either directly through conductive nanowires or via electron mediators while the protons pass through a PEM to the cathode. The electrons travel through an external circuit, generating electricity, and eventually combine with protons and oxygen, the reaction on cathode and anode along with overall reaction is shown in Figure 6.5. MFCs offer several advantages, such as sustainable energy production, wastewater treatment, and bioremediation, as they utilize organic waste as fuel [38]. They operate at room temperature, are eco-friendly, and can reduce greenhouse gas emissions. However, MFCs face limitations, including low power output, slow electron transfer rates, and high costs associated with electrode materials and membranes [39]. Additionally, maintaining stable microbial communities is challenging. MFCs are widely used in remote environmental sensors and biosensors in wastewater treatment plants for monitoring pollutants [40]. The detail discussion of fuel cell along with advantage, disadvantage, limitation, and application is shown in Table 6.2.

## 6.3 Design and Development Considerations for Fuel Cells

Fuel cell design must address miniaturization, fuel storage, durability, and safety. Each type – PEMFCs, DMFCs, SOFCs, and MFCs – has specific requirements that impact efficiency and performance in various applications are shown in Table 6.3.

### 6.3.1 Applications of Portable and Wearable Fuel Cells

Portable and wearable fuel cells have established as a revolutionary energy solution, offering efficient and sustainable power for various applications [49].

Beside, there are some other benefits such as compact design, high energy density, and reduced environmental impact making them ideal for consumer application in field of electronics and healthcare [50]. The medical devices, military and defense applications, and remote and emergency power scenarios are shown in Table 6.4.

### 6.3.2 Market Trends and Future Prospects

It has been seen that the global fuel cell market is experiencing rapid growth, with its valuation expected to rise from USD 4.78 billion in 2023 to USD 62.17 billion by 2034, driven by a remarkable CAGR of 27% [58], as shown in Figure 6.6. This expansion is fuelled by the increasing demand for portable and wearable fuel cells across various industries including consumer electronics, healthcare, and defense [59]. Portable fuel cells are an alternative to conventional batteries, powering smartphones, laptops, and smart watches, while in healthcare, they are being integrated into implantable biosensors and pacemakers to provide long-lasting energy solutions. The military sector also benefits from these advancements, with fuel cells being used to power soldier-worn

**Table 6.2:** Types of fuel cell along with advantage, disadvantage, limitation, and application.

| Fuel cell type | Advantages | Disadvantages | Limitations | Applications |
|---|---|---|---|---|
| **PEMFCs** | – High efficiency, quick start-up, compact design, operates at low temperature (60–80 °C) [41], and zero emissions (only water as by-product) | – Requires high-purity hydrogen<br>– Expensive platinum catalyst<br>– Sensitive to CO poisoning, complex water management | – Limited durability and high cost of materials<br>– Infrastructure for hydrogen storage is lacking | – Electric vehicles<br>– Portable power sources, backup power systems, and residential applications |
| **DMFCs** | – Uses liquid methanol as fuel (easier storage)<br>– Compact and lightweight<br>– Operates at low temperature (60–130 °C)[42] | – Low efficiency compared to PEMFC<br>– Methanol is toxic and flammable<br>– Expensive catalyst<br>– Methanol crossover reduces performance | – Limited lifespan<br>– Slow reaction kinetics<br>– Carbon emissions from methanol | – Portable electronics<br>– Military applications<br>– Small-scale power generation |
| **SOFCs** | – High efficiency<br>– Operates on various fuels (natural gas, biogas, and hydrogen) [43]<br>– Long lifespan<br>– High operating temperature (800–1,000 °C) allows internal reforming | – High operating temperature leads to material degradation<br>– Slow start-up time<br>– Complex thermal management | – Fragile ceramic materials<br>– Requires expensive components<br>– Limited portability | – Stationary power generation<br>– Industrial applications<br>– Combined heat and power (CHP) systems |
| **MFCs** | – Converts organic waste into electricity<br>– Sustainable and eco-friendly<br>– Operates at ambient temperature [44]<br>– Can be used in wastewater treatment | – Low power density<br>– Slow microbial growth<br>– Requires complex biological control | – Limited scalability<br>– Sensitive to environmental conditions<br>– Long start-up time | – Wastewater treatment in plants<br>– Remote sensing<br>– Bioenergy production |

**Table 6.3:** Design and development considerations for fuel cells.

| Fuel cell type | Miniaturization challenges | Fuel storage and delivery | Durability and lifespan | Safety concerns |
|---|---|---|---|---|
| **PEMFCs** | – High proton-conductive, thin, flexible membranes (e.g., Nafion)<br>– Micro-scale flow channels for gas distribution and water management [45]<br>– Nanostructured catalysts for higher power density | – High-pressure tanks, metal hydrides, and cryogenic storage.<br>– On-demand hydrogen from hydrocarbons/ammonia<br>– Optimized gas diffusion layers for precise fuel delivery | – Perfluorinated membranes resist chemical degradation<br>– Platinum alloys and protective layers prevent catalyst poisoning<br>– Efficient cooling systems prevent thermal stress | – Leak detection sensors and explosion-proof enclosures<br>– Proper electrical insulation prevents hazards<br>– Protective casings for portable applications |
| **DMFCs** | – Microfluidics for methanol and water management<br>– Ultra-thin proton-conductive membranes<br>– Nanostructured catalysts enhance reaction rates | – Compact, refillable methanol cartridges<br>– Controlled fuel injection via micro-pumps<br>– Vapor management systems prevent clogging | – Sulfonate polymers reduce methanol crossover<br>– Platinum-ruthenium alloys resist poisoning<br>– Anticorrosive coatings enhance durability | – Leak-proof containers and ventilation systems<br>– Thermal management prevents overheating<br>– Sensor-based shutdown mechanisms |
| **SOFCs** | – Thin ceramic electrolyte with high ionic conductivity<br>– Miniature interconnects for electron flow<br>– High-temperature materials for efficiency | – Uses natural gas/biogas reforming for hydrogen<br>– Heat exchangers for waste heat recovery[46]<br>– Sealing systems prevent gas leakage | – YSZ ceramics resist thermal cycling<br>– Protective coatings prevent oxidation and corrosion<br>– Nickel catalysts resist coking and sulfur poisoning | – Insulated enclosures prevent burns<br>– CO and hydrogen leak sensors<br>– Robust materials handle thermal expansion |
| **MFCs** | – Optimized microbial biofilms for electron transfer<br>– Compact carbon-based electrodes<br>– Integrated sensors monitor microbial activity | – Uses organic waste (e.g., wastewater and agricultural waste)<br>– Controlled nutrient supply for microbial growth<br>– Surface management prevents electrode fouling[47] | – Microbial community diversity ensures stability[48]<br>– Corrosion-resistant electrodes (graphite/stainless steel)<br>– Bio-fouling prevention for membrane integrity | – Pathogen control prevents contamination<br>– Electrical isolation ensures safe handling<br>– Secure waste disposal for residual biomass |

**Table 6.4:** Applications of portable and wearable fuel cells.

| Name of portable or wearable fuel cells | Applications |
|---|---|
| **Remote and emergency power** | – Fuel cells provide electricity during natural disasters, power outages, and for off-grid applications<br>– PEMFCs and SOFCs are efficient and have rapid start-up capabilities<br>– Can use locally available fuels (e.g., biogas and ethanol) for energy independence [51]<br>– Portable fuel cells benefit hikers, campers, and outdoor enthusiasts<br>– Silent operation and low emissions make them ideal for sensitive environments<br>– Used in remote research stations and isolated communities for sustainable energy |
| **Consumer electronics** | – Fuel cells are integrated into smartphones, laptops, and smart watches to enhance battery life and enable faster charging [52]<br>– PEMFCs and DMFCs are suitable due to their lightweight nature and high energy output<br>– DMFCs provide continuous power, eliminating frequent recharging<br>– Minimal heat generation and silent operation enhance user comfort<br>– Compact design allows integration into slim modern gadgets [53] |
| **Healthcare and medical devices** | – Fuel cells power implantable biosensors, pacemakers, and other medical devices [54]<br>– MFCs and biofuel cells use organic substrates (e.g., glucose and oxygen) from bodily fluids for power. Reduces the need for invasive battery replacement [55]<br>– PEMFCs are used in portable diagnostic equipment and oxygen concentrators<br>– Provides stable and uninterrupted power for medical monitoring and treatments |
| **Military and defense** | – Wearable fuel cells provide reliable power for communication devices, surveillance equipment, and soldier systems [56]<br>– SOFCs and DMFCs offer long-duration power and operate on multiple fuels (e.g., methanol and hydrogen).<br>– Lightweight and durable, reducing the burden of heavy battery packs<br>– Low infrared signature enhances stealth operations [57]<br>– Powers advanced sensors, GPS, and night-vision devices |

systems, communication devices, and surveillance equipment in remote operations. Leading companies, such as Ballard Power Systems, Intelligent Energy, and Plug Power, are at the forefront of fuel cell commercialization, and introducing hydrogen-powered chargers and compact fuel cell generators [60]. Other players, like Lilliputian Systems and Horizon Fuel Cell Technologies, are focused on developing innovative fuel cell solutions for consumer electronics and portable applications. Future advancements in the sector include flexible fuel cells for wearable, biofuel-based systems that convert organic waste into energy, and hybrid fuel cell solutions integrated with renewable energy sources. These innovations aim to increase energy efficiency, reduce costs, and improve the sustainability of fuel cell technology.

Despite promising advancements, the commercialization of fuel cells faces several challenges. High production costs, particularly due to expensive platinum-based catalysts and advanced membranes, remain a major barrier [61]. Additionally, regulatory hurdles related to hydrogen storage, transportation, and safety standards require further refinement. Public perception and safety concerns regarding hydrogen handling also pose obstacles to widespread adoption. Addressing these challenges through continued research, policy support, and public awareness initiatives will be critical in ensuring the successful integration of portable and wearable fuel cells into mainstream applications.

## 6.3.3 Sustainability and Environmental Impact

Fuel cells offer significant green energy benefits by producing electricity through electrochemical reactions rather than combustion, thereby reducing greenhouse gas emissions and air pollutants [62]. As they primarily utilize hydrogen as a fuel, the only byproduct is water, making them a cleaner alternative to fossil fuel-based power sources. This characteristic makes fuel cells ideal for applications where environmental sustainability is important. The comprehensive life cycle analysis of fuel cells evaluates the environmental impact from fuel production to waste disposal [63]. While these processes can be energy-intensive, advancements in recycling technologies and the development of alternative catalysts are helping to reduce the carbon footprint [64]. During operation, fuel cells have minimal emissions, but life management as recycling of components and proper handling of spent materials, is essential to minimize environmental impact. For instance, hydrogen can be produced through water electrolysis powered by solar or wind energy, creating a fully green energy cycle [65]

Additionally, biofuel-based systems can utilize organic waste to generate hydrogen, which reduce dependence on fossil fuels and contributing to circular economy initiatives. Hybrid energy solutions that combine fuel cells with solar panels or battery storage systems can provide continuous and reliable power in off-grid and remote areas. However, challenges such as the energy-intensive production of hydrogen and the need for infrastructure for hydrogen storage and distribution must be addressed to fully realize the environmental benefits of fuel cells [66].

## 6.3.4 Challenges for Development of Fuel Cell

Portable and wearable fuel cell technologies have emerged as promising alternatives to traditional battery-powered systems because of their lightweight nature, high energy density, and environmentally friendly operation [67], as shown in Table 6.5.

**Table 6.5:** Comprehensive comparison of various portable and wearable fuel cells, evaluated across key parameters.

| Types | Power density ($W/cm^2$) | Fuel storage compatibility | Weight | Environmental issues | Applications |
|---|---|---|---|---|---|
| **PEMFCs** | 0.3–1.2 | Hydrogen gas | Light to moderate | Low emissions (water as by-product) and challenges with hydrogen storage and infrastructure | Medical implants and biosensors |
| **DMFCs** | 0.1–0.3 | Liquid methanol | Light to moderate | Methanol is toxic and flammable and requires careful handling and disposal | Consumer electronics, military equipment, and battery chargers |
| **SOFCs** | 0.5–2.0 | Hydrogen and hydrocarbon fuels | Heavier due to ceramic components | High operating temperatures; emits $CO_2$ when using hydrocarbon fuels and material degradation concerns | Auxiliary power units, remote power generation |
| **MFCs** | 0.001–0.01 | Glucose, organic matter | Very light | Biodegradable, minimal emissions, and low power output limits applicability | |

Fuel cells, primarily utilizing hydrogen, methanol, or biofuels, offer continuous power supply with quick refueling, making them suitable for various applications, including military operations, healthcare devices, consumer electronics, and emergency backup power. The growing demand for miniaturized, lightweight energy sources has led to advancements in PEMFCs, DMFCs, and MFCs, which provide extended operation times compared to conventional lithium-ion batteries [68]. Wearable fuel cells, integrated into textile or compact devices, are particularly gaining attention for powering biosensors, smart textiles, and medical implants, ensuring uninterrupted operation without frequent recharging. In military applications, soldiers require lightweight power solutions for communication and navigation devices, making fuel cells a viable alternative to heavy battery packs [69]. Similarly, in healthcare, wearable biofuel cells using glucose or lactate as fuel sources are being developed to power implants such as pacemakers, glucose monitors, and neuron stimulators, reducing the need for invasive surgeries for battery replacements. Despite their benefits, the widespread adoption of portable and wearable fuel cells faces contests including high production costs, fuel storage issues, and the need for durable, flexible materials to enhance integration into wearable devices [70]. Research is focused on improving catalyst efficiency, membrane durability, and miniaturization to make fuel cells more commercially viable. Recent developments include flexible, stretchable fuel cells that can conform to body movements without performance degradation, enhancing their usability in next-generation wearable [71]. Additionally, hybrid systems combining fuel cells with supercapacitors or batteries are being explored to provide stable power output and quick response times. The push for sustainability has also led to interest in biofuel cells that utilize organic fuels derived from sweat, urine, or ambient air, enabling self-sustaining energy generation for low-power devices [72]. Governments and private companies are investing in research to overcome technical limitations and make these technologies accessible for real-world applications [73]. As fuel cell technology advances, improvements in fuel storage, system miniaturization, and cost reduction will drive their widespread adoption in portable and wearable electronics, contributing to the shift towards sustainable, on-demand energy solutions for modern-day applications [74].

## 6.4 Conclusion

Portable and wearable fuel cell technologies have considered as a revolutionary solution to report the growing energy demands of modern electronics while overcoming the limitations of conventional batteries. With advancements in PEMFCs, DMFCs, and MFCs, these technologies offer high energy density, longer operational time, and eco-friendly alternatives for powering consumer electronics, healthcare devices, and military equipment. However, challenges such as miniaturization, fuel storage, durability, and safety remain critical for their widespread adoption. Recent innovations in flexible materials, biofuel integration, and hybrid energy systems are flagging the way for more efficient and cost-effective designs. With continuous research and development, portable and wearable fuel cells have the potential to revolutionize the future of sustainable energy solutions, offering reliable and long-lasting power for next-generation devices while contributing to global efforts towards green and clean energy transitions.

## References

[1] Geetha, M., Safamariyam, E., Roshan, S. and Synumol, P., 2025. Biofuel cells: a novel innovation. In *Biofuel Cells and Energy Generation* (pp. 1–26). Woodhead Publishing: Cambridge, UK.

[2] Jang, J.K., Kan, J., Bretschger, O., Gorby, Y.A., Hsu, L., Kim, B.H. and Nealson, K.H., 2013. Electricity generation by microbial fuel cell using microorganisms as catalyst in cathode. *Journal of microbiology and biotechnology*, *23*(12), pp.1765–1773.

[3] Aquino Neto, S. and De Andrade, A.R., 2013. New energy sources: the enzymatic biofuel cell. *Journal of the Brazilian Chemical Society*, *24*, pp.1891–1912.

[4] Meshram, P., Pandey, B.D. and Mankhand, T.R., 2014. Extraction of lithium from primary and secondary sources by pre-treatment, leaching and separation: A comprehensive review. *Hydrometallurgy*, *150*, pp.192–208.

[5] Panwar, N.L., Kothari, R. and Tyagi, V.V., 2012. Thermo chemical conversion of biomass–Eco friendly energy routes. *Renewable and Sustainable Energy Reviews*, *16*(4), pp.1801–1816.

[6] Makol, N., Gupta, P., Mital, M. and Syal, M., 2024. Green Hydrogen: Understanding Techno-Economic Viability for Industries. *OIDA International Journal of Sustainable Development*, *17*(02), pp.33–42.

[7] Hannan, M.A., Abbas, S.M., Samad, S.A. and Hussain, A., 2011. Modulation techniques for biomedical implanted devices and their challenges. *Sensors*, *12*(1), pp.297–319.

[8] Fan, X., Liu, B., Ding, J., Deng, Y., Han, X., Hu, W. and Zhong, C., 2020. Flexible and wearable power sources for next-generation wearable electronics. *Batteries & Supercaps*, *3*(12), pp.1262–1274.

[9] Richards, G.A., McMillian, M.M., Gemmen, R.S., Rogers, W.A. and Cully, S.R., 2001. Issues for low-emission, fuel-flexible power systems. *Progress in Energy and Combustion Science*, *27*(2), pp.141–169.

[10] Xu, Q., Zhang, F., Xu, L., Leung, P., Yang, C. and Li, H., 2017. The applications and prospect of fuel cells in medical field: A review. *Renewable and Sustainable Energy Reviews*, *67*, pp.574–580.

[11] Chakraborty, I., Olsson, R.T., Andersson, R.L. and Pandey, A., 2024. Glucose-based biofuel cells and their applications in medical implants: a review. *Heliyon*. 10(13), e33615.

[12] Ramasubramanian, B., Sundarrajan, S., Rao, R.P., Reddy, M.V., Chellappan, V. and Ramakrishna, S., 2022. Novel low-carbon energy solutions for powering emerging wearables, smart textiles, and medical devices. *Energy & Environmental Science*, *15*(12), pp.4928–4981.

[13] Mekhilef, S., Saidur, R. and Safari, A., 2012. Comparative study of different fuel cell technologies. *Renewable and Sustainable Energy Reviews*, *16*(1), pp.981–989.

[14] Yin, J., Li, X., Yu, J., Zhang, Z., Zhou, J. and Guo, W., 2014. Generating electricity by moving a droplet of ionic liquid along graphene. *Nature nanotechnology*, *9*(5), pp.378–383.

[15] Kassakian, J.G. and Jahns, T.M., 2013. Evolving and emerging applications of power electronics in systems. *IEEE Journal of Emerging and Selected Topics in Power Electronics*, *1*(2), pp.47–58.

[16] Torres, C.I., Lee, H.S. and Rittmann, B.E., 2008. Carbonate species as OH− carriers for decreasing the pH gradient between cathode and anode in biological fuel cells. *Environmental science & technology*, *42*(23), pp.8773–8777.

[17] Williams, R.H., Larson, E.D., Katofsky, R.E. and Chen, J., 1995. Methanol and hydrogen from biomass for transportation. *Energy for Sustainable Development*, *1*(5), pp.18–34.

[18] An, L., Zhao, T.S., Chen, R. and Wu, Q.X., 2011. A novel direct ethanol fuel cell with high power density. *Journal of power sources*, *196*(15), pp.6219–6222.

[19] Lin, L., Zhang, L., Luo, Y., Luo, J., Chen, C. and Jiang, L., 2022. Highly-integrated and cost-efficient ammonia-fueled fuel cell system for efficient power generation: A comprehensive system optimization and techno-economic analysis. *Energy Conversion and Management*, *251*, p.114917.

[20] Wee, J.H., 2007. Applications of proton exchange membrane fuel cell systems. *Renewable and sustainable energy reviews*, *11*(8), pp.1720–1738.

[21] Lindorfer, J., Rosenfeld, D.C. and Böhm, H., 2020. Fuel cells: Energy conversion technology. In *Future energy* (pp. 495–517). Elsevier: Amsterdam, The Netherlands.

[22] Scofield, M.E., Liu, H. and Wong, S.S., 2015. A concise guide to sustainable PEMFCs: recent advances in improving both oxygen reduction catalysts and proton exchange membranes. *Chemical Society Reviews*, *44*(16), pp.5836–5860.

[23] Chalk, S.G. and Miller, J.F., 2006. Key challenges and recent progress in batteries, fuel cells, and hydrogen storage for clean energy systems. *Journal of Power Sources*, *159*(1), pp.73–80.

[24] Potter, S., Davis, D., Born, M., Bayer, M., Howell, J. and Mankins, J., 2008. *Wireless power transmission options for space solar power* (No. MSFC-2135).

[25] Min, B., Cheng, S. and Logan, B.E., 2005. Electricity generation using membrane and salt bridge microbial fuel cells. *Water research*, *39*(9), pp.1675–1686.

[26] Dai, S., Li, M., Li, H., Shi, Y., Zhang, H., Wang, D., Xiang, K., Zou, J. and Luo, G., 2025. Advancements in Electrocatalytic Methanol Oxidation: Catalyst Design, Reaction Mechanisms, and Renewable Energy Applications. *ChemSusChem*, *18*, p.e202402767.

[27] Andersson, J. and Grönkvist, S., 2019. Large-scale storage of hydrogen. *International journal of hydrogen energy*, *44*(23), pp.11901–11919.

[28] Cayuela, M.L., Velthof, G.L., Mondini, C., Sinicco, T. and Van Groenigen, J.W., 2010. Nitrous oxide and carbon dioxide emissions during initial decomposition of animal by-products applied as fertilisers to soils. *Geoderma*, *157*(3–4), pp.235–242.

[29] Ducey, R.A., Feickert, C.A., Weaver, W., Menicucci, D., Abdallah, T., Akhil, A. and Balog, R., 2006. Control dynamics of adaptive and scalable power and energy systems for military micro grids.

[30] Huang, Z., Lu, L., Jiang, D., Xing, D. and Ren, Z.J., 2017. Electrochemical hythane production for renewable energy storage and biogas upgrading. *Applied energy*, *187*, pp.595–600.

[31] Schröder, U., 2007. Anodic electron transfer mechanisms in microbial fuel cells and their energy efficiency. *Physical Chemistry Chemical Physics*, *9*(21), pp.2619–2629.

[32] Cimenti, M. and Hill, J.M., 2009. Direct utilization of liquid fuels in SOFC for portable applications: challenges for the selection of alternative anodes. *Energies*, *2*(2), pp.377–410.

[33] Stambouli, A. Boudghene, and Enrico Traversa. "Solid oxide fuel cells (SOFCs): a review of an environmentally clean and efficient source of energy." *Renewable and sustainable energy reviews* 6, no. 5 (2002): 433–455.

[34] Schäfer, A.W., Barrett, S.R., Doyme, K., Dray, L.M., Gnadt, A.R., Self, R., O'Sullivan, A., Synodinos, A.P. and Torija, A.J., 2019. Technological, economic and environmental prospects of all-electric aircraft. *Nature Energy*, *4*(2), pp.160–166.

[35] Oh, S., Min, B. and Logan, B.E., 2004. Cathode performance as a factor in electricity generation in microbial fuel cells. *Environmental science & technology*, *38*(18), pp.4900–4904.

[36] Obileke, K., Onyeaka, H., Meyer, E.L. and Nwokolo, N., 2021. Microbial fuel cells, a renewable energy technology for bio-electricity generation: A mini-review. *Electrochemistry Communications*, *125*, p.107003.

[37] Gallert, C. and Winter, J., 2005. Bacterial metabolism in wastewater treatment systems. *Environmental Biotechnology. Concepts and Applications*.

[38] Gude, V.G., 2016. Wastewater treatment in microbial fuel cells–an overview. *Journal of cleaner production*, *122*, pp.287–307.

[39] Qiao, Y., Bao, S.J. and Li, C.M., 2010. Electrocatalysis in microbial fuel cells – from electrode material to direct electrochemistry. *Energy & Environmental Science*, *3*(5), pp.544–553.

[40] Do, M.H., Ngo, H.H., Guo, W., Chang, S.W., Nguyen, D.D., Liu, Y., Varjani, S. and Kumar, M., 2020. Microbial fuel cell-based biosensor for online monitoring wastewater quality: a critical review. *Science of The Total Environment*, *712*, p.135612.

[41] Amamou, A.A., Kelouwani, S., Boulon, L. and Agbossou, K., 2016. A comprehensive review of solutions and strategies for cold start of automotive proton exchange membrane fuel cells. *Ieee Access*, *4*, pp.4989–5002.

[42] Joghee, P., Malik, J.N., Pylypenko, S. and O'Hayre, R., 2015. A review on direct methanol fuel cells–In the perspective of energy and sustainability. *MRS Energy & Sustainability*, *2*, p.E3.

[43] Abdelkareem, M.A., Tanveer, W.H., Sayed, E.T., Assad, M.E.H., Allagui, A. and Cha, S.W., 2019. On the technical challenges affecting the performance of direct internal reforming biogas solid oxide fuel cells. *Renewable and Sustainable Energy Reviews*, *101*, pp.361–375.

[44] Hamedani, E.A., Abasalt, A. and Talebi, S., 2024. Application of microbial fuel cells in wastewater treatment and green energy production: a comprehensive review of technology fundamentals and challenges. *Fuel*, *370*, p.131855.

[45] Jamalpour, S., Shamsabadi, A., Arab, K., Tohidian, M., Tamsilian, Y., Hooshyari, K. and Rahmani, S., 2024. Two-Dimensional Nanomaterials-Based Polymer Nanocomposites for Fuel Cell Applications. *Two-Dimensional Nanomaterials-Based Polymer Nanocomposites: Processing, Properties and Applications*, pp.465–508.

[46] Pieta, I.S., Donazzi, A. and Resini, C., 2018. Fuel processing for solid oxide fuel cells. *Modeling, Design, Construction, and Operation of Power Generators with Solid Oxide Fuel Cells: From Single Cell to Complete Power System*, pp.97–141.

[47] Logan, B.E. and Regan, J.M., 2006. Microbial fuel cells – challenges and applications. *Environmental science & technology, 40*(17), pp.5172–5180.

[48] Briones, A. and Raskin, L., 2003. Diversity and dynamics of microbial communities in engineered environments and their implications for process stability. *Current opinion in biotechnology, 14*(3), pp.270–276.

[49] Abbas, S.Z., Beddu, S., Kamal, N.L.M., Rafatullah, M. and Mohamad, D., 2024. A review on recent advancements in wearable microbial fuel cells. *Journal of Environmental Chemical Engineering*, 12, p.112977.

[50] Mosa, A.S.M., Yoo, I. and Sheets, L., 2012. A systematic review of healthcare applications for smartphones. *BMC medical informatics and decision making, 12*, pp.1–31.

[51] Li, J., Cheng, J., Zhang, Y., Chen, Z., Nasr, M., Farghali, M., Rooney, D.W., Yap, P.S. and Osman, A.I., 2024. Advancements in solid oxide fuel cell technology: bridging performance gaps for enhanced environmental sustainability. *Advanced Energy and Sustainability Research, 5*(11), p.2400132.

[52] Pramanik, P.K.D., Sinhababu, N., Mukherjee, B., Padmanaban, S., Maity, A., Upadhyaya, B.K., Holm-Nielsen, J.B. and Choudhury, P., 2019. Power consumption analysis, measurement, management, and issues: A state-of-the-art review of smartphone battery and energy usage. *ieee Access, 7*, pp.182113–182172.

[53] Kamarudin, S.K., Achmad, F. and Daud, W.R.W., 2009. Overview on the application of direct methanol fuel cell (DMFC) for portable electronic devices. *International Journal of hydrogen energy, 34*(16), pp.6902–6916.

[54] Nishaa, V., Spoorthi, B.V., Soumya, B.T. and Vishwesh, S.D., 2022. Powering implantable medical devices with biological fuel cells. *ECS Transactions, 107*(1), p.19197.

[55] Sharma, V., Sharma, M., Gaur, R., Gupta, Y. and Rimza, S., 2025. Applications of biofuel cells. In *Biofuel Cells and Energy Generation* (pp. 429–452). Woodhead Publishing: Amsterdam, The Netherlands.

[56] Gao, M., Wang, P., Jiang, L., Wang, B., Yao, Y., Liu, S., Chu, D., Cheng, W. and Lu, Y., 2021. Power generation for wearable systems. *Energy & Environmental Science, 14*(4), pp.2114–2157.

[57] Ivanova, K. and Gallasch, G.E., 2016. Analysis of Emerging Technologies and Trends for ADF Combat Service Support 2016. *DST Group Report DST-Group-GD-0946, Defence Science and Technology Group Land Division, Edinburgh SA, 5111.*

[58] Itani, K. and De Bernardinis, A., 2023. Review on new-generation batteries technologies: trends and future directions. *Energies, 16*(22), p.7530.

[59] Gao, M., Wang, P., Jiang, L., Wang, B., Yao, Y., Liu, S., Chu, D., Cheng, W. and Lu, Y., 2021. Power generation for wearable systems. *Energy & Environmental Science, 14*(4), pp.2114–2157.

[60] Dolan, C., Gangi, J., Homann, Q., Fink, V. and Kopasz, J., 2020. *2019 Fuel Cell Technologies Market Report* (No. ANL-20/58). Argonne National Lab.(ANL), Argonne, IL (United States).

[61] Hughes, A.E., Haque, N., Northey, S.A. and Giddey, S., 2021. Platinum group metals: A review of resources, production and usage with a focus on catalysts. *Resources, 10*(9), p.93.

[62] Brouwer, J., 2010. On the role of fuel cells and hydrogen in a more sustainable and renewable energy future. *Current Applied Physics, 10*(2), pp.S9–S1

[63] Abdelkareem, M.A., Elsaid, K., Wilberforce, T., Kamil, M., Sayed, E.T. and Olabi, A., 2021. Environmental aspects of fuel cells: A review. *Science of The Total Environment, 752*, p.141803.

[64] Ferella, F., Innocenzi, V. and Maggiore, F., 2016. Oil refining spent catalysts: A review of possible recycling technologies. *Resources, Conservation and Recycling, 108*, pp.10–20.

[65] Hassan, N.S., Jalil, A.A., Rajendran, S., Khusnun, N.F., Bahari, M.B., Johari, A., Kamaruddin, M.J. and Ismail, M., 2024. Recent review and evaluation of green hydrogen production via water electrolysis for a sustainable and clean energy society. *International Journal of Hydrogen Energy, 52*, pp.420–441.

[66] Chalk, S.G. and Miller, J.F., 2006. Key challenges and recent progress in batteries, fuel cells, and hydrogen storage for clean energy systems. *Journal of Power Sources*, *159*(1), pp.73–80.

[67] Pistoia, G., 2008. Battery operated devices and systems: From portable electronics to industrial products. The Netherlands: Elsevier, Amsterdam.

[68] Huang, L., Li, T., Liu, C., Quan, X., Chen, L., Wang, A. and Chen, G., 2013. Synergetic interactions improve cobalt leaching from lithium cobalt oxide in microbial fuel cells. *Bioresource technology*, *128*, pp.539–546.

[69] Singh, S., Sharma, P.K., Moon, S.Y. and Park, J.H., 2024. Advanced lightweight encryption algorithms for IoT devices: survey, challenges and solutions. *Journal of Ambient Intelligence and Humanized Computing*, 15, pp.1–18.

[70] Song, C., 2002. Fuel processing for low-temperature and high-temperature fuel cells: Challenges, and opportunities for sustainable development in the 21st century. *Catalysis today*, *77*(1–2), pp.17–49.

[71] Chalk, S.G. and Miller, J.F., 2006. Key challenges and recent progress in batteries, fuel cells, and hydrogen storage for clean energy systems. *Journal of Power Sources*, *159*(1), pp.73–80.

[72] Zou, Y. and Li, Z., 2024. Self-Powered Sensors for Wearable Detections. *Portable and Wearable Sensing Systems: Techniques, Fabrication, and Biochemical Detection*, pp.207–232.

[73] Kaufman, D.R., Patel, V.L., Hilliman, C., Morin, P.C., Pevzner, J., Weinstock, R.S., Goland, R., Shea, S. and Starren, J., 2003. Usability in the real world: assessing medical information technologies in patients' homes. *Journal of biomedical informatics*, *36*(1–2), pp.45–60.

[74] Wagih, M., Balocchi, L., Benassi, F., Carvalho, N.B., Chiao, J.C., Correia, R., Costanzo, A., Cui, Y., Georgiadou, D., Gouveia, C. and Grosinger, J., 2022. Microwave-enabled wearables: Underpinning technologies, integration platforms, and next-generation roadmap. *IEEE Journal of Microwaves*, *3*(1), pp.193–226.

Mradul Kumar, Amit Kumar Rathoure

# 7 Fuel Cell Electric Vehicles (FCEVs): Recent Trends and Challenges

**Abstract:** Fuel cell electric vehicles (FCEVs) are becoming increasingly popular as a clean and efficient transport solution compared to internal combustion and battery electric vehicles (BEVs) because they have high energy efficiency, quick refueling time, and no emissions. This chapter discusses the development and principle of operation of fuel cells, which produce electricity by the electrochemical reaction of hydrogen and oxygen. It summarizes major types of fuel cells, such as proton exchange membrane fuel cells, solid oxide fuel cells, alkaline fuel cells, direct methanol fuel cells, and microbial fuel cells, mentioning their mechanisms, advantages, and drawbacks. The challenges of high manufacturing costs, inferior hydrogen infrastructure, storage, and intricate energy management systems are also mentioned. The relative strengths and weaknesses of each technology are illustrated with a comparison to BEVs. The application of energy storage devices, such as batteries and supercapacitors, to enhance FCEV performance is discussed together with the latest developments in artificial intelligence integration and hybrid energy systems. This chapter also gives the directions for the future in terms of the necessity for green hydrogen production, infrastructure creation, and facilitation policies to ensure mass adoption of FCEVs for clean transportation.

**Keywords:** Fuel cell electric vehicles, combustion, battery, emissions, electrochemical, hydrogen, clean transportation

## 7.1 Introduction

Fuel cells were first introduced in the nineteenth century when Sir William Grove, a Welsh scientist, first demonstrated the basic principle of a fuel cell in 1839. From 1950 to 1960, fuel cell technology gained popular attention, particularly its use in space programs [1]. The Gemini V spacecraft was the first to use fuel cells in 1965. Since then, the development of fuel cells has been a continuous pursuit, with researchers and engineers working to improve their efficiency, reliability, and cost-effectiveness. Fuel cells work by using a chemical reaction between hydrogen and oxygen to generate electricity, with water being the only by-product. What makes them stand out from traditional combustion engines is their ability to deliver cleaner energy much more efficiently, often reaching efficiency levels as high as 79% [2]. Due to their significantly higher energy efficiency – often surpassing 60% – fuel cell vehicles are particularly alluring. By contrast, conventional engines lag well behind, with diesel engines com-

 | https://doi.org/10.1515/9783112181072-007

ing in at only 4% and gasoline engines at an average of about 22% [3]. Fuel cells provide a clean and effective means of powering electric vehicles, which is why they are such a promising alternative. Fuel cell electric vehicles (FCEVs) are starting to look like a viable alternative to both conventional engines and battery-powered EVs because of their special blend of powerful energy performance, fuel efficiency, long driving range, and environmental advantages. They are a clever way to reduce greenhouse gas emissions in the transportation sector because they produce electricity by reacting hydrogen and oxygen, with the only by-product being water.

The transportation sector is a significant contributor to greenhouse gas emissions, accounting for 20–25% of global energy consumption and carbon dioxide emissions [4]. Looking for other environmentally friendly substitutes to replace conventional energy sources for vehicles is urgently required as the demand for personal transportation keeps rising. Fuel cells can be used as an energy source to minimize the environmental impact of the transportation sector. Also, the contribution of fuel cell energy vehicles in reducing the greenhouse effect is worth considering. Fuel cells can achieve around 60% efficiency, which makes them an energy-efficient option than conventional engines running over conventional fuel sources, which operate at 22% or even 4% efficiency [5]. Figure 7.1 shows the global distribution of FCEVs, with the highest adoption in countries like Japan, South Korea, the United States, and parts of Europe, driven by strong hydrogen infrastructure and supportive policies FCEVs are a more eco-friendly option for the transportation sector due to their higher efficiency, which reduces energy consumption and emissions.

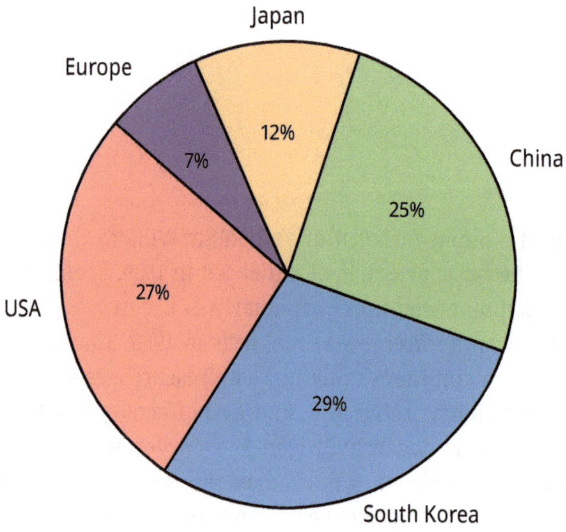


**Figure 7.1:** Distribution of FCEVs around the world (adapted after [6]).

Fuel cells possess enormous potential, but FCEVs still face many challenges as they have a short driving range, high cost, and lack of refueling stations. Fuel cells and batteries cannot replace internal combustion engines as of now. Expensive hydrogen production and risks involved in storing it safely. These challenges are need to be addressed for the further adoption of fuel cell technology, as it possesses a massive potential for the future [7].

Development in fuel cell technology is still in its early phase and is focused upon making this technology low cost, adaptive, and productive. These developments are necessary for fuel cells to make it a truly sustainable energy source. Some developments have been made in this field like the making of different types of fuel cells, the enhancement of membrane materials, and the exploration of innovative research methods, all of which will further advance the technology. Some developments in membrane materials for proton exchange membrane fuel cells (PEMFCs) are made like, they are now more durable, chemically stable, and capable of conducting protons due to innovations like metal-organic frameworks, polymer blends, and hybrid membranes. Polybenzimidazole (PBI)-based membranes, among others, had shown improved performance at elevated temperatures. Techniques such as cross-linking and the use of nanofillers further enhanced membrane durability, while acid-base membranes, including those based on PBI, have shown better performance at high temperatures.

## 7.2 Classifications of Fuel Cells

### 7.2.1 Proton Exchange Membrane Fuel Cells

PEMFCs use a polymer membrane to move protons from the anode to the cathode. These fuel cells have a higher power output and can operate at low temperatures which is a favorable condition for vehicles and mobile power systems, that's is why they are highly recommended to use in portable electronic devices and transportation systems [8]. Hydrogen gas is supplied to PEMFCs, which is converted into protons and electrons at the anode. In this process of operation, electrons move through the external circuit to generate electricity, whereas protons pass through the polymer membrane to reach the cathode, water is the by-product of their reaction with oxygen [9].

PEMFCs are most popular due to their high efficiency, very low emissions, and quick startup time. They are compact and lightweight, which makes them an ideal choice for use in cars and portable electronic devices [10]. However, there are some limitations of PEMFCs that can affect the working of these fuel cells as they are expensive and sensitive to impurities. Researchers are actively working to reduce costs and boost efficiency of these fuel cells. The research is mainly focused upon developing better system integration and control strategies also upon researching novel materials like improved electrode designs, advanced membranes, and substitute catalysts [11, 12].

### 7.2.2 Solid Oxide Fuel Cells

Solid oxide fuel cell (SOFC) is an electrochemical device that operates at high temperatures and uses solid ceramic electrolyte to move oxygen ions. SOFC uses fuels like hydrogen or natural gas and produce electricity, water, and heat, which can also be reused. SOFCs are efficient, low in emissions, different fuels could also be used like biogas, and last for a longer duration. As they operate on high temperature, which may leads to more stress, material degradation and low life of the fuel cells, so research is going on to improve materials and system design [13].

### 7.2.3 Alkaline Fuel Cells

Alkaline fuel cells (AFCs) are the type of fuel cells that carry ions from the anode to the cathode using an electrolyte like potassium hydroxide (KOH), which is alkaline in nature, as well as AFCs are sensitive to carbon dioxide and other contaminants which restrict their performance. AFCs are usually utilized in applications like spaceships and submarines during early days, because of their great efficiency and long lifespan. The anode of AFCs receives hydrogen gas, which is then divided into protons and electrons. While the electrons move through an external circuit to produce electricity, the protons are carried to the cathode via the alkaline electrolyte. Protons and oxygen mix at the cathode to create water [14].

AFCs need high-purity hydrogen and oxygen to function effectively, and can function comparatively at low temperature, ranging between 70 and 90 °C, when compared to other fuel cells. Fuel impurities may lead to electrolyte to deteriorate and do not perform well. AFCs have a long lifespan, great efficiency, and quick reaction times, among other benefits. Because they use low-cost materials and electrolyte can easily be recycled, they may also be reasonably priced. They do have drawbacks too, though, such as their sensitivity to contaminants, carbon dioxide, and requirement for highly pure reactants.

### 7.2.4 Direct Methanol Fuel Cells

Direct methanol fuel cells (DMFCs) use methanol as the fuel and the proton exchange membrane (PEM) as the electrolyte. Due to the features like easy to handle and store, high energy density, and could be used in portable electronic devices. These fuel cells are actively researched among the parallelly available options. At the anode, methanol is passed to DMFCs, where the catalyst is oxidized to yield protons, electrons, and carbon dioxide. Whereas the electrons move through the outer circuit to produce current, PEM carries the protons to the cathode. At cathode, water is formed and oxygen is decreased. The overall reaction of DMFC is given below:

$$CH_3OH + 1.5\ O_2 \rightarrow CO_2 + 2H_2O$$

Discussing over the benefit side, DMFCs possess a high energy density, low emissions, and an easy handling and storage are some of the benefits of DMFCs. Because of their simple and easy design, they convert fuel into hydrogen without any help of an external reformer [15]. There are some drawbacks of DMFCs that can affect their performance like low reaction times and comparatively low efficiency. Research is going on to solve these challenges and to work on improving affordability, performance, and its scalability. The major focus of these research is to develop better system integration, smarter control strategies, and the development of new membrane materials and catalyst-like use of non-noble catalyst as they are cheap and equally effective too [16]. One approach would be to make catalyst enhancement used in DMFCs to speed up the methanol oxidation reaction and improve overall efficiency. We can also tune the fuel cell's design, such as optimizing the flow channels and membrane electrode assembly to boost performance parallelly keeping the costs down [17].

## 7.2.5 Microbial Fuel Cells

Microbial fuel cells (MFCs) are the devices that use the catabolic activities of the microorganism to produce electricity and organic materials. These fuel cells show great promise for things like treating sewage and wastewater efficiently [18], producing sustainable electricity [19], it could also be used as a sensor to analyze pollutants, and control and monitor processes [18]. In MFCs, the bacteria are floating in a liquid or attached to an electrode. These bacteria feed upon materials like wastewater or food waste, they release electrons. These electrons travel to the anode and then move through an outer circuit to the cathode and generate electricity. At cathode, the electrons react with oxygen to form water. The overall process can be summed up in a simple following reaction:

$$\text{Organic matter} + O_2 \rightarrow CO_2 + H_2O + \text{electricity}$$

Their design is also quite simple, which means they don't need expensive catalysts or highly purified reactants. However, there are some challenges faced by MFCs, like slow reaction rates, low efficiency, and practical application of these fuel cells on a larger scale poses substantial challenges. To fix these challenges, research is going on to improve performance and reduce its costs. However, a better system design, smarter control techniques, and the development of new materials, such as improved electrodes and membranes are some of the main focused areas. Another possible approach is to enhance the performance of the microbes itself, for example, by using mixed microbial cultures to improve electron transfer or by applying genetically modified bacteria.

## 7.3 Working Principle of Fuel cells

Fuel cells are amazing electrochemical devices that can directly transform the energy stored in fuels, like hydrogen or natural gas, into usable electrical energy through an electrochemical process. The very core of this conversion process is the fuel cell, which generates heat, water, and electricity by combining an oxidant (generally atmospheric oxygen) with a fuel (usually hydrogen) in a controlled atmosphere [20]. The electrochemical reaction between the fuel (usually hydrogen) and the oxidant (usually oxygen from the air) inside the fuel cell forms the basis of a fuel cell's basic operation [21]. An oxidant, usually oxygen, is reduced at the cathode, and a fuel, like hydrogen, is oxidized at the anode in this reaction. The ions can move between the anode and cathode to complete the electrical circuit through the electrolyte, a substance that permits the flow of ions but not electrons. The reaction between hydrogen and oxygen in a fuel cell can be summed up as the production of electrical energy and water. After being delivered to the anode, the hydrogen fuel undergoes an oxidation reaction that releases electrons and creates protons. Detailed working of fuel cell is discussed in Figure 7.2.

Reaction taking place at anode: $2H_2 \rightarrow 4H^+ + 4e^-$

Reaction taking place at cathode: $O_2 + 4e^- + 4H^+ \rightarrow 2H_2O$

Overall cell reaction: $2H_2 + O_2 \rightarrow 2H_2O(l) + \text{electricity} + \text{heat}$

## 7.4 Key Challenges

FCEVs are a reliable alternative to traditional cars and other vehicles, but there are still some major challenges which are on its way for widespread adoption. These include high production costs of hydrogen, limited hydrogen refueling stations and infrastructure, and the complexity of managing energy systems and its storage. To make FCEVs a more practical option for sustainable transportation, it's important to tackle these issues. The following sections take a closer look at the key challenges FCEVs currently face.

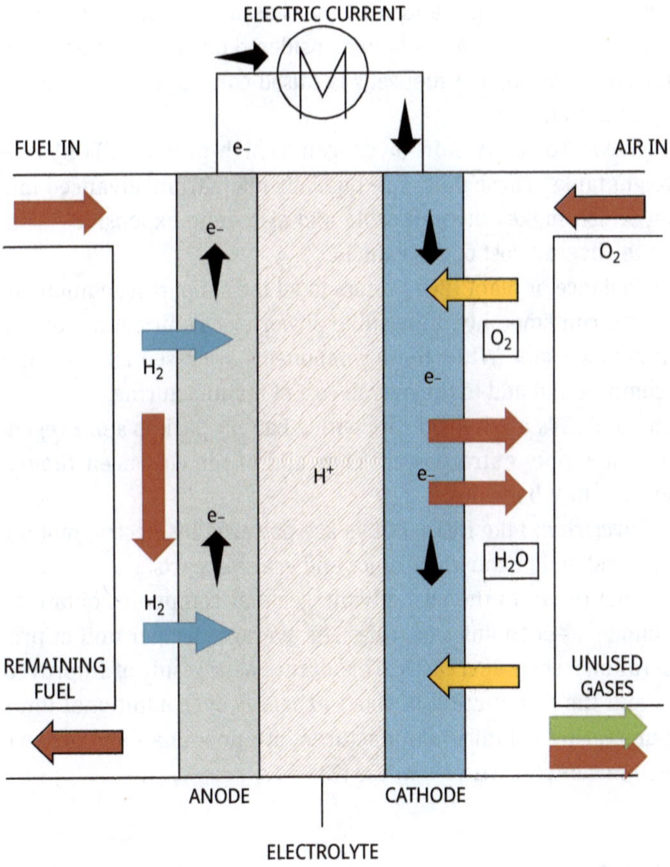


**Figure 7.2:** Schematic representation of the actual working of a fuel cell.

## 7.4.1 High Production Costs

*Manufacturing Costs*: Right now, FCEVs are more expensive to produce than traditional internal combustion engine vehicles (ICEVs) and often even battery electric vehicles (BEVs). However, as technology continues to improve and production scales up, these costs are expected to come down. Below is a breakdown of the key factors that contribute to the overall cost of producing FCEVs.

### 7.4.1.1 Principal Cost Elements

*Fuel Cell Stack*: The heart of every FCEV is the fuel cell stack – the part responsible for converting hydrogen into electricity. It's made up of several high-cost components, like bipolar plates, advanced membranes, and platinum-based catalysts. Because of these expensive materials, the fuel cell stack is often considered one of the most costly parts of the vehicle. The cost of producing fuel cells is raised considerably by the use of pricey elements like platinum [22].

*Hydrogen Storage System*: To safely store hydrogen at high pressure, FCEVs require strong yet lightweight tanks. These tanks are typically made from advanced materials like carbon fiber, which makes them durable and also quite expensive – contributing significantly to the overall cost of the vehicle.

*Balance of Plant*: The balance of plant (BOP) refers to all the supporting equipment that helps a fuel cell system run smoothly, things like power electronics, humidifiers, cooling systems, and air compressors. While these components are essential, they also make the system more complex and add to the overall cost of manufacturing.

*Battery*: The majority of FCEVs are hybrid cars with a battery pack to store regenerative braking energy and supply extra power. They add to the cost even though they are usually smaller than BEV batteries.

*Electric Motor and Powertrain*: Like BEVs, FCEVs are powered by electric motors, which are accompanied by related power electronics and gearbox parts.

*Economies of Scale*: This refers to the cost advantages that companies obtain as their output increases, known as economies of scale. The average cost per unit of production falls as the operation's scale increases. This occurs as a result of improved operational efficiencies and the distribution of fixed expenses over additional units due to FCEVs' low manufacturing volume, manufacturers are now unable to achieve economies of scale, which makes cost worries worse [23].

## 7.4.2 Hydrogen infrastructure

For FCEVs to be widely accepted, the hydrogen infrastructure is essential. It includes all of the equipment and facilities required to manufacture, store, transport, and fill up these cars with hydrogen fuel. To fully utilize FCEVs as a clean transportation medium, a strong hydrogen infrastructure must be built.

### 7.4.2.1 Hydrogen Production

There are several ways to produce hydrogen. Some of the most popular and economical techniques at the moment are as follows:

*Steam Methane Reforming (SMR)*: It creates carbon dioxide as a by-product (sometimes known as "grey hydrogen").

*Electrolysis*: This is the process of separating water into hydrogen and oxygen using electricity. This process is referred to as "green hydrogen" and produces very little emissions if it is powered by renewable energy sources.

*Alternative Techniques*: Alternative techniques, such as thermochemical water splitting, biomass gasification, coal gasification, and biological processes, are also being investigated.

*Refueling Stations*: The practicality of FCEVs is limited by the lack of hydrogen infrastructure filling stations, as buyers are reluctant to purchase cars without readily available refueling options. As of the end of 2023, there were approximately 900 hydrogen refueling stations worldwide, with the majority located in Asia (Japan, South Korea, and China) and North America (primarily California). Europe is also expanding its hydrogen refueling network. However, the distribution remains limited compared to gasoline or electric charging infrastructure. In 2024, around 125 new hydrogen refueling stations were opened worldwide, 42 in Europe, around 30 in China, 25 in South Korea, 8 in Japan and 13 in North America.

*Investment Requirements*: A major obstacle to progress is the significant amount of money needed to establish a complete hydrogen infrastructure [23]. The cost assumptions used in the station analysis are summarized in Table 7.1.

**Table 7.1:** Cost assumptions used in station analysis [24].

| | $CH_2$ stations | | | $LH_2$ stations | | |
|---|---|---|---|---|---|---|
| | 2023 | 2030 | 2040 | 2023 | 2030 | 2040 |
| Average new station capacity (tons/day) | 1.5 | 1.5 | 2.5 | 4 | 5 | 10 |
| Capital cost ($1,000/kg/day of capacity) | $4.7 | $3.6 | $3.0 | $2.7 | $2.4 | $1.8 |
| Capital cost per kg produced at 80% utilization | $2.1 | $1.6 | $1.3 | $1.1 | $1.1 | $0.8 |
| Fixed OPEX at 80% utilization ($/kg) | $0.9 | $0.8 | $0.7 | $0.7 | $0.7 | $0.6 |
| Variable operating cost ($/kg) | $0.8 | $0.7 | $0.5 | $0.6 | $0.5 | $0.4 |

### 7.4.3 Hydrogen Storage Challenges

*Storage Technology*: Because it directly affects the vehicle's weight, range, and overall performance, on-board storage technology is a crucial component of FCEVs. Despite having a large mass energy content, hydrogen has a low volumetric energy density, which makes compact and effective storage a major difficulty. Because existing technologies can be cumbersome and costly, efficiently storing hydrogen continues to be a major concern [23].

The most often utilized storage mechanism in commercial FCEVs at the moment is compressed hydrogen gas storage, high pressures (usually 350 or 700 bar) are used to compress hydrogen gas, which is then kept in tanks made specifically for this purpose. This is a relatively mature technology, allowing for fast refueling times (similar to gasoline vehicles); however, on the disadvantage side, it requires high pressures, leading to heavy and bulky tanks.

Liquid hydrogen storage is another method of hydrogen storage in its liquid state at cryogenic temperatures (−253 °C or 20 K) [25]. One of the advantages is that it has higher volumetric energy density compared to compressed gas, potentially allowing for longer driving ranges with smaller tanks. Talking about its disadvantages, it maintains extremely low temperatures, leading to energy losses due to boil-off. The liquefaction process is energy-intensive. This technology is not very much common in passenger FCEVs but can be used for heavy-duty and aerospace applications.

*Safety Issues*: Due to high flammability of hydrogen, there are safety issues that need strict safety regulations in infrastructure and vehicle design [6].

### 7.4.4 Complexities of Technology

*Energy Management Systems*: To get the best performance, FCEVs rely on advanced energy management systems. While these systems are crucial, they can also make the vehicle design more complex and add to the overall cost [26].

*Cold Start Problems*: FCEVs can struggle to start in cold weather, which may affect their performance and reliability, especially in regions with harsh winters [6].

Despite these challenges, many researchers believe that ongoing technological improvements and greater investment in hydrogen infrastructure could help overcome these hurdles over time. As the demand for zero-emission vehicles continues to grow, FCEVs may become a strong complement to (BEVs), especially for uses that need longer driving ranges and faster refueling.

## 7.5 Comparative Analysis with Other Vehicle Technologies

FCEVs and BEVs are emerging as appealing alternatives to traditional gasoline-powered cars, offering the potential to cut down greenhouse gas emissions and reduce our reliance on fossil fuels. This chapter explores how these two innovative technologies stack up, comparing their performance, environmental impact, and cost considerations.

Thanks to recent improvements in battery technology and powertrain design, BEVs have come a long way. Many of them can now travel over 150 miles on a single charge and offer acceleration that rivals traditional gasoline-powered cars [27]. BEVs also offer major environmental benefits, since they produce no direct emissions while driving, helping to improve air quality, especially in urban areas. However, how eco-friendly they truly are depending a lot on how the electricity used to charge them is to be generated. Their environmental impact is much lower when that power comes from renewable sources like wind or solar energy[28].

FCEVs run on hydrogen, which is converted into electricity through a chemical reaction inside the fuel cell. Like BEVs, they produce no direct emissions, and they also offer some extra perks such as longer driving ranges and faster refueling times. That said, the high cost of fuel cell technology and the lack of widespread infrastructure for producing, transporting, and storing hydrogen have been major barriers to their broader adoption [29].

### 7.5.1 FCEVs Versus BEVs

There are certain benefits and drawbacks for each technology when comparing FCEVs and BEVs as shown in Table 7.2. Though they differ greatly in terms of infrastructure requirements, operational effectiveness, and market dynamics, both are essential to the shift to zero-emission transportation. An organized comparison of the main features of FCEVs and BEVs can be found in Table 7.2.

**Table 7.2:** Comparative analyses of FCEVs and BEVs [30].

| S. no. | Feature | BEVs | FCEVs |
|---|---|---|---|
| 1 | Energy source | Lithium-ion batteries | Hydrogen fuel cell |
| 2 | Performance and efficiency | Lower operational cost and higher efficiency | Longer driving ranges and quicker refueling times. |
| 3 | Infrastructure requirements | Highly dependent upon charging stations | Require hydrogen refueling stations, which are limited in number, hence impacting accessibility [1] |
| 4 | Efficiency | 73% | 22% |
| 5 | Environmental impact | The environmental impact depends upon the source of electricity | The method of hydrogen production plays a key role in deciding its environmental impact |
| 6 | Weight | Reduced powertrain weight in comparison to BEVs' massive battery packs, which could enable larger cars to carry more cargo | The weight of battery packs can affect payload and possibly range |
| 7 | Recycling | Fuel cell components can be recycled easily, for example, platinum | Battery recycling is complex and still in a development phase |

### 7.5.2 Sustainability Factors and Environmental Impact

Because of their promise for improved sustainability and a smaller environmental effect than conventional ICEVs, FCEVs have become a viable alternative. High energy efficiency, minimal direct emissions, and the capacity to produce hydrogen from renewable energy sources are just a few benefits of fuel cell technology [31].

When compared to conventional gasoline-powered vehicles, FCEVs have the potential to lower greenhouse gas emissions, which is one of its primary sustainability features. Clean hydrogen production techniques are crucial since the life cycle assessment of fuel cell vehicles has revealed that the usage stage frequently has the highest energy consumption and environmental effect [31, 28]. The environmental advantages of fuel cell vehicles will increase as the electrical grid and hydrogen production processes depend more on renewable energy sources [28].

Additionally, by diversifying the transportation energy mix, fuel cell vehicles help lessen the industry's reliance on finite fossil fuels [28]. By boosting resilience and energy security, this can improve the transportation system's overall sustainability. However, there are still a number of obstacles preventing the widespread use of FCEVs, including the high initial cost in comparison to traditional gasoline vehicles and the scarcity of infrastructure for hydrogen refueling.

## 7.6 Energy Storage Systems for Fuel Cell-Based Electric Vehicles (FCEVs)

The energy storage system (ESS) is the backbone of FCEVs. Specific energy, energy density, working temperature, life expectancy, and cost are the most crucial variables to take into account while selecting the optimal storage media for FCEVs and fuel cell hybrid electric vehicles (FCHEVs). Figure 7.3(a) shows different ESSs used in FCHEVs and FCEVs, including fuel cells, batteries, and supercapacitors for efficient energy management and Figure 7.3(b) illustrates the drivetrain configuration of an FCEV, highlighting key components such as the hydrogen tank, fuel cell stack, power electronics, and electric motor used for vehicle propulsion. Nowadays, fuel cells are the most common source of electricity for FCEVs. One of these sources has a significant energy supply capability and is referred to as the prime energy source (PES) [32]. In comparison, the rechargeable power storage arrangement (RPSA) has a high power capacity and the ability to reverse [33, 34]. A prime source is used to sustain a constant power stream, whereas RPSA is employed for reformative slowing and transitory power streams. When it comes to PES cold starts and unexpected high-energy demands, like speeding up an automobile, the RPSA bears a major duty. Batteries and fuel cells have better energy densities but lower power densities, per published studies [35, 36]. Batteries and fuel cells are hence appropriate PES. Different kinds of flywheels and capacitors, on the other hand, have a low energy density, a high-power concentration, and a quick discard rate, suggesting that they can produce a lot of power but that it would only survive a short while. They can be used as RPSA because of their characteristics. The energy sources and storage systems used by FCEVs and FCHEVs are described in the subsections below.

## 7.6.1 Power Storage Sources

A reliable ESS plays an important role in FCEVs and FCHEVs in terms of power supply and braking energy recovery. Minimizing hydrogen consumption, protecting the fuel cell from the rapid dynamic load, storing braking energy and maintaining power over the load within the prescribed limits can be achieved with ESS [37]. Each vehicle application has a different selection of ESS and places its own requirements on them. Battery and supercapacitors are the two which widely used ESS in the FCHEV.

A battery is a device that converts chemical energy into electrical energy through electrochemical processes and fulfills several functions. It not only supplies the vehicle with electricity and helps to recover braking energy, but can also act as a primary energy supplier to cover the vehicle's power requirements [38]. In this scenario, the fuel cell serves as a constant energy source that replenishes the battery's state of charge to the initial level, especially in the case of Plug-in Fuel Cell Hybrid Electric Vehicle (PFCHEV) and Fuel Cell Range-Extended Electric Vehicle (FCREEV) [39]. Improving performance, increasing longevity, cutting down on charging time, and lowering prices are the main goals of battery development. Li-ion, Ni-MH, and lead acid batteries are all commonly used; however, because of their high specific energy and energy density, Ni-MH and Li-ion batteries are preferred in battery-powered vehicles [40]. Battery characteristics like capacity, cycle life, cost, and charging rate have greatly improved over decades of research and commercial use, satisfying a sizable percentage of EV needs [41, 42].

In several studies, a supercapacitor has also been selected as the vehicle's ESS. Due to the porous carbon electrodes' large specific active surface area, they have a significantly higher energy storage density than conventional capacitors, which can reach up to 12 W h/kg [43]. Supercapacitors have a long lifespan, low maintenance costs, a large operating temperature range, a strong tolerance to overcharge and over temperature, and comparatively lower prices [44]. In addition to ultra-fast charging and discharging, the supercapacitor can handle relatively higher power demands without shortening its lifespan. In contrast to the new quick lithium-ion battery, which can be charged to 70% capacity in a matter of minutes, a supercapacitor can take anywhere from 1 to 10 s to charge. As a result, the supercapacitor is a great option for FCHEV's peak-load provider [45, 46]. In FCHEVs, the battery and the supercapacitor can be operated separately as an ESS. Another promising approach is to combine batteries with supercapacitors to create a multi-source energy storage system (MESS). Research has shown that this kind

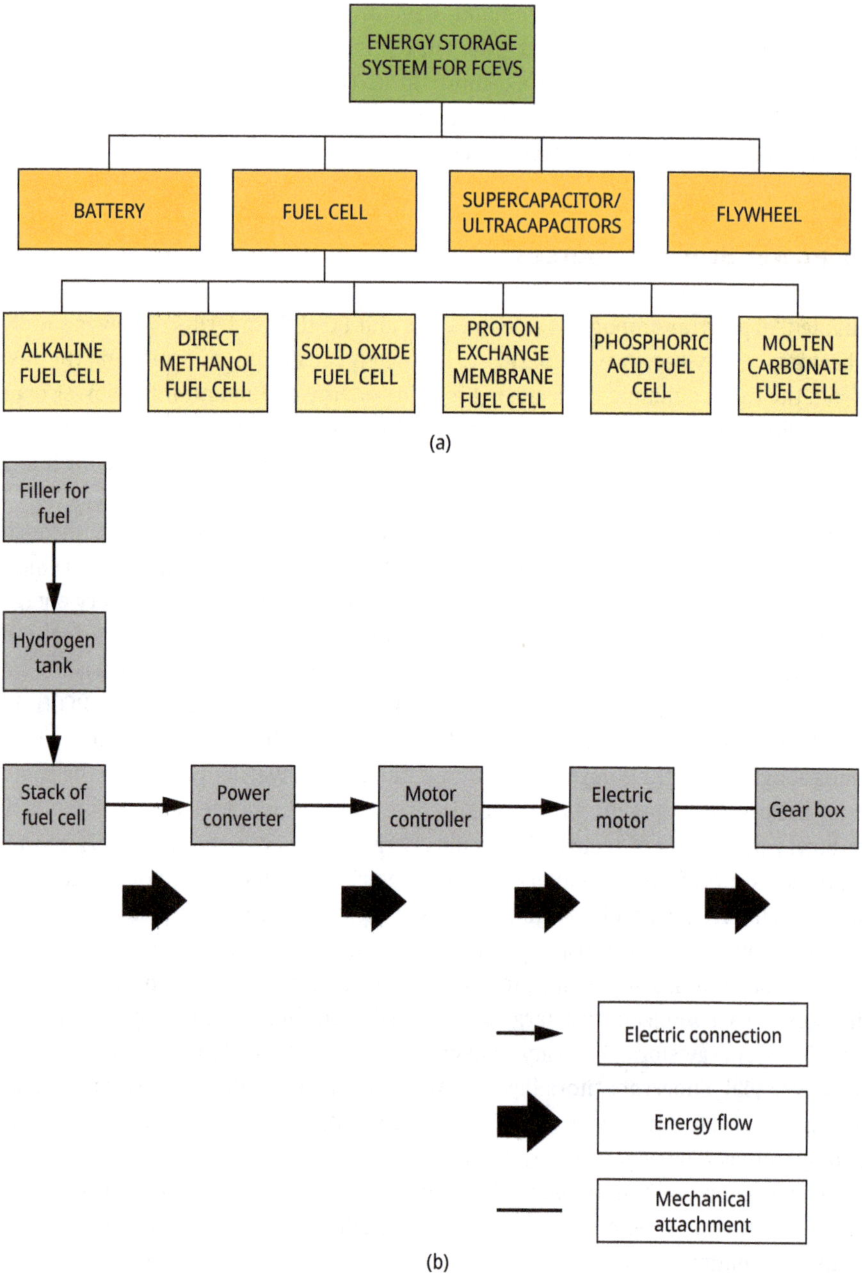


**Figure 7.3:** (a) Different FCHEV and FCEV energy storage systems. (b) FCEV drivetrain configuration.

of hybrid setup is a strong alternative, as it can meet the demands for both high energy and high power – making it well-suited for FCHEVs.

According to research cited in [47, 48] and [49], the battery functions as the primary ESS and the supercapacitor as the source of peak power. Additionally, the supercapacitor serves as the only ESS in studies like [50, 51], providing kinetic energy to the propulsion system. This involves recovering energy during braking or deceleration and delivering high-torque output during acceleration or high-speed operation [52]. A comparison of the three-battery and supercapacitor types is shown in Table 7.3 [53–55].

**Table 7.3:** A comparison of three-battery chemistries and supercapacitor technologies [56–59].

| Categories | Lead-acid battery | Li-ion battery | Ni-MH battery | Supercapacitors |
|---|---|---|---|---|
| Energy density (Wh/kg) | 30–40 | 100–120 | 60–80 | 4–15 |
| Power density (W/kg) | 200–300 | 600–2,000 | 800–1,500 | 1,000–10,000 |
| Life cycle (times) | 300–400 | >1,000 | >1,000 | >100,000 |
| Efficiency (%) | 75 | 90 | 75 | 85–98<br>Fast rates of charge and discharge combined with a remarkably long lifespan |
| Strength | Reduced expenses, quick release, and effective recycling | Greater energy density, higher voltage, reduced self-discharge, lightweight design, longer cycle life, and no memory effect | Increased capacity for energy storage, quick charging and discharging, and longer lifespan | |
| Weaknesses | Decreased efficacy in cold climates and a shorter lifespan | Reduced lifespan in high temperatures, rigorous security requirements, and a tight prohibition against overcharging and over-discharging | High production costs, significant self-discharge, and the need for cooling systems | Restricted energy density |

## 7.7 Conclusion and Future Scope

FCEVs offer one of the most promising routes to sustainable and zero-emission mobility, providing benefits of high efficiency, fast refueling, and extended driving ranges. Numerous key challenges remain, however, including high production costs, minimal hydrogen infrastructure, and safety and regulation issues. Some of the latest studies have concentrated on hybrid energy management schemes that combine fuel cells and batteries to maximize system efficiency and dependability. In addition to that, artificial intelligence (AI) integration has revealed potential toward optimizing fuel cell performance, facilitating predictive maintenance, and controlling energy flow in real time.

Infrastructure development is the major bottleneck in scaling up FCEV adoption. The existing hydrogen distribution network is small and mostly catering to large industrial consumers, thus the imperative for widespread policy support, investment in public-private partnerships, and streamlined regulatory schemes for scaling up refueling stations. At the same time, the shift toward green hydrogen production through the use of renewable energy sources such as solar, wind, and hydro via electrolysis is vital in bringing down the carbon intensity and bringing sustainability to FCEVs in the long term.

Forward, the future potential of FCEVs is in the development of cutting-edge technologies like light-weighting materials, cost-effective thermal management, modular fuel cell stacks, and AI-based control systems. There is also major potential in breaking new ground in alternative hydrogen production technology, including biomass gasification and biological pathways, to lower cost and diversify supply. Strategic efforts, such as government-initiated demonstration projects, consumer incentives, and educational outreach, can expedite public acceptance and commercialization. In addition, global collaboration on standardization and harmonization with overarching renewable energy objectives will be critical to bringing about a hydrogen-based transportation infrastructure. Through concerted research efforts, policy assistance, and technological advancement, FCEVs are poised to be a keystone of clean and resilient mobility in the near term.

## References


[1] D. Novak, H.D. 2023. Mohammadian, Challenges of a sustainable energy and vehicle-related value chain for BEVs and FCEVs through the 5th wave theory, Progress Energy Fuels (PEF) J, 12, 3308.

[2] Y. Haseli, 2018. Maximum conversion efficiency of hydrogen fuel cells, International Journal of Hydrogen Energy, 43, 9015–9021.

[3] Y. Manoharan, S.E. Hosseini, B. Butler, H. Alzhahrani, B.T.F. Senior, T. Ashuri, J. Krohn, 2019. Hydrogen fuel cell vehicles; current status and future prospect, Applied Sciences, 9, 2296.

[4] D.M. Nguyen, M.A. Kishk, M.-S. Alouini, 2022. Toward sustainable transportation: Accelerating vehicle electrification with dynamic charging deployment, IEEE Transactions on Vehicular Technology, 71, 9283–9296.

[5] C. Pistidda, 2021. Solid-state hydrogen storage for a decarbonized society, Hydrogen, 2, 428–443.

[6] M. Waseem, M. Amir, G.S. Lakshmi, S. Harivardhagini, M. Ahmad, 2023. Fuel cell-based hybrid electric vehicles: An integrated review of current status, key challenges, recommended policies, and future prospects, Green Energy and Intelligent Transportation, 2, 100121.

[7] P. Prem, P. Sivaraman, J. Sakthi Suriya Raj, M. Jagabar Sathik, D. Almakhles, 2020. Fast charging converter and control algorithm for solar PV battery and electrical grid integrated electric vehicle charging station, Automatika: časopis za automatiku, mjerenje, elektroniku, računarstvo i komunikacije, 61, 614–625.

[8] S.M.R. Niya, M. Hoorfar, 2013. Study of proton exchange membrane fuel cells using electrochemical impedance spectroscopy technique–A review, Journal of Power Sources, 240, 281–293.

[9] X. Yuan, H. Wang, J.C. Sun, J. Zhang, 2007. AC impedance technique in PEM fuel cell diagnosis – A review, International Journal of Hydrogen Energy, 32, 4365–4380.

[10] R. Palanisamy, R. Sahasrabuddhe, M. Hiteshkumar, J. Puranik, 2021. A new energy regeneration system for A BLDC motor driven electric vehicle, International Journal of Electrical and Computer Engineering (IJECE), 11, 2986.

[11] Y. Zhang, H. Liu, J. Li, M. Sumner, C. Xia, 2019. DC–DC Boost Converter With a Wide Input Range and High Voltage Gain for Fuel Cell Vehicles, IEEE Transactions on Power Electronics, 34, 4100–4111.

[12] R. Palanisamy, V. Sugavanam, K. Vakesan, S. Ramamurthi, S. Kuppusamy, U. Sengamalai, T. Thentral, 2022. Implementation of multicarrier PWM based 7-level Z-source cascaded H-bridge inverter, International Journal of Power Electronics and Drive Systems (IJPEDS), 13, 322.

[13] J. Li, J. Liu, 2019. A Novel Buck–Boost Converter With Low Electric Stress on Components, IEEE Transactions on Industrial Electronics, 66, 2703–2713.

[14] X. Liu, H. Li, Z. Wang, 2014. A Fuel Cell Power Conditioning System With Low-Frequency Ripple-Free Input Current Using a Control-Oriented Power Pulsation Decoupling Strategy, IEEE Transactions on Power Electronics, 29, 159–169.

[15] X. Zhao, L. Zhang, R. Born, J.S. Lai, 2016. Solution of input double-line frequency ripple rejection for high-efficiency high-power density string inverter in photovoltaic application, 2016 IEEE Applied Power Electronics Conference and Exposition (APEC), pp. 1148–1154 IEEE: USA.

[16] K. Bhunia, J.H. Seok, M. Perumalsamy, K.S. Bejigo, V. Elumalai, S.U. Lee, S.-j. Kim, 2024. Atomically dispersed dual iron and manganese anchored nitrogen doped reduced graphene as robust cathode catalyst for direct methanol fuel cell and aluminium air battery, Nano Energy, 129, 109966.

[17] R. Palanisamy, R. Ravindran, N. Gupta, G. Sardana, I. Sekar, V. Marthanda, S. Kuppusamy, 2023. Simulation of 3D-space vector modulation for neutral point clamped inverters, Bulletin of Electrical Engineering and Informatics, 12, 719–726.

[18] P. Jalili, A. Ala, P. Nazari, B. Jalili, D.D. Ganji, 2024. A comprehensive review of microbial fuel cells considering materials, methods, structures, and microorganisms, Heliyon, 10, e25439.

[19] J.R. Kim, Y. Zuo, J.M. Regan, B.E. Logan, 2008. Analysis of ammonia loss mechanisms in microbial fuel cells treating animal wastewater, Biotechnology and bioengineering, 99, 1120–1127.

[20] B. B, 2020. A Review of Design and Implementation of Hydrogen Fuel Cell with Modified Proton Exchange Membrane, International Journal of Engineering Research and, V9.

[21] B. Huang, Y. Qi, M. Murshed, 2010. Solid Oxide Fuel Cell: Perspective of Dynamic Modeling and Control, IFAC Proceedings Volumes, 43, 133–146.

[22] T. Rahman, M.S. Miah, T.F. Karim, M.S. Hossain Lipu, A.M. Fuad, Z.U. Islam, M.M.N. Ali, M.N. Shakib, S. Sahrani, M.R. Sarker, 2024. Empowering Fuel Cell Electric Vehicles Towards Sustainable Transportation: An Analytical Assessment, Emerging Energy Management, Key Issues, and Future Research Opportunities, World Electric Vehicle Journal, 15, 484.

[23] M.S. Munsi, R.P. Joshi, 2024. Comprehensive Analysis of Fuel Cell Electric Vehicles: Challenges, Powertrain Configurations, and Energy Management Systems, IEEE Access, 145459–145482.
[24] L. Fulton, 2024. U.D.I.H.S. Team, California FCEV and Hydrogen Refueling Station Deployment: Requirements and Costs to 2050.
[25] T. Zhang, J. Uratani, Y. Huang, L. Xu, S. Griffiths, Y. Ding, 2023. Hydrogen liquefaction and storage: Recent progress and perspectives, Renewable and Sustainable Energy Reviews, 176, 113204.
[26] K. Nithin, V. Manoj, B. Mohith, 2023. Fuel cell hybrid electric vehicle: A review on current status, key challenges and future prospects, Int. J. Sci. Res. Eng. Manag., 7, 1–11.
[27] M. DeLuchi, Q. Wang, D. Sperling, 1989. Electric vehicles: performance, life-cycle costs, emissions, and recharging requirements, Transportation Research Part A: General, 23, 255–278.
[28] A. Ajanovic, R. Haas, 2019. Economic and environmental prospects for battery electric-and fuel cell vehicles: a review, Fuel cells, 19, 515–529.
[29] Q. Hassan, I.D.J. Azzawi, A.Z. Sameen, H.M. Salman, 2023. Hydrogen Fuel Cell Vehicles: Opportunities and Challenges, Sustainability, 15, 11501.
[30] A. Parikh, M. Shah, M. Prajapati, 2023. Fuelling the sustainable future: a comparative analysis between battery electrical vehicles (BEV) and fuel cell electrical vehicles (FCEV), Environmental Science and Pollution Research, 30, 57236–57252.
[31] I. Dincer, M.A. Rosen, 2011. Sustainability aspects of hydrogen and fuel cell systems, Energy for sustainable development, 15, 137–146.
[32] S.B. Rao, Y.P. Kumar, M. Amir, F. Ahmad, 2022. An adaptive neuro-fuzzy control strategy for improved power quality in multi-microgrid clusters, IEEE Access, 10, 128007–128021.
[33] D.M. Bellur, M.K. Kazimierczuk, 2007. DC-DC converters for electric vehicle applications, 2007 Electrical Insulation Conference and Electrical Manufacturing Expo, IEEE: Piscataway, NJ, USA, pp.286–293.
[34] X. Zhou, B. Sheng, W. Liu, Y. Chen, L. Wang, Y.-F. Liu, P.C. Sen, 2021. A high-efficiency high-power-density on-board low-voltage DC–DC converter for electric vehicles application, IEEE Transactions on Power Electronics, 36, 12781–12794.
[35] T. Selmi, A. Khadhraoui, A. Cherif, 2022. Fuel cell–based electric vehicles technologies and challenges, Environmental Science and Pollution Research, 29, 78121–78131.
[36] H.S. Das, C.W. Tan, A.H.M. Yatim, 2017. Fuel cell hybrid electric vehicles: A review on power conditioning units and topologies, Renewable and Sustainable Energy Reviews, 76, 268–291.
[37] I. Aschilean, M. Varlam, M. Culcer, M. Iliescu, M. Raceanu, A. Enache, M.S. Raboaca, G. Rasoi, C. Filote, 2018. Hybrid electric powertrain with fuel cells for a series vehicle, Energies, 11, 1294.
[38] N. Sulaiman, M. Hannan, A. Mohamed, E. Majlan, W.W. Daud, 2015. A review on energy management system for fuel cell hybrid electric vehicle: Issues and challenges, Renewable and Sustainable Energy Reviews, 52, 802–814.
[39] J. Deng, C. Bae, J. Marcicki, A. Masias, T. Miller, 2018. Safety modelling and testing of lithium-ion batteries in electrified vehicles, Nature Energy, 3, 261–266.
[40] J.P. Trovão, M.A. Silva, C.H. Antunes, M.R. Dubois, 2017. Stability enhancement of the motor drive DC input voltage of an electric vehicle using on-board hybrid energy storage systems, Applied Energy, 205, 244–259.
[41] O. Veneri, C. Capasso, S. Patalano, 2018. Experimental investigation into the effectiveness of a super-capacitor based hybrid energy storage system for urban commercial vehicles, Applied Energy, 227, 312–323.
[42] D. Deng, 2015. Li-ion batteries: basics, progress, and challenges, Energy Science & Engineering, 3, 385–418.
[43] C. Capasso, D. Lauria, O. Veneri, 2018. Experimental evaluation of model-based control strategies of sodium-nickel chloride battery plus supercapacitor hybrid storage systems for urban electric vehicles, Applied energy, 228, 2478–2489.

[44] A.F. Burke, 2007. Batteries and ultracapacitors for electric, hybrid, and fuel cell vehicles, Proceedings of the IEEE, 95, 806–820.
[45] J. Ruan, P.D. Walker, N. Zhang, J. Wu, 2017. An investigation of hybrid energy storage system in multi-speed electric vehicle, Energy, 140, 291–306.
[46] A. Tahri, H. El Fadil, F. Belhaj, K. Gaouzi, A. Rachid, F. Giri, F. Chaoui, 2018. Management of fuel cell power and supercapacitor state-of-charge for electric vehicles, Electric Power Systems Research, 160, 89–98.
[47] K. Kaya, Y. Hames, 2019. Two new control strategies: For hydrogen fuel saving and extend the life cycle in the hydrogen fuel cell vehicles, International Journal of Hydrogen Energy, 44, 18967–18980.
[48] Y. Wang, Z. Sun, Z. Chen, 2019. Rule-based energy management strategy of a lithium-ion battery, supercapacitor and PEM fuel cell system, Energy Procedia, 158, 2555–2560.
[49] H. Jiang, L. Xu, J. Li, Z. Hu, M. Ouyang, 2019. Energy management and component sizing for a fuel cell/battery/supercapacitor hybrid powertrain based on two-dimensional optimization algorithms, Energy, 177, 386–396.
[50] M. Bougrine, A. Benalia, E. Delaleau, M. Benbouzid, 2018. Minimum time current controller design for two-interleaved bidirectional converter: Application to hybrid fuel cell/supercapacitor vehicles, international journal of hydrogen energy, 43, 11593–11605.
[51] H. Fathabadi, 2018. Fuel cell hybrid electric vehicle (FCHEV): Novel fuel cell/SC hybrid power generation system, Energy Conversion and Management, 156, 192–201.
[52] X. Lü, Y. Qu, Y. Wang, C. Qin, G. Liu, 2018. A comprehensive review on hybrid power system for PEMFC-HEV: Issues and strategies, Energy conversion and Management, 171, 1273–1291.
[53] H. Wang, Q. Wang, B. Hu, 2017. A review of developments in energy storage systems for hybrid excavators, Automation in Construction, 80, 1–10.
[54] A. Eftekhari, 2018. Metrics for fast supercapacitors as energy storage devices, ACS Publications: Washington, DC, USA.
[55] F. Odeim, J. Roes, A. Heinzel, 2015. Power management optimization of an experimental fuel cell/battery/supercapacitor hybrid system, Energies, 8, 6302–6327.
[56] A. Townsend, I.N. Jiya, C. Martinson, D. Bessarabov, R. Gouws, 2020. A comprehensive review of energy sources for unmanned aerial vehicles, their shortfalls and opportunities for improvements, Heliyon, 6, e25439.
[57] Z.P. Cano, D. Banham, S. Ye, A. Hintennach, J. Lu, M. Fowler, Z. Chen, 2018. Batteries and fuel cells for emerging electric vehicle markets, Nature energy, 3, 279–289.
[58] Z. Qi, G.M. Koenig, 2017. Flow battery systems with solid electroactive materials, Journal of Vacuum Science & Technology B, 35, e25439.
[59] W. Guan, X. Huang, 2021. A modular active balancing circuit for redox flow battery applied in energy storage system, IEEE Access, 9, 127548–127558.

Chetan Chauhan

# 8 Environmental Impact of Green Hydrogen Production

**Abstract:** Green hydrogen production has become a sustainable, promising alternative to fossil fuels, offering a pathway toward carbon neutrality and environmental sustainability. This chapter explores the environmental impact of green hydrogen production, assessing both its benefits and associated challenges. By utilizing renewable energy sources such as solar, wind, and hydropower, green hydrogen minimizes greenhouse gas emissions and reduces dependency on conventional energy sources. However, the scalability of hydrogen technologies presents challenges, including land and water resource constraints, energy-intensive electrolysis processes, and the environmental footprint of material extraction for electrolyzers and storage systems. This chapter critically examines these aspects, integrating recent advancements in catalytic processes, electrolysis efficiency, and policy frameworks that support the hydrogen economy. Furthermore, the discussion highlights the role of green hydrogen in achieving global climate goals, emphasizing the need for life cycle assessments, circular economy approaches, and sustainable infrastructure development. By addressing these concerns, this chapter provides a comprehensive understanding of how green hydrogen can contribute to a cleaner and more sustainable energy future.

**Keywords:** Green hydrogen, greenhouse, gas emissions, renewable energy integration, sustainable, development goals, environmental metrics

## 8.1 Introduction

Global energy landscape is undergoing a transformative shift toward environmental sustainability, driven by the urgent need to mitigate climate change and reduce greenhouse gas (GHG) emissions [1, 2]. Amid such an energy transition, green hydrogen has gained considerable attention as a sustainable and versatile energy carrier capable of supporting a carbon-neutral future [3, 4]. Unlike blue or gray hydrogen, which are derived from natural gas and fossil fuels, respectively, green hydrogen is produced through the electrolysis of water powered entirely by renewable energy resources such as wind, solar, and hydropower [5]. This approach not only avoids direct $CO_2$ emissions but also offers synergies with the integration of renewable energy into the grid, energy storage, and the decarbonization of various sectors including transport and heavy industry [6, 7]. The environmental appeal of green hydrogen lies in its potential to serve as a sustainable alternative to fossil-based fuels and natural gas, thereby reducing the environmental footprint during energy generation and con-

 | https://doi.org/10.1515/9783112181072-008

sumption [3, 8]. As countries align with the goals of the Paris Agreement and strive toward net-zero emissions, green hydrogen is emerging as a central component of national and international energy strategies [9, 10]. Its application extends across multiple domains, including power generation, industrial feedstocks, mobility, and heating, making it a crucial pillar in the development of a diversified and resilient energy system [2, 11]. However, while green hydrogen holds promise for environmental sustainability, its production, distribution, and storage are not without environmental implications [8, 12]. The electrolysis process, although emission-free at the point of operation, is energy-intensive and dependent on the accessibility of low-cost renewable electricity [5, 7]. As of 2025, global hydrogen production is gradually increasing, with green hydrogen contributing only 0.3–0.5%. Green hydrogen offers near-zero $CO_2$ emissions and is projected to grow significantly by 2030. However, large-scale deployment raises concerns related to resource use, as well as the environmental costs associated with manufacturing electrolyzers, particularly the sourcing and processing of critical materials like platinum group metals and rare earth elements [13]. Furthermore, the infrastructure required for hydrogen transport and storage must be evaluated through a life cycle perspective to ensure true environmental gains [5, 14]. In light of such complexities, this chapter aims to provide a critical analysis of the environmental impact of green hydrogen production (Table 8.1). It explores both the ecological benefits and the potential trade-offs associated with scaling up green hydrogen technologies. Emphasis is placed on recent advancements in electrolysis efficiency, innovations in catalyst development, and upcoming regulatory frameworks and policies that support the transition to a hydrogen-based economy [11, 14]. Moreover, the chapter underscores the importance of LCA (life cycle assessment) methodologies, circular economy principles, and integrated resource planning to optimize the environmental performance of green hydrogen systems [5, 12]. By synthesizing current scientific understanding and policy discourse, this introduction sets the stage for a comprehensive exploration of how green hydrogen can be harnessed responsibly to contribute to global sustainability goals. The chapter advocates for a balanced approach that not only accelerates technological development but also safeguards environmental integrity in the pursuit of a cleaner, greener future.

## 8.2 Environmental Benefits of Green Hydrogen Production

Green hydrogen, produced via water electrolysis powered through renewable energy sources, offers multiple environmental benefits that are critical in the context of global efforts to combat climate change and transition to sustainable energy systems. The following sections describe the major environmental advantages of green hydrogen, supported by recent research and literature (Figure 8.1).

**Table 8.1:** Comparative overview of the global status of total hydrogen and green hydrogen production.

| Parameter | Total hydrogen production | Green hydrogen production | Remarks | References |
|---|---|---|---|---|
| **Annual production (2025)** | ~95–100 million tons (Mt) | ~250,000–500,000 tons | Green $H_2$ is ~0.3–0.5% of total | [1, 3, 15, 16] |
| **Main production method** | SMR (76%), Coal Gasification (23%) | Electrolysis using renewable electricity (wind/solar/hydro) | Green $H_2$ uses renewable energy sources | [1–3, 5, 7] |
| **$CO_2$ emissions (annual)** | ~900 million tons $CO_2$ | Near-zero emissions | Major environmental benefit of Green $H_2$ | [3, 8, 12] |
| **Share in global $H_2$ market** | ~99.5% (fossil based hydrogen) | ~0.5% | Green $H_2$ still at pilot/early scale | [4, 10, 12, 15] |
| **Production cost (2024)** | ~$1–2/kg (fossil-based, excluding CCS) | ~$3–6/kg | Green $H_2$ cost declining rapidly | [18–21] |
| **Future target (2030)** | – | 10–15 Mt/year (~10–15% of future hydrogen market) | Ambitious growth trajectory | [6, 12, 17, 16] |
| **Installed capacity (electrolyzers)** | Mostly SMR/coal gasification plants | ~3.5 GW (2024) globally | Electrolyzer deployment scaling rapidly | [4, 7, 10, 16] |

## 8.2.1 Reduction in GHGs (Greenhouse Gas Emissions)

One of the most evident environmental advantages of green hydrogen is its **zero-emission** nature, using renewable energy resources. Unlike gray hydrogen (obtained from natural gas via steam methane reforming), which emits approximately 9–12 kg of $CO_2$ per kg of hydrogen, green hydrogen emits no direct greenhouse gases during its production or end-use stages [1, 2]. This makes it a pivotal technology in reducing the carbon footprint during application spectra in several sectors, including energy and industry [2, 22]. According to the IRENA (International Renewable Energy Agency), widespread adoption of green hydrogen production could reduce annual $CO_2$ emissions by up to 830 million tons, particularly if it replaces fossil-derived hydrogen and fossil fuels in industrial and transportation applications [2, 9]. LCAs have also demonstrated that green hydrogen significantly reduces GHG emissions compared to conventional fuels, especially when electrolysis comes from wind or solar resources [5, 8].

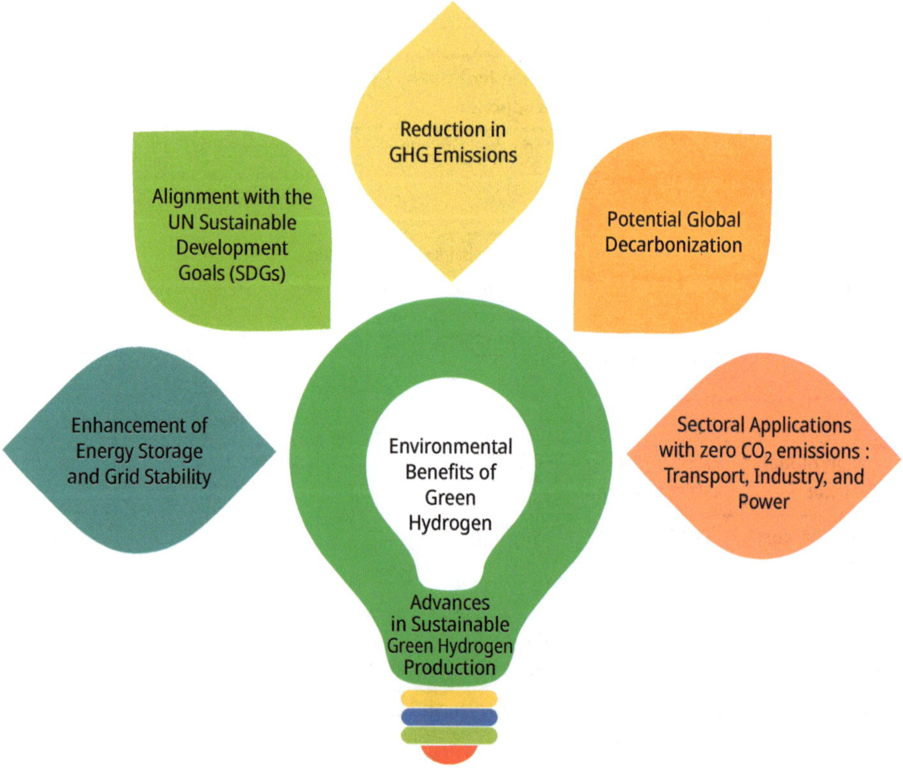


**Figure 8.1:** Environmental benefits of green hydrogen.

### 8.2.2 Potential Global Decarbonization in Sectoral Applications

Certain industrial processes and transport modes, such as **steelmaking, cement production, ammonia synthesis, aviation, and long-haul trucking**, are considered "hard-to-abate" due to their high energy intensity demands and limited alternatives to fossil fuels. Green hydrogen offers a viable pathway to decarbonize such sectors [13]. Hydrogen-powered trains and buses are already operational in parts of Europe and Asia [23]. The Hydrogen Council estimates that hydrogen could help abate 20% of total $CO_2$ emissions by 2050 when deployed across various sectors in combination with electrification and energy efficiency measures across various sectors in combination with electrification and energy efficiency measures [3, 15]. For instance, in steel production, refining, and fertilizer production, hydrogen can act as a reducing agent, enabling the shift from carbon-intensive methods like coal-based reduction in steel to direct hydrogen reduction instead of $CO_2$ [3]. Pilot projects such as HYBRIT in Sweden have shown the feasibility of steel production using green hydrogen, with the potential to cut emissions by 90% [24]. Similarly, in ammonia production – a process traditionally reliant on hydrogen from natural gas – green hydrogen can eliminate the associated $CO_2$ emissions, contributing to cleaner agricultural supply chains [25]. Hydrogen fuel cells offer a zero-emission solution in the transport sector, for heavy-duty vehicles and even including trains, ships, and even aircraft, where battery-based electrification is less feasible due to weight and energy density constraints [11].

### 8.2.3 Enhancement of Energy Storage and Grid Stability

Beyond emissions reductions, green hydrogen plays a vital role in supporting the renewable energy integration and grid stability. Electrolyzers can convert surplus into hydrogen, which can be stored over long periods and converted back into electricity through turbines or fuel cells during peak demand or when supply is low, or support the integration of intermittent solar and wind power [1, 19, 26, 27]. This capacity makes hydrogen a critical energy carrier and enabler of seasonal energy storage, addressing the intermittency challenges of VRE (variable renewable energy) sources. Furthermore, hydrogen production can be explored to balance electricity grids, reduce curtailment of excess renewable power, and provide ancillary services such as frequency regulation [4]. Additionally, hydrogen infrastructure can support sector coupling, linking electricity, heating, transport, and industry in an integrated energy system, which enhances overall efficiency and sustainability [14]. It can also be used in gas turbines for power generation with zero $CO_2$ emissions [28].

### 8.2.4 Alignment with the UN SDGs (Sustainable Development Goals)

Green hydrogen contributes significantly to several of the UN SDGs (Sustainable Development Goals), particularly SDG 7 (Affordable and Clean Energy), SDG 9 (Industry, Innovation, and Infrastructure), and SDG 13 (Climate Action) [17, 29]. Green hydrogen supports energy diversification and the transition to renewable sources, helping to ensure accessibility to clean and sustainable energy [3]. Hydrogen enables the development of sustainable industrial processes, including green steel, cement, and ammonia production [5]. By replacing fossil fuels, green hydrogen drastically reduces GHG emissions and supports commitments underlined by the Paris Agreement to lower global warming [12, 30].

## 8.3 Environmental Challenges and Concerns

While green hydrogen holds substantial promise for lowering the GHG emissions and enabling a clean and sustainable energy future, it also brings several environmental and sustainability challenges that must be addressed to ensure truly green outcomes. This section outlines key concerns related to resource consumption, energy intensity, critical material use, and end-of-life management, supported by current research and data [1–5, 7, 12, 29, 30] (Figure 8.2).

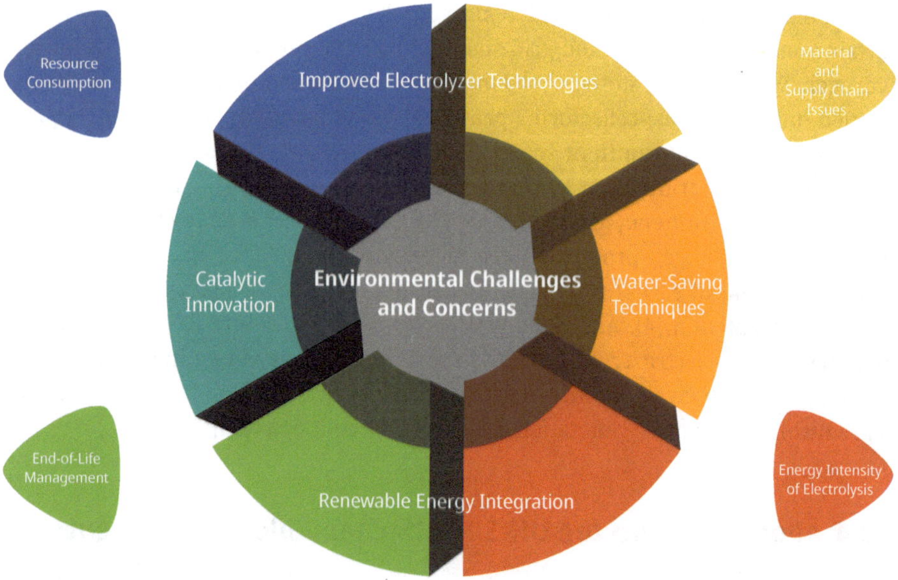


**Figure 8.2:** Environmental challenges and concerns in the advanced era of hydrogen production technology.

### 8.3.1 Resource Consumption

A critical input in green hydrogen production is water via electrolysis. For every kilogram of hydrogen produced, approximately **9–12 L of deionized or purified water** are required [31, 32]. While this may appear modest in comparison to water use in agriculture or thermal power generation, at **large-scale deployment**, especially in water-stressed regions, water sourcing and purification can pose significant environmental and socio-economic challenges [33, 34]. In addition, large-scale green hydrogen production using **renewable energy sources** like solar and wind demands **extensive land** for installation of photovoltaic (PV) panels or wind turbines. This could result in **habitat fragmentation, biodiversity loss,** and **land-use conflicts**, especially if not carefully planned [20]. Integrating hydrogen projects with existing infrastructure or deploying them on degraded or non-arable lands may help mitigate such impacts [6, 8].

### 8.3.2 Energy Intensity of Electrolysis

Electrolysis is inherently energy-intensive, where 1 kg of green hydrogen production requires nearly 50–55 kWh of electricity [20]. Although powered by renewable energy, the **conversion** losses remain substantial, leading to low round-trip efficiency (typically less than 40%) when hydrogen is stored and later converted back into electricity [21]. Ongoing research seeks to improve electrolyzer efficiency, particularly through the development of SOECs (solid oxide electrolysis cells) and AEM (anion exchange membranc) tcchnologies, which promise lower operating costs and better energy performance [35, 36]. However, the energy input remains a limiting factor, particularly when considering hydrogen's role in seasonal or long-term energy storage.

### 8.3.3 Material and Supply Chain Issues

The large-scale production of electrolyzers and fuel cells relies on expensive materials, such as PGMs (platinum-group metals) like platinum, iridium, and ruthenium, particularly in PEM (proton exchange membrane) electrolyzers [37]. Mining and refining these materials involve significant environmental impacts, including land degradation, toxic waste generation, and high energy consumption, especially when sourced unsustainably. Moreover, geopolitical and supply chain vulnerabilities related to rare earth elements and critical minerals can hinder large-scale deployment and drive-up costs, necessitating innovations in material substitution, recycling, and ethical sourcing strategies [38, 39].

### 8.3.4 End-of-Life Management

At present, there is limited infrastructure and policy focus on managing the EoL (end-of-life) phase of electrolyzers, hydrogen storage systems, and associated components. Without proper recycling or disposal strategies, these systems could contribute to growing electronic and material waste [40]. Implementing circular economy principles, such as designing for disassembly, reusing precious metals, and establishing closed-loop supply chains – can significantly reduce environmental footprints and enhance sustainability [41]. However, global efforts toward establishing standardized EoL protocols for hydrogen technologies are still in nascent stages and require urgent attention.

## 8.4 Advances in Sustainable Hydrogen Production

Sustainable hydrogen production via water electrolysis powered by renewable energy sources is a promising pathway to achieving a clean, low-carbon energy future [2, 30]. Significant and evident advancements in technology and its integration are paving the way for cost-effective, and environmentally friendly green hydrogen production [5, 21, 42]. Below are the key developments driving this transformation:

### 8.4.1 Catalytic Innovations

One of the most critical components in the electrolysis process is the catalyst, which facilitates the HER (hydrogen evolution reaction) and the OER (oxygen evolution reaction). Traditionally, noble metals, including platinum and iridium, have been explored due to their superior catalytic activities. However, their scarcity and high cost have driven intensive research into alternative materials. In the current scenario, the recent innovations have focused on non-precious metal catalysts, and transition metal-based entities such as nickel, cobalt, and iron phosphides, sulfides, and nitrides. These materials demonstrate promising catalytic performance and durability in both acidic and alkaline media [43, 44]. Additionally, biomimetic catalysts inspired by natural enzymes like hydrogenases are emerging as a viable option for sustainable hydrogen production due to their high catalytic activity and selectivity under mild conditions [45]. Advanced nanostructuring techniques and dopant engineering are further enhancing the surface area and electronic properties of these catalysts, thereby improving their overall efficiency [46]. These efforts significantly lower the reliance on rare elements, enabling the economically viable production of sustainable hydrogen.

### 8.4.2 Improved Electrolyzer Technologies

Electrolyzers play a vital role in the production of green hydrogen, and innovations in this domain are enhancing both performance and sustainability. Two leading types of next-generation electrolyzers are PEM electrolyzers and SOECs. PEM electrolyzers offer high current densities, rapid response times, and compact designs, ideal for coupling with variable resources. Recent improvements include the development of membranes with higher conductivity and durability, as well as cost-effective bipolar plates [42, 47]. SOECs operate at high temperatures (600–850 °C), which allows for greater electrical efficiency as thermal energy assists in driving the electrochemical reaction. Advances in ceramic materials and thermal management systems are improving the long-term stability and reducing the degradation of SOECs [48, 49].

### 8.4.3 Renewable Energy Integration

Integrating green hydrogen production directly with solar PV systems and wind farms can significantly enhance energy efficiency and sustainability. Such coupling allows for real-time utilization of excess renewable energy, which is otherwise not possible due to grid limitations [2, 20]. Hybrid systems that use smart inverters and energy management software optimize hydrogen production by matching the variability of renewable output with the electrolyzer's operating conditions. This approach minimizes energy losses and reduces the dependency on external grid infrastructure. Additionally, co-location of electrolyzers at renewable energy sites reduces transmission and distribution losses, leading to a more efficient overall system [21, 45]. Demonstration projects in regions like Australia, Germany, and the Middle East are validating the feasibility of such integrated systems on commercial scales [16, 19].

### 8.4.4 Water-Saving Techniques

A significant limitation of conventional water electrolysis is its dependency on high-purity freshwater, which poses a challenge in arid and water-scarce regions. Therefore, current research is intentionally focused on designing systems that can explore non-potable water sources, such as sea, brackish, and wastewater [7, 44]. Recent advances include the use of membrane-free electrolyzers and electrochemical water purification pre-treatment stages, which enable the use of seawater by selectively removing impurities that would otherwise poison the catalysts or reduce efficiency [42, 43]. Additionally, seawater electrolysis technologies incorporating corrosion-resistant electrodes and selective AEMs are making progress. These systems aim to reduce chlorine evolution, a major competing reaction in saline water electrolysis, thereby enhancing safety and efficiency [44, 48]. These collective innovations represent a major stride toward making green hydrogen a scalable and accessible solution for decarbonizing various broad spectrum applications, including transportation and industry, electricity generation, and storage [6, 22].

## 8.5 Policy and Regulatory Frameworks

The successful scaling and adoption of sustainable hydrogen technologies depend not only on technological advancements but also on the framing of robust **policy and regulatory guidelines**. Governments across the globe are realizing the role of green hydrogen in achieving decarbonization targets and, therefore, formulating strategies, subsidies, and incentives to foster its sustainable development [1, 2, 6]. Sustainable approach-based frameworks are vital for reducing investment risks, supporting infrastructure development, and encouraging private sector participation [17, 50].

### 8.5.1 Importance of Policy Support

In the current scenario, the green hydrogen production, being in the early stages of commercial deployment, faces challenges such as high investment and operational costs compared to traditional hydrogen derived from fossil fuels and natural gas (commonly known as gray hydrogen and blue hydrogen) [2, 6, 18]. To bridge these economic gaps, supportive policy framework interventions are essential. Such measures may include capital subsidies and production incentives to offset the initial investment costs, tax breaks and exemptions to reduce operational expenses, public-private partnerships for infrastructure development, including pipelines, hydrogen refueling stations, and storage systems. Regulatory standards and certification schemes need to be implemented to ensure the environmental integrity of green hydrogen production [6, 15, 17]. Such measures accelerate the market uptake of green hydrogen technologies to create a favorable climate for sustainable investment [9, 18].

### 8.5.2 European Union's Hydrogen Strategy

The EU (European Union) has been at the forefront of integrating sustainable hydrogen production technologies into its climate and energy policy framework. European Commission launched the EU Hydrogen Strategy in July 2020, which outlines a comprehensive roadmap for green hydrogen deployment across the continent [16, 41]. Key features include a phased approach to scale green hydrogen production, with a target of installing 40 GW of sustainable hydrogen electrolyzers in the EU by 2030, investment of up to €470 billion by 2050 in green hydrogen infrastructure and technologies, establishment of the European Clean Hydrogen Alliance to foster collaboration among industry, government, and academia to further development of a hydrogen certification system to ensure traceability and sustainability [16, 41]. The EU strategy also emphasizes cross-border infrastructure and trade, positioning the EU as a global leader in hydrogen production and innovation [16, 41].

### 8.5.3 India's National Green Hydrogen Mission

India, a major emitter of GHGs, has also acknowledged the strategic importance of green hydrogen. In January 2023, the NGHM (National Green Hydrogen Mission) was approved by the Union Cabinet with an initial outlay of approximately USD 2.4 billion. Key components of the mission include: production target of 5 million metric tons of green hydrogen per annum by 2030, with associated renewable energy capacity of 125 GW, financial incentives under the SIGHT (Strategic Interventions for Green Hydrogen Transition) program to support domestic manufacturing of electrolyzers and production of green hydrogen, support for pilot projects in hard-to-abate sectors such as fertilizers, steel, and heavy transport, development of green hydrogen hubs, policies for land allocation, and support for port infrastructure to enable hydrogen exports and establishment of a single-window clearance mechanism to streamline project approvals. India aims to become a global hub for green hydrogen production to ensure energy security, job creation, export competitiveness, and sustainability [47].

### 8.5.4 Global Trends in Hydrogen Policy

According to the IEA (International Energy Agency), over 40 countries have announced national hydrogen strategies or are in the process of developing one [16]. Beyond the EU and India, several other countries are integrating hydrogen into their national energy strategies:
- Japan and South Korea have long-term hydrogen roadmaps focusing on transport and fuel cell technologies [16].
- Australia is investing in large-scale hydrogen projects aimed at exports, particularly to Asian markets [16].
- USA, under the Infrastructure Investment and Jobs Act (2021), allocated $9.5 billion for clean hydrogen, including regional clean hydrogen hubs and electrolyzer R&D [16, 17].

These policies serve as a catalyst for international collaboration, standards harmonization, and market creation [51, 52]. Robust policy and regulatory frameworks are essential enablers of the green hydrogen economy. From funding and incentives to infrastructure and standardization, these policies reduce uncertainties and drive investment into hydrogen technologies [6, 16, 18]. With strategic initiatives like the EU Hydrogen Strategy and India's National Green Hydrogen Mission, governments are laying the foundation for a resilient, low-carbon hydrogen ecosystem that aligns with global climate goals [9, 14, 16].

## 8.6 Life Cycle Assessment (LCA) and Environmental Metrics in Green Hydrogen Production

The growing global interest in green hydrogen as a sustainable energy carrier prompts deeper scrutiny of its overall environmental performance. While green hydrogen, produced through the electrolysis of water via renewable energy, is touted as a zero-carbon fuel at the point of use, a comprehensive evaluation of its life cycle impacts is essential to verify its sustainability credentials. This is where LCA and associated environmental metrics play a critical role. LCA is a standardized methodology used to quantify the environmental impacts associated with all stages of a product's life, from resource extraction (cradle) through production, distribution, use, and end-of-life disposal (grave) [5]. In the context of green hydrogen, LCA captures the environmental effects of renewable electricity generation (e.g., solar and wind), electrolyzer manufacturing and operation, water use and purification, hydrogen storage, compression, and distribution and end-use applications (mobility, industry, etc.) [2]. LCA ensures a holistic and scientifically grounded assessment of sustainability by identifying emissions, resource consumption, and ecological effects throughout the hydrogen supply chain [9].

**Figure 8.3:** Key environmental metrics and challenges in LCA for green hydrogen.

### 8.6.1 Key Environmental Metrics for Green Hydrogen

Several quantitative indicators are commonly used in LCA studies to assess the sustainability of green hydrogen. Among them, the most important are GWP (global warming potential), water footprint, and EROI (energy return on investment). GWP measures the climate impact of a process over a given time horizon (usually 100 years), expressed in terms of $CO_2$-equivalents (kg $CO_2$-eq) per unit of product (Figure 8.3). For green hydrogen, GWP reflects the emissions associated with the manufacture and deployment of renewable energy systems, electrolyzers, and balance-of-plant components [7]. Typical GWP values for various hydrogen production pathways are listed in Table 8.2. Although green hydrogen emits no $CO_2$ at the point of use, emissions from upstream stages, such as solar panel or wind turbine production, can contribute to the total GWP [1].

**Table 8.2:** Typical GWP values for blue, gray, and green hydrogen production pathways [60].

| Type of hydrogen | Method of production | GWP range (kg $CO_2$-eq/kg $H_2$) |
|---|---|---|
| Type of hydrogen | SMR with carbon capture and storage | ~3–6 |
| | Electrolysis via renewable energy | 0.5–4 |
| | SMR (steam methane reforming) | ~10–12 |

Electrolysis of water requires about 9 L of water to produce 1 kg of hydrogen. The water footprint encompasses not only this direct use but also indirect water usage in renewable energy generation (e.g., cooling systems in solar thermal plants), manufacturing of components like PV panels and electrolyzers, and water treatment and purification [26]. In water-scarce regions, the source of water becomes a sustainability constraint. Therefore, research is increasingly focusing on alternatives such as seawater electrolysis and electrolysis of treated wastewater [7]. EROI refers to the ratio of usable energy output to the total energy input required for production and delivery. High EROI values of green hydrogen indicate its better energy efficiency and sustainability. Green hydrogen's EROI can vary depending on the efficiency of electrolyzers, type of renewable energy source, and location-specific conditions. Typical EROI for green hydrogen varies from 1.3 to 4, lower than traditional fuels due to energy losses in various steps, including electrolysis, compression, and storage [53]. Efficiency improvements in PEM and solid oxide electrolyzers are expected to improve this metric [5].

### 8.6.2 Role of LCA in Policy, Certification, and Industry

Regulatory bodies including EU across the globe are increasingly relying on LCA to define thresholds for sustainable hydrogen production (e.g., <3.4 kg $CO_2$-eq/kg $H_2$ in

EU taxonomy), design certification schemes for green hydrogen (e.g., CertifHy in Europe), evaluate subsidies and carbon credits based on environmental performance, and guide investment decisions for public and private stakeholders [6, 15]. For instance, the European Commission and countries such as Germany, Japan, and India are adopting LCA-based criteria for green hydrogen incentives under national hydrogen missions [11]. While LCA provides valuable insights, its application to green hydrogen is challenged by data uncertainties in emerging technologies, regional variability in electricity mix and water availability, allocation issues (e.g., in co-products like oxygen), and dynamic system boundaries as the hydrogen production economy evolves [12]. Future research and development with standardization efforts need to be explored to enhance the comparability and robustness of LCA studies across different geographies and technologies. LCA and environmental metrics such as GWP, water footprint, and EROI are critical for ensuring that green hydrogen lives up to its potential as a sustainable energy solution. These tools provide a rigorous basis for decision-making, technology development, and policy formation. As global efforts intensify to decarbonize energy systems, transparent and science-based environmental assessments will be key to the responsible deployment of green hydrogen technologies [3].

## 8.6.3 Challenges in Applying LCA to Hydrogen

LCA is a vital tool for evaluating the environmental sustainability of hydrogen production, particularly green hydrogen. Its practical application faces several challenges that affect the accuracy, comparability, and reliability of results (Figure 8.3) [5, 7], [8].

### 8.6.3.1 Data Uncertainties in Emerging Technologies

Green hydrogen technologies, such as high-efficiency electrolyzers (PEM, SOEC) and renewable-powered electrolysis systems, are still developing. Limited availability of empirical performance data, reliance on modeling assumptions, and rapid technological changes lead to significant uncertainties in LCA results [5, 7, 37]. For instance, the actual lifespan, efficiency degradation, and maintenance needs of electrolyzers can vary widely, affecting the assessment outcomes [54].

### 8.6.3.2 Regional Variability in Electricity Mix and Water Availability

LCA outcomes are highly sensitive to the geographic context, mostly the carbon intensity of electricity used and the availability of water resources. For example, hydrogen produced via electrolysis via grid electricity in a coal-dominant sector may have a higher GWP than gray hydrogen from natural gas [2, 8]. Additionally, water scarcity in arid regions raises sustainability concerns about water bodies use, influencing largely the water footprint metric [7].

#### 8.6.3.3 Allocation Issues

Electrolysis not only produces hydrogen but also oxygen as a co-product. How emissions and resource use are allocated between these outputs can significantly influence the LCA results. At different allocations, the methods like mass-based, energy-based, or economic can yield different environmental impact scores, creating inconsistency in assessments [55].

#### 8.6.3.4 Dynamic System Boundaries

As the hydrogen economy evolves, the system boundaries of LCA studies must adapt. For example, integration with carbon capture, smart grids, or hydrogen carriers (e.g., ammonia) introduces new stages and components into the hydrogen value chain. Moreover, future changes in renewable technology, policy interventions, or infrastructure development can alter LCA assumptions over time [12, 19, 56].

#### 8.6.3.5 Need for Standardization and Methodological Harmonization

The lack of standardized LCA frameworks and functional units across studies leads to challenges in comparing results. Differences in modeling approaches, databases, impact assessment methods (e.g., ReCiPe vs. TRACI), and time horizons further complicate meta-analysis or policy formulation based on LCA data. Therefore, harmonization and international guidelines are needed to ensure transparency and reproducibility [5, 7, 57].

## 8.7 Future Prospects and Sustainable Infrastructure in Green Hydrogen Production

The environmental impact of sustainable hydrogen production can be significantly reduced through the intervention of innovative infrastructure strategies. Decentralized production, advanced

storage and transport technologies, and carefully planned scalability are essential to ensure that the green hydrogen economy is not only clean at the point of use but sustainable across its full life cycle (Figure 8.4) [12, 58].

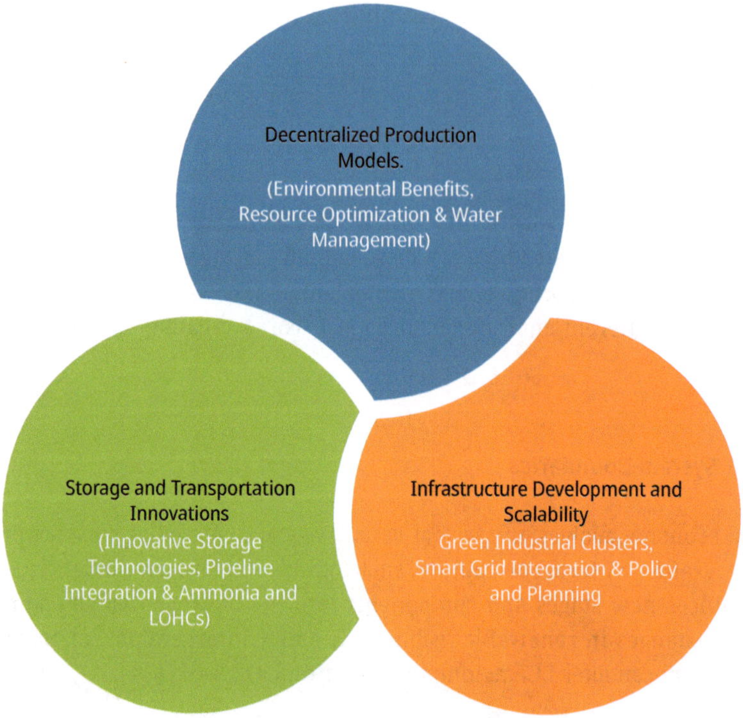


**Figure 8.4:** Future prospects and sustainable infrastructure in green hydrogen production.

## 8.7.1 Decentralized Production Models

Decentralized or localized green hydrogen production – especially through small- to medium-scale electrolysis units co-located with renewable energy sources – presents significant environmental and economic advantages. Distributed systems reduce the need for long-distance hydrogen transport, thereby lowering transmission-related energy losses and carbon emissions [59]. Co-location with solar or wind farms enables direct utilization of surplus renewable energy, minimizing curtailment and enhancing grid stability [37]. Some decentralized systems are exploring the use of non-potable or recycled water sources to address regional water stress issues [60]. Moreover, modular electrolyzer designs are making it easier to implement decentralized hydrogen hubs in remote or rural regions [4, 10].

### 8.7.2 Storage and Transportation Innovations

Efficient and sustainable hydrogen transportation and storage are vital for minimizing life cycle emissions and enabling the widespread adoption of green hydrogen. Solid-state hydrogen storage materials (e.g., metal hydrides) and cryo-compressed hydrogen tanks are under development to enhance safety, reduce energy losses, and increase volumetric energy density [16]. Repurposing existing natural gas pipelines for hydrogen blends or pure hydrogen transport offers a cost-effective and lower-impact alternative to new infrastructure [3]. Hydrogen can also be converted into ammonia or liquid organic hydrogen carriers for easier storage and long-distance shipping. However, the reconversion process must be optimized to minimize energy penalties and emissions [61]. These innovations are essential for improving the EROI and reducing the GWP of green hydrogen systems [23, 62].

### 8.7.3 Infrastructure Development and Scalability

Scaling up of green hydrogen production and application infrastructure requires coordinated development of production plants, refueling stations, storage facilities, and distribution networks, all while ensuring a low environmental footprint. European countries like Germany and the Netherlands are developing hydrogen valleys or industrial hubs that concentrate green hydrogen production and use in specific regions. These hubs benefit from economies of scale, shared infrastructure, and reduced logistics-related emissions [63, 64]. Integrating green hydrogen systems with smart energy grids enhances demand-response capabilities, improving efficiency and reducing wasteful oversupply of energy [13]. Long-term infrastructure planning considering environmental metrics like land use, water footprint, and lifecycle emissions is key to avoiding unintended ecological impacts during scale-up [12, 58].

## 8.8 Conclusion

Sustainable hydrogen production stands at the forefront of the global clean energy transition, offering a transformative pathway to decarbonize hard-to-abate sectors, reduce GHG emissions, and for the effective integration of renewable energy systems. Its alignment with the United Nation's SDGs underscores its potential as a key driving force of sustainable development and climate resilience. Nevertheless, harnessing the full environmental benefits of green hydrogen necessitates a balanced approach that addresses its inherent challenges – such as intensive water and land requirements, high energy inputs for electrolysis, reliance on critical raw materials, and the lack of comprehensive end-of-life management. Technological advancements in catalyst development, electrolyzer performance, and system integration with renewables are gradually mitigating these concerns, moving the sector toward greater efficiency and sustainability. To ensure green hydrogen truly fulfills its promise as a clean and sustainable energy vector, a life cycle-based, circular economy framework must be adopted – guided by forward-thinking policies, robust environmental assessments, and continuous innovation. Green hydrogen can significantly contribute to building a low-carbon, environmentally sound, and resilient global energy future through global strategic planning and responsible implementation.

## References


[1] Araújo, O.Q. and de Medeiros, J.L., 2023. *Hydrogen in the energy transition: some roles, issues, and questions.* Clean Technologies and Environmental Policy, 25(7), pp.2071–2073.

[2] Staffell, I., Scamman, D., Abad, A.V., Balcombe, P., Dodds, P.E., Ekins, P., Shah, N. and Ward, K.R., 2019. *The role of hydrogen and fuel cells in the global energy system.* Energy & Environmental Science, 12(2), pp.463–491.

[3] Elegbeleye, I., Oguntona, O. and Elegbeleye, F., 2025. *Green Hydrogen: Pathway to Net Zero Green House Gas Emission and Global Climate Change Mitigation.* Hydrogen, 6(2), p.29.

[4] Mokrzycki, E. and Gawlik, L., 2024. *The Development of a Green Hydrogen Economy.* Energies, 17(13), p.3165.

[5] Bhandari, R., Trudewind, C.A. and Zapp, P., 2014. *Life cycle assessment of hydrogen production via electrolysis–a review.* Journal of Cleaner Production, 85, pp.151–163.

[6] Bianco, E. and Blanco, H., 2020. *Green hydrogen: a guide to policy making.*

[7] Hassan, N.S., Jalil, A.A., Rajendran, S., Khusnun, N.F., Bahari, M.B., Johari, A., Kamaruddin, M.J. and Ismail, M., 2024. *Recent review and evaluation of green hydrogen production via water electrolysis for a sustainable and clean energy society.* International Journal of Hydrogen Energy, 52, pp.420–441.

[8] Suleman, F., Dincer, I. and Agelin-Chaab, M., 2015. *Environmental impact assessment and comparison of some hydrogen production options.* International Journal of Hydrogen Energy, 40(21), pp.6976–6987.

[9] Birol, F., 2019. *The future of hydrogen: seizing today's opportunities.* IEA Report prepared for the G20, p.442.

[10] Bains, P., Bennett, S., Collina, L., Connelly, E., Delmastro, C., Evangelopoulou, S., Fajardy, M., Shari, B.E., Moumouni, Y., Ohunakin, O.S., Blechinger, P., Madougou, S. and Rabani, A., 2024. *Exploring the role of green hydrogen for distributed energy access planning towards net-zero emissions in Nigeria.* Sustainable Energy Research, 11(1), p.16.

[11] Johnravindar, D. and Selvakumar, R., 2024. *Applications of Green Hydrogen with Commercial Feasibility: Identifying Gaps, Perspectives, and Bottlenecks.* In *Green Hydrogen Economy for Environmental Sustainability. Volume 2: Applications, Challenges, and Policies* (pp. 1–30). American Chemical Society.

[12] Wei, S., Sacchi, R., Tukker, A., Suh, S. and Steubing, B., 2024. *Future environmental impacts of global hydrogen production.* Energy & Environmental Science, 17(6), pp.2157–2172.

[13] Vogl, V., Åhman, M. and Nilsson, L.J., 2018. *Assessment of hydrogen direct reduction for fossil-free steelmaking.* Journal of Cleaner Production, 203, pp.736–745.

[14] Nour, S. and Arbussà, A., 2024. *Green Hydrogen: Revolutionizing the Energy Sector and Facilitating Cross-Sector Collaboration.* Available at SSRN 5249405.

[15] Council, H., 2022. *Hydrogen insights 2022.*

[16] International Energy Agency (IEA), 2024. *Global Hydrogen Review 2024.*

[17] Council, H., 2024. *Scaling Hydrogen Financing for Development.*

[18] Shirizadeh, B., Ailleret, A., Guillon, A., Bovari, E., El Khatib, N., Douguet, S., Issa, C.B., Brauer, J. and Trüby, J., 2023. *Towards a resilient and cost-competitive clean hydrogen economy: the future is green.* Energy & Environmental Science, 16(12), pp.6094–6109.

[19] Zun, M.T. and McLellan, B.C., 2023. *Cost projection of global green hydrogen production scenarios.* Hydrogen, 4(4), pp.932–960.

[20] Glenk, G. and Reichelstein, S., 2019. *Economics of converting renewable power to hydrogen.* Nature Energy, 4(3), pp.216–222.

[21] Schmidt, O., Gambhir, A., Staffell, I., Hawkes, A., Nelson, J. and Few, S., 2017. *Future cost and performance of water electrolysis: An expert elicitation study.* International Journal of Hydrogen Energy, 42(52), pp.30470–30492.

[22] Abdalla, A.M., Hossain, S., Nisfindy, O.B., Azad, A.T., Dawood, M. and Azad, A.K., 2018. Hydrogen production, storage, transportation and key challenges with applications: a review. *Energy Conversion and Management*, 165, pp.602–627.

[23] Giakoumakis, G. and Sidiras, D., 2025. Production and Storage of Hydrogen from Biomass and Other Sources: Technologies and Policies. Energies, 18(3), p.650.

[24] Pei, A., Jones, A., Skreiberg, Ø. and Wang, L., 2020. Hydrogen use in steel production: State-of-the-art and outlook. *Energy Reports*, 6, pp.524–532

[25] Rouwenhorst, K.H., Van der Ham, A.G., Mul, G. and Kersten, S.R., 2019. Islanded ammonia power systems: Technology review & conceptual process design. Renewable and Sustainable Energy Reviews, 114, p.109339.

[26] Wolfram, P., Kyle, P., Fuhrman, J., O'Rourke, P. and McJeon, H., 2024. The hydrogen economy can reduce costs of climate change mitigation by up to 22%. *One Earth*, *7*(5), pp.885–895.

[27] Boettcher, S.W., 2024. Introduction to Green Hydrogen. Chemical Reviews, 124(23), pp.13095–13098.

[28] Ozowe, W., Ikevuje, A.H., Ogbu, A.D. and Esiri, A.E., 2023. Hydrogen production and utilization in oil and gas facilities.

[29] Aljohani, B.S., Aljohani, K., Kandasamy, M., Vellaiyan, S. and Devarajan, Y., 2025. Enhancing combustion performance and emission reduction in ammonium hydroxide-diesel blends: A comparative study of metal-based and carbon-based nanocatalysts for hydrogen separation. International Journal of Hydrogen Energy, 100, pp.646–657.

[30] Turner, J.A., 2004. Sustainable hydrogen production. Science, 305(5686), pp.972–974.

[31] Rouwenhorst, K.H., Van der Ham, A.G., Mul, G. and Kersten, S.R., 2019. Islanded ammonia power systems: Technology review & conceptual process design. Renewable and Sustainable Energy Reviews, 114, p.109339.

[32] Dodds, P.E., Staffell, I., Hawkes, A.D., Li, F., Grünewald, P., McDowall, W. and Ekins, P., 2015. Hydrogen and fuel cell technologies for heating: A review. International journal of hydrogen energy, 40(5), pp.2065–2083.

[33] Osman, A.I., Mehta, N., Elgarahy, A.M., Hefny, M., Al-Hinai, A., Al-Muhtaseb, A.A.H. and Rooney, D.W., 2022. Hydrogen production, storage, utilisation and environmental impacts: a review. Environmental Chemistry Letters, pp.1–36.

[34] Carmo, Marcelo, David L. Fritz, Jürgen Mergel, and Detlef Stolten. "A comprehensive review on PEM water electrolysis." International journal of hydrogen energy 38, no. 12 (2013): 4901–4934.

[35] Palmer, G., Floyd, J., Palmer, G. and Floyd, J., 2020. Hydrogen as an energy carrier. Energy Storage and Civilization: A Systems Approach, pp.125–137.

[36] Sarmah, M.K., Singh, T.P., Kalita, P. and Dewan, A., 2023. Sustainable hydrogen generation and storage–a review. RSC advances, 13(36), pp.25253–25275.

[37] Revinova, S., Lazanyuk, I., Ratner, S. and Gomonov, K., 2023. Forecasting development of green hydrogen production technologies using component-based learning curves. *Energies*, 16(11), p.4338.

[38] UNFCCC, C., 2021. Nationally determined contributions under the Paris agreement. Synthesis report by the secretariat.

[39] Citaristi, I., 2022. International energy agency – iea. In The Europa directory of international organizations 2022 (pp. 701–702). Routledge.

[40] Van Hulst, N., 2019. The clean hydrogen future has already begun.

[41] Press, E., Van de Graaf, T., Blanco, H., Bianco, E., Tsang, W., Ferroukhi, R. and Gielen, D., 2022. Geopolitics of the energy transformation: the hydrogen factor.

[42] Carmo, Marcelo, David L. Fritz, Jürgen Mergel, and Detlef Stolten. "A comprehensive review on PEM water electrolysis." International journal of hydrogen energy 38, no. 12 (2013): 4901–4934.

[43] Zhang, X., Wang, J., Zong, K., Chen, Z., Yang, X., Yang, L., Wang, X. and Chen, Z., 2025. Recent advances in non-noble metal-based electrocatalysts for hybrid water electrolysis systems. Carbon Energy, p.e679.

[44] Wang, G., Xiang, T., Ren, X., Zhang, L. and Chen, C., 2024. Transition metal-based electrocatalysts for hydrogen production from seawater: A review. International Journal of Hydrogen Energy, 73, pp.775–790.

[45] Maukonen, J., 2023. Techno-economic analysis of dynamic operation of water electrolysis.

[46] Ozbilen, A., Dincer, I. and Rosen, M.A., 2013. Comparative environmental impact and efficiency assessment of selected hydrogen production methods. Environmental Impact Assessment Review, 42, pp.1–9.

[47] Gouws, S. and Mackay, J., 2024. Production of green hydrogen through PEM water electrolysis. Pure and Applied Chemistry, 96(10), pp.1383–1401.

[48] Buttler, A. and Spliethoff, H., 2018. Current status of water electrolysis for energy storage, grid balancing and sector coupling via power-to-gas and power-to-liquids: A review. Renewable and Sustainable Energy Reviews, 82, pp.2440–2454.

[49] Lahrichi, A., El Issmaeli, Y., Kalanur, S.S. and Pollet, B.G., 2024. Advancements, strategies, and prospects of solid oxide electrolysis cells (SOECs): Towards enhanced performance and large-scale sustainable hydrogen production. Journal of Energy Chemistry.

[50] Wald, M.L., 2004. Questions about a hydrogen economy. Scientific American, 290(5), pp.66–73.

[51] Mayyas, A., Steward, D. and Mann, M., 2019. The case for recycling: Overview and challenges in the material supply chain for automotive li-ion batteries. Sustainable materials and technologies, 19, p. e00087.

[52] Gouws, S. and Mackay, J., 2024. Production of green hydrogen through PEM water electrolysis. Pure and Applied Chemistry, 96(10), pp.1383–1401.

[53] Boettcher, S.W., 2024. Introduction to Green Hydrogen. Chemical Reviews, 124(23), pp.13095–13098.

[54] Citaristi, I., 2022. International energy agency – iea. In The Europa directory of international organizations 2022 (pp. 701–702). Routledge.

[55] Schwartzkopff, J., 2022. The future role of gas in a climate-neutral Europe. Report Based on the Discussions of an Expert Group Convened by the Heinrich-Böll-Stiftung European Union and Environmental Action Germany (Deutsche Umwelthilfe), 6.

[56] Corbeau, A.S., 2025. Europe's Long Journey to Renewable Hydrogen. Fletcher F. World Aff., 49, p.19.

[57] Eh, C.L., Tiong, A.N., Kansedo, J., Lim, C.H., How, B.S. and Ng, W.P.Q., 2022. Circular hydrogen economy and its challenges. Chemical Engineering Transactions, 94, pp.1273–1278.

[58] Noyan, O.F., Hasan, M.M. and Pala, N., 2023. A global review of the hydrogen energy eco-system. Energies, 16(3), p.1484.

[59] Otaki, T. and Shaw, R., 2024. Global hydrogen energy: Potentials and challenges. Energy, Sustainability and Resilience: A Futuristic Vision from Asia, pp.127–138.

[60] Khan, M.A., Al-Attas, T., Roy, S., Rahman, M.M., Ghaffour, N., Thangadurai, V., Larter, S., Hu, J., Ajayan, P.M. and Kibria, M.G., 2021. Seawater electrolysis for hydrogen production: a solution looking for a problem?. Energy & Environmental Science, 14(9), pp.4831–4839.

[61] Lebrouhi, B.E., Djoupo, J.J., Lamrani, B., Benabdelaziz, K. and Kousksou, T., 2022. Global hydrogen development-A technological and geopolitical overview. International Journal of Hydrogen Energy, 47(11), pp.7016–7048.
[62] Abdalla, A.M., Hossain, S., Nisfindy, O.B., Azad, A.T., Dawood, M. and Azad, A.K., 2018. Hydrogen production, storage, transportation and key challenges with applications: A review. Energy conversion and management, 165, pp.602–627.
[63] Ryding, S.O., 1999. ISO 14042 Environmental management• Life cycle assessment• life cycle impact assessment. The International Journal of life cycle assessment, 4(6), pp.307–307.
[64] GoElegbeleye, I., Oguntona, O. and Elegbeleye, F., 2025. Green Hydrogen: Pathway to Net Zero Green House Gas Emission and Global Climate Change Mitigation. Hydrogen, 6(2), p.29.uy, A., Kotani, M., Le Marois, J.B. and Levi, P., 2023. Global hydrogen review 2023.

Nishant Singh, Virendra Kumar, Sanjeev Kumar, Devendra Singh

# 9 Low-Friction Materials and Coatings for Sustainable Hydrogen Systems

**Abstract:** Hydrogen becomes more important energy carrier for sustainable systems as the global energy scene moves toward low-carbon solutions. Particularly in fuel cells, compressors, storage systems, and pipelines, the great usage of hydrogen technologies provides significant tribological challenges resulting from hydrogen embrittlement, increased diffusivity, and material deterioration. With low-friction materials and surface coatings meant to increase the efficiency, durability, and sustainability of hydrogen infrastructure, this work explores their advancement and use. Analyzed are advanced materials with friction-reducing properties and resistance to hydrogen-induced degradation: diamond-like carbon, transition metal dichalcogenides, and creative nanocomposites. Surface engineering techniques include atomic layer deposition and plasma-enhanced coating systems that offer exact control of surface properties in hydrogen-rich environments are also investigated in this work. Minimizing energy losses, lowering maintenance needs, and extending component lifetime elements vital for the economic and environmental sustainability of hydrogen technologies – rely on low-friction systems. The outcomes underline the need of multidisciplinary approaches combining hydrogen engineering, tribology, and materials research to hasten the use of robust, sustainable hydrogen systems.

**Keywords:** Low-friction materials, surface coatings, hydrogen systems, sustainable energy, tribology, hydrogen embrittlement, wear resistance

## 9.1 Introduction

Often including dynamic components like compressors, pumps, valves, and fuel cells, hydrogen systems cover production, storage, and use. The strict purity criteria for hydrogen, which remove the option of utilizing oil-based lubricants [1], make the functioning of these components in dry-running conditions sometimes indispensable. For tribological design, this poses significant challenges since lack of traditional lubrication can aggravate friction and wear. Within mechanical systems, friction causes a direct energy loss. Minimizing friction between mating components [2] greatly affects the efficiency of both the compressor and motor in hydrogen compressors. Low-friction coatings can help to significantly lower energy usage, therefore improving the general efficiency of hydrogen infrastructure. Factors influencing the performance of materials are shown in Figure 9.1.

 | https://doi.org/10.1515/9783112181072-009

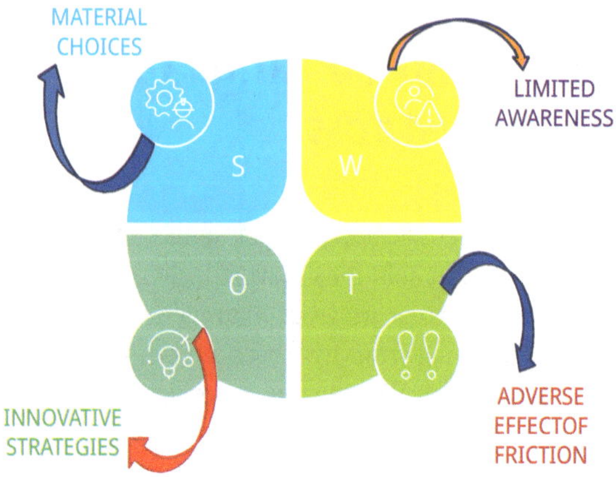


**Figure 9.1:** Enhancing mechanical performance of materials.

Degradation of components may result in early failure and expensive interruptions in operations. This aspect is especially vital in hydrogen systems, where dependability is essential for guaranteeing safety and avoiding leaks. Low-friction coatings have the potential to prolong the lifespan of essential components, minimizing the need for maintenance and replacement, which in turn improves the overall reliability of the system [3]. The incompatibility of traditional lubricants with high-purity hydrogen highlights the need for materials that can function efficiently in dry-running environments. Materials and coatings that provide self-lubrication and sustain low friction without the need for external lubrication are crucial for various hydrogen applications.

## 9.2 Low-Friction Coating Technologies for Hydrogen Systems

Various coating technologies have surfaced as effective options for minimizing friction and wear in hydrogen environments. The coatings are engineered to deliver a low coefficient of friction (COF), exceptional wear resistance, and longevity tailored to the unique operating conditions of hydrogen systems. Two-dimensional materials, including graphene and molybdenum disulfide ($MoS_2$), have attracted considerable interest because of their layered architecture and exceptionally low shear strength. The materials demonstrate a high specific surface area, substantial in-plane strength, and minimal layer-layer interaction, leading to exceptionally low friction and wear resistance. Lincong Liu et al. [4] conducted a comprehensive review of the latest developments in the scientific exploration and industrial utilization of antifriction, anti-wear, and lubrication properties of 2D materials. They provided a summary of fundamental nanoscale friction mechanisms, encompassing interfacial friction and surface friction. Yanbao Guo et al. [5] examined recent advancements in 2D materials within the field of tribology, providing a summary of well-established nanoscale friction mechanisms, including surface friction and interlayer friction. John F. Curry et al. [6] illustrated the significance of microstructure in the friction and oxidation behavior of $MoS_2$, showing that ordered $MoS_2$ exhibited remarkable resistance to both atomic and molecular oxygen, maintaining its low-friction properties even under extreme oxidative conditions. Diamond-like carbon (DLC) coatings are amorphous materials recognized for their exceptional hardness, minimal COF, and outstanding wear resistance [7]. DLC coatings can be customized for particular applications by adjusting their composition and structure, such as the ratio of sp3 (diamond-like) to sp2 (graphitic) carbon bonding [8]. Hydrogenated DLC (a-C:H) films exhibit remarkably low-friction coefficients in high vacuum or inert environments, a phenomenon referred to as "superlubricity." The presence of hydrogen in the film is thought to be essential for the establishment and maintenance of this superlow-friction regime. Lei Sun et al. [11] discovered that a-C:H films applied to 9Cr18Mo steel through programmable high-power pulse magnetron sputtering demonstrated exceptionally low friction when interacting with different mating materials in a dry air environment. The optimal friction performance was observed in self-mating tribosystems comprising a-C:H films on steel surfaces paired with a-C:H-coated steel balls, which can be ascribed to their reduced electron affinities. The addition of metals like Ag or Ti to DLC films can improve their mechanical properties and resistance to nanoscratching [12]. The nanocomposite films demonstrated improved resistance and low-friction coefficients, rendering them ideal for protective coatings and solid lubricant applications. The capability to manipulate the dimensions and density of metallic nanoparticles paves the way for customizing coating efficacy. The addition of silicon to DLC films can further improve their tribological properties and thermal stability. A study assessing the use of a graded hydrogen amorphous DLC film with silicon on cylinder liners indicated a potential reduction in the COF by as much as 19% [13]. Engine tests revealed a reduction in friction mean effective pressure (FMEP) of up to 12% and a slight decrease in brake-specific fuel consumption (BSFC) of up to 2.5% within the low engine speed range. Molybdenum disulfide ($MoS_2$) is a well-known solid lubricant that provides low friction under both ambient and vacuum conditions [14]. The TiN/$MoS_2$ bilayer coatings for applications requiring extremely high hardness, ultra-low COF, and high durability at temperatures lower than 60 K. However, $MoS_2$ coatings typically have poor wear resistance and low ability to withstand higher contact pressures [15].

## 9.3 Factors Influencing the Performance of Low-Friction Materials in Hydrogen Environments

The performance of low-friction materials and coatings in hydrogen systems is influenced by a variety of factors, the factors affecting the degradation of coatings performance is illustrated in Figure 9.2:

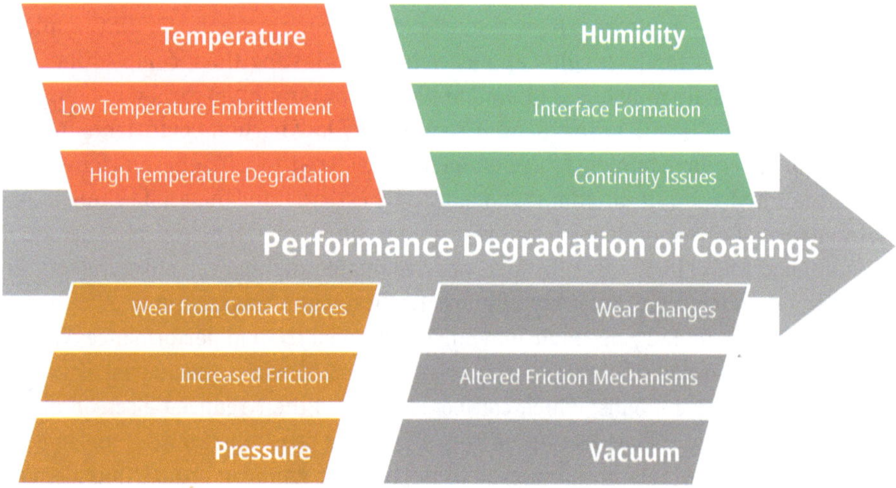


**Figure 9.2:** Factors affecting performance degradation of coatings.

- Temperature: The operating temperature can significantly affect the stability and performance of coatings. High temperatures can lead to the degradation of some materials, while low temperatures can cause embrittlement.
- Pressure: High-pressure environments can increase friction and wear due to increased contact forces.
- Humidity: Relative humidity can play a critical role in the formation and continuity of low-frictional interfaces, as demonstrated by Kuriyagawa et al. [16] for hydrogenated carbon nitride coatings.
- Vacuum: Vacuum environments present unique challenges for tribological systems, as the absence of air and moisture can alter friction and wear mechanisms [17].

## 9.4 Material Properties

- Hardness: High hardness is generally desirable for wear resistance.
- Elastic Modulus: The elastic modulus affects the contact pressure and deformation behavior of the material.

- Surface Roughness: A smooth surface finish can reduce friction and wear.
- Adhesion Strength: Good adhesion to the substrate is essential for preventing coating delamination.
- Chemical Inertness: Resistance to chemical reactions with hydrogen or other environmental components is crucial for long-term stability.

## 9.5 Tribological Parameters

- Sliding Speed: The sliding speed affects the frictional heating and wear rate.
- Contact Load: The contact load influences the contact pressure and the extent of deformation.
- Mating Material: The choice of mating material can significantly impact the friction and wear behavior of the coating.

## 9.6 Friction and Wear Characteristics of Various Materials in Hydrogen Energy Systems

The application of a hydrogen-free diamond-like carbon (H-free DLC) coating to piston rings in a mass-production gasoline engine reportedly cut friction by 18%. Since DLC is most effective in thin oil film conditions common near the piston top dead center (TDC), the friction reducing impact was more noticeable when the H-free DLC was applied to the top ring. For O-ring uses in hydrogen energy systems, Hiroyoshi Tanaka et al. [18] looked examined the friction and wear properties of different rubber materials. Under hydrogen atmosphere, many rubber materials such as silicon rubber, EPDM rubber, fluorinated rubber, and natural rubber-based materials were assessed. The materials performance in hydrogen system is revealed in Figure 9.3. Depending on its type, surroundings gasses, and running conditions, every material displayed different friction and wear performances. Whereas a fluorinated rubber with carbon black filler showed low COF [18], a silicone rubber shown low wear and high, stable COF in hydrogen. Promising technologies for the generation of sustainable energy are hydrogen fuel cells [19]. Improving energy economy and extending operational lifetime in hydrogen fuel cells depend on optimizing temperature management. Hong-Soo Kim et al. [20] investigated how silver nanoparticles might be used on stainless steel electrodes to enhance their electrocatalytic performance for the hydrogen evolution reaction (HER). Annick Vanhulsel et al. [21] evaluated the possibility of highly hydrogenated DLC coatings for solid lubricant applications in space, discovering that the test environment had a decisive effect on wear rate, with the coatings able of giving ultra-low-friction values.

Michele Ferrari et al. investigated superhydrophobic coatings from recyclable materials for protection and friction reduction in marine environments [23].

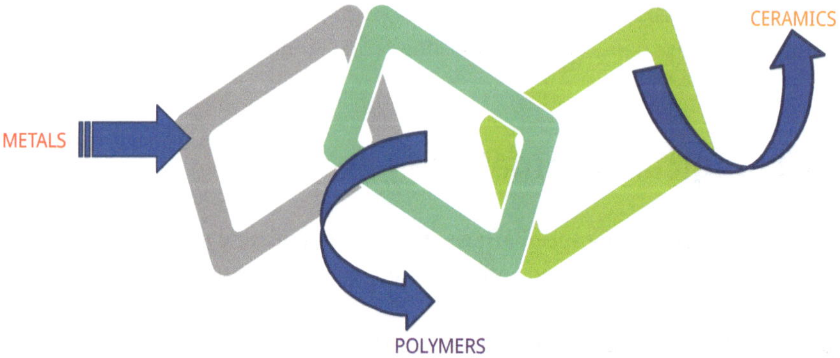


**Figure 9.3:** Materials performance in hydrogen system.

# 9.7 How Do Low-Friction Coatings Prolong Hydrogen Fuel Cell Life?

Low-friction coating methods are important for making hydrogen fuel cell systems last longer since they greatly reduce friction and make parts last longer. Near-frictionless carbon (NFC) and solid lubricants like graphene are examples of coatings that are made to work in hydrogen settings, where regular lubricants may not be enough. This document explains how these technologies work and what benefits they will always have.

## 9.7.1 Ways to Reduce Wear and Lowering Friction

NFC coatings have cut the frictional force by four times and the wear rate by more than two orders of magnitude in thrust bearings for air compressors. Solid lubricants like polytetrafluoroethylene (PTFE), graphite, and $MoS_2$ work well to reduce friction in hydrogen gas and liquid environments. They are a reliable alternative to liquid lubricants [24]. Graphene coatings have superlubricity, which makes worn surfaces last much longer in hydrogen applications [25].

### 9.7.2 Lasting Benefits

Fuel cell parts last longer because they are better at resisting scuffing and wear [26]. Less friction means less parasitic energy loss, which makes the whole system work better. When hybrid vehicles use low-friction coatings, they use less fuel and release less pollution. Low-friction coatings have many advantages, but there are still questions about how well they work in different environments. This shows that further study is needed to make them more useful in hydrogen fuel cell systems.

## 9.8 Essential Traits of Low-Friction Coating Technologies That Optimize Hydrogen Flow in Hydrogen Systems

To make hydrogen systems work better, we need low-friction coating technology. The coatings are designed to lower friction and make parts last longer when they are in hydrogen environments. The chemical makeup, structural properties, and manufacturing techniques of these technologies are all important for their success in hydrogen applications.

### 9.8.1 Chemical Composition

- Films of Hydrogenated Carbon: Fluorine-doped hydrogenated carbon films are great at reducing wear in hydrogen systems because they have very low-friction coefficients (about 0.01) and very high hardness (around 11.04 GPa) [27].
- Complex Thin Films: Carbon-silicon-hydrogen films with 30 wt% hydrogen are chemically stable and have low friction, which are important for parts in hydrogen systems [28].

### 9.8.2 Structural Properties

Multilayered coatings like TiAlN and TiMoN make it much harder for hydrogen to get through. They can even get as low as 1/100 of what is seen in untreated surfaces. This happens because their crystal structures are so thin and they have many grain boundaries that can trap hydrogen [29]. Coatings with columnar crystal structures have better hydrogen barrier characteristics, which make hydrogen flow more efficiently [30]. Low-friction coatings make hydrogen flow more efficiently, but there are still problems with their long-term stability and performance in diverse working situations.

More research is needed to improve these coatings so they can be used in more ways in the hydrogen economy.

## 9.9 Challenges in Low-Friction Materials and Coatings for Hydrogen Systems

Many difficulties still exist even if low-friction materials and coatings for hydrogen systems have shown impressive advancements. Ensuring the coatings' long-lasting stability and endurance under very demanding situations is a great difficulty. Further research is needed to produce coatings competent of surviving high temperatures, pressures, and corrosive environments for extended periods of time:

- The cost of coating materials and deposition techniques could prevent their more general use. Improving hydrogen technologies' economic viability depends on developing reasonably priced coating solutions. A major difficulty is raising coatings' manufacturing to meet the needs of a large hydrogen infrastructure demand. Broad acceptance depends on the development of scalable deposition techniques and production processes.
- Development of more effective coatings depends on a deeper awareness of the basic tribo-mechanisms existing in hydrogen settings. This includes investigating, at the nanoscale level, how hydrogen affects friction, wear, and material deterioration. Guaranteeing quality and dependability depends on establishing consistent testing techniques to evaluate low-friction materials in hydrogen settings.

## 9.10 The Positive and Negative Aspects About Low-Friction Coatings in Hydrogen Systems and How to Get the Most Out of Them for Diverse Uses

There are pros and cons to using low-friction coatings in hydrogen systems, and each one can be tailored to fit different needs. These coatings make seals and parts last longer and work better in hydrogen settings by reducing friction and wear. They help keep the systems gas-tight. However, there are problems that need to be solved, including as the material's sensitivity to hydrogen and the need for very particular application circumstances.

### 9.10.1 Benefits of Low-Friction Coatings

Better seal efficiency low-friction coatings make seals in hydrogen storage systems much better at keeping gas in. This reduces the amount of hydrogen that escapes and extends the life of the seal by reducing wear [31]:

- Lower Friction Coefficient: Coatings like DLC and solid lubricants (like PTFE and $MoS_2$) have low-friction coefficients, which are important in hydrogen environments where regular lubricants may not work well [32, 33].
- Stability at High Temperatures: Some coatings, including those with boron nitride and graphene, can keep their low friction even at very high temperatures. This makes them good for use in situations where there are a lot of stresses [34].

### 9.10.2 Problems with Low-Friction Coatings

- How Quickly the Material Reacts: According to Gradt and Theiler [33], some low-friction materials may stop working or break down when they are in hydrogen settings. As a result, it is very important to carefully choose and assess these materials.
- Complicated Application Requirements: Low-friction coatings often need certain climatic conditions to be made, which can make the manufacturing process more difficult.

Choosing the right materials for low-friction coatings is important for getting the greatest performance. You need to think about the environment around the coating and the needs of the application. According to Keunecke et al. [35], CrC/a-C:H coatings can make things more resistant to wear and work well with lubricants like MoDTC. On the contrary, low-friction coatings have a number of important benefits. However, their usefulness can be restricted by the environment and the materials used to make them. This shows how important it is to keep looking at ways to make hydrogen systems more durable.

## 9.11 Future Research Directions

Despite these advancements, significant research gaps remain, including the need for improved multi-material powder bed fusion techniques, numerical modeling of friction stir welding, understanding of friction-induced vibration and noise, optimization of friction materials for seismic isolation, computational support for emerging technology education, the effects of electric fields on nanoscale friction, standardization in soft robotics, sustainable welding technologies, molecular sieving using graphene oxide membranes, principles of green tribology, low-friction materials for seismic dampers, liquid super repellent surfaces, friction in two-dimensional materials, and the use of cork façades in architecture [36]:

- Self-Healing Coatings: Developing coatings that can self-repair damage caused by wear or corrosion would significantly extend their lifespan.
- Adaptive Coatings: Coatings that can adapt their properties in response to changing operating conditions could provide optimal performance under a wider range of conditions.
- Bio-inspired Materials: Exploring bio-inspired materials and designs could lead to new and innovative solutions for low-friction tribology.

The following areas still need more research: multi material powder bed fusion techniques, numerical modeling of friction stir welding, understanding of friction-induced vibration and noise, optimization of friction materials for seismic isolation, computational empowerment for emerging technologies education, standardizing in soft robotics, environmentally friendly welding technologies, molecular sieving through graphene oxide membranes, principles of green tribology, low-friction materials for seismic dampers, liquid super repellent surfaces, friction in 2D materials, cork façade in architecture. Once these voids are closed, we can proceed with the evolution of much enhanced materials.

## 9.12 Conclusion

Advancing sustainable hydrogen systems depends much on low-friction materials and coatings. These materials are important since they help to reduce energy losses, increase durability, and enable dry-running conditions for a spectrum of hydrogen uses. Though there has been significant progress in this field, some challenges still exist like improving durability, cutting costs, and raising manufacturing capacity. Unlocking the whole promise of hydrogen as a clean and sustainable energy carrier requires constant research and invention in new materials, advanced deposition techniques, and a thorough understanding of tribo-mechanisms. Particularly as society's attention on climate change grows, hydrogen energy is poised to become ever more important in the search of a green and low-carbon planet. Low-friction materials' development is driven by the increasing demand for improved performance in a variety of uses, sustainability, and energy economy. Minimizing friction and wear advantages innovative technologies including inorganic-organic hybrids, 2D materials, DLC coatings, hydrogels, and biomimetic materials. In surface engineering, thin layer deposition, surface texturing, and electropolymerization all greatly enhance material tribological characteristics. Further choices for lowering friction and wear are provided by lubrication techniques including solid lubrication, additive inclusion, and use of green tribology concepts.

## References


[1] Schbel, M., Langela, Marc, Koch, Thomas, and Pllinger, A. 2023. "High-Strength PPS-Polymer Composites for Hydrogen High-Pressure Applications". Key Engineering Materials. https://doi.org/10.4028/p-3B1Yxv

[2] Dhere, N. and Pai, Anil. NaN. "Development of TiN/MoS". None. https://doi.org/10.1115/isec2004-65050

[3] Linek, Tomasz. 2018. "Effects of Applying WC/C Protective Coating on Structural Elements Working in Cavitation Environment". IntechOpen. https://doi.org/10.5772/intechopen.80719

[4] Liu, Lincong, Zhou, Ming, Jin, Long, Li, Liangchuan, Mo, Y., Su, Guoshi, Li, Xiao, Zhu, Hongwei, and Tian, Yu. 2019. "Recent advances in friction and lubrication of graphene and other 2D materials: Mechanisms and applications". Friction. https://doi.org/10.1007/S40544-019-0268-4

[5] Guo, Yanbao, Zhou, Xuanli, Lee, Kyungju, Yoon, H. C., Xu, Quan, and Wang, Deguo. 2021. "Recent development in friction of 2D materials: from mechanisms to applications". Nanotechnology. https://doi.org/10.1088/1361-6528/abfa52

[6] Curry, John F., Wilson, M., Luftman, H. S., Strandwitz, Nicholas C., Argibay, Nicolas, Chandross, Michael, Sidebottom, Mark A., and Krick, Brandon A. 2017. "Impact of Microstructure on $MoS_2$ Oxidation and Friction". American Chemical Society. https://doi.org/10.1021/acsami.7b06917

[7] 2012. "Caracterizao e desempenho de um filme de carbono amorfo hidrogenado tipo diamante (a-C:H) dopado com silcio, aplicado em camisa de cilindro de motor combusto interna". None. https://doi.org/10.11606/D.85.2012.TDE-10122012-125406

[8] Kolawole, Funsho Olaitan, Kolawole, Shola Kolade, Varela, Luis Bernardo, Owa, Adebayo Felix, Ramrez, Marco A., and Tschiptschin, Andr Paulo. 2020. "Diamond-Like Carbon (DLC) Coatings for Automobile Applications". IntechOpen. https://doi.org/10.5772/intechopen.95063

[9] Ohtake, Naoto, et al. 2021. "Properties and Classification of Diamond-Like Carbon Films". Multidisciplinary Digital Publishing Institute. https://doi.org/10.3390/ma14020315

[10] Rusanov, A., Fontaine, J., Martin, J., Mogne, T., and Nevshupa, R. 2008. "Gas desorption during friction of amorphous carbon films". None. https://doi.org/10.1088/1742-6596/100/8/082050

[11] Sun, Lei, Jia, Q., Zhang, Bin, Gao, Kaixiong, Tan, Xin, Lai, Zhenguo, and Zhang, Junyan. 2021. "Achieving ultralow friction of a C:H film grown on 9Cr18Mo steel for industrial application via programmable high power pulse magnetron sputtering". Surface and Interface Analysis. https://doi.org/10.1002/sia.7022

[12] Constantinou, Marios, Avgeropoulos, Apostolos, Moschovas, Dimitrios, Varotsis, Constantinos, Patsalas, P., Kelires, P. C., and Constantinides, Georgios. 2018. "Metal (Ag/Ti)-Containing Hydrogenated Amorphous Carbon Nanocomposite Films with Enhanced Nanoscratch Resistance: Hybrid PECVD/PVD System and Microstructural Characteristics". Multidisciplinary Digital Publishing Institute. https://doi.org/10.3390/nano8040209

[13] Hrtwig, Fabian, Lorenz, L., Makowski, S., Krause, M., Habenicht, C., and Lasagni, A. 2022. "Low-Friction of ta-C Coatings Paired with Brass and Other Materials under Vacuum and Atmospheric Conditions". Materials. https://doi.org/10.3390/ma15072534

[14] Hrtwig, Fabian. 2022. "Triboscopic Insights into Long-Term Tribologic Behavior of Brass/ta-C in High Vacuum". None. https://doi.org/10.26799/cp-sict-plasmatech-tribology-2022/1

[15] Berman, Diana, Erdemir, Ali, and Sumant, Anirudha V. 2014. "Graphene: a new emerging lubricant". Elsevier BV. https://doi.org/10.1016/j.mattod.2013.12.003

[16] Kuriyagawa, Kazuya, Tamagawa, Tatsuki, and Adachi, Koshi. 2024. “Impact of Relative Humidity on the Formation of Low-Frictional Interface and its Continuity in Tribological Systems with Hydrogenated Carbon Nitride Coatings”. Tribology letter. https://doi.org/10.1007/s11249-024-01899-5

[17] Higuchi, Tsuyoshi, Mabuchi, Y., Ichihara, H., Murata, T., and Moronuki, Mmasaki. 2012. “Development of Hydrogen-Free Diamond-Like Carbon Coating for Piston Rings”. None. https://doi.org/10.4271/2012-01-1327

[18] Tanaka, Hiroyoshi, Norikyo, Suguru, Aoyagi, Ayako, Tanaka, Hiroyoshi, Morita, T., Sawae, Yoshinori, and Sugimura, J. 2024. “Friction and wear properties of several rubber materials for high pressure O-ring”. None. https://doi.org/10.61319/bcwrtc88

[19] Aikins, Selasi Agbale, Avevor, James, and Enyejo, L. A. 2024. “Optimizing Thermal Management in Hydrogen Fuel Cells for Smart HVAC Systems and Sustainable Building Energy Solutions”. None. https://doi.org/10.38124/ijsrmt.v3i9.351

[20] Kim, Hong-Soo, Kim, Hwapyong, Flores, Monica Claire, Jung, Gyusuck, and In, Su-il. 2021. “Stable Surface Technology for HER Electrodes”. Catalysts. https://doi.org/10.3390/CATAL11060693

[21] Vanhulsel, Annick, Jacobs, R. N., Acker, Karel Van, Roberts, E., Velasco, Francisco, Sherrington, Ian, and Gaillard, Lionel. 2005. “Development of Highly Hydrogenated DLC Coatings for Solid Lubricant Applications in Space”. None. https://doi.org/10.1115/wtc2005-63450

[22] Kolhe, M. 2019. “The 5th International Symposium on Hydrogen Energy, Renewable Energy and Materials (HEREM 2019)”. None. https://doi.org/10.1088/1757-899x/605/1/011001

[23] Ferrari, Michele, Benedetti, A., and Cirisano, Francesca. 2019. “Superhydrophobic Coatings from Recyclable Materials for Protection in a Real Sea Environment”. Multidisciplinary Digital Publishing Institute. https://doi.org/10.3390/coatings9050303

[24] Gradt, T., & Theiler, G. (2011). Tribological Behaviour of Solid Lubricants in Hydrogen Environment. *Tribology Online*, *6*(2), 117–122. https://doi.org/10.2474/TROL.6.117

[25] Sumant, A. V., Berman, D., & Erdemir, A. (2014). *Low friction wear resistant graphene films*. https://www.freepatentsonline.com/y2018/0223208

[26] Fenske, G. R., Woodford, J., Erdemir, A., & Garland, N. L. 2003. IV.H.9 Low-Friction Coatings and Materials for Fuel Cell Air Compressors. https://www.hydrogen.energy.gov/pdfs/progress04/ivh9_fenske.

[27] Li, W., Zhang, B., Zhou, Y., Li, Q., & Zhang, J. 2013. Ultra-low friction of fluorine-doped hydrogenated carbon film with curved graphitic structure. Surface and Interface Analysis, 45(8), 1233–1237. https://doi.org/10.1002/SIA.5260

[28] Gregoire, T. C., Ducarroir, M., Scordo, S., & Boher, C. 1995. Low friction carbon@-silicon@-hydrogen@ complex thin film – produced by plasma-CVD e.g. on metal pinion or bearing part.

[29] Tamura, M. (2019). Hydrogen Permeation of Multi-Layered-Coatings. Advanced Materials Research, 1152, 9–18. https://doi.org/10.4028/WWW.SCIENTIFIC.NET/AMR.1152.9

[30] Cooke, K. E., & Sun, H. (2013). Physical vapour deposition coatings for demanding applications in the new hydrogen economy. Transactions of The Institute of Metal Finishing, 91(6), 294–297. https://doi.org/10.1179/0020296713Z.000000000145.

[31] Lev, L. C., Podorashi, D. A., Lukitsch, M. J., Michler, T., & Xiao, X. 2010. Improved seal performance for hydrogen storage and supply systems. https://patents.google.com/patent/WO2011008898A3/en

[32] Takahisa, K., Nosaka, M., Tohyama, M., & Suzuki, M. 2016. Low-friction coating production method and sliding method. https://www.freepatentsonline.com/y2017/0362711.html

[33] Gradt, T., & Theiler, G. (2011). Tribological Behaviour of Solid Lubricants in Hydrogen Environment. Tribology Online, 6(2), 117–122. https://doi.org/10.2474/TROL.6.117

[34] Sumant, A. V., & Ayyagari, V. A. 2019. Low friction coatings. https://patents.google.com/patent/US20210094067A1/en

[35] Keunecke, M., Bewilogua, K., Becker, J., Gies, A., & Grischke, M. 2012. CrC/a-C:H coatings for highly loaded, low friction applications under formulated oil lubrication. Surface & Coatings Technology, 207, 270–278. https://doi.org/10.1016/J.SURFCOAT.2012.06.085
[36] Farrar-Gaines, D., Maisano, A. J., Freeman, A. W., Xia, Z., LaBarre, E., Watson, G., & Dawib, Y. 2021. Development of an environmentally-durable, low-friction coating. MRS Communications, 11(3), 1–5. https://doi.org/10.1557/S43579-021-00017-Z

Nishant K. Singh, Sanjeev Kumar, Vijay Singh

# 10 Environmental Impact of Green Hydrogen Production

**Abstract:** Hydrogen minimizes fossil fuel use and transforms energy architectures, promoting sustainable development. This study reviews green hydrogen's domestic and worldwide development, stressing its growing importance in the global energy sector. The talk then shifts to green hydrogen generating technologies and their novel approaches. The pros and cons of green hydrogen generating systems are examined to assess their efficacy and potential. Future hydrogen production may include water electrolysis, indicating a shift. This conversation discusses green hydrogen technology's development hurdles and application prospects, including its integration potential. Green hydrogen is expected to help achieve sustainable global energy development and a cleaner, safer energy future.

**Keywords:** Green hydrogen, hydrogen, generating technologies, efficacy, sustainable energy, water electrolysis

## 10.1 Introduction

Rising worldwide population and fast economic growth have driven energy demand to explode, harm the environment, and cause climate change. With these problems, to move to a sustainable, low-carbon energy system, we have to quickly develop better, more efficient, long-term substitutes for fossil fuels. One is the creation of hydrogen from sustainable sources. Green hydrogen energy has become a worldwide focus during this transition because of its efficiency and environmental friendliness [1]. Particularly in the transportation and heavy industries sectors, hydrogen ($H_2$) is expected to be a necessary component in the global change toward a net-zero energy system [2]. With estimates showing a possible sixfold increase by 2050, the increasing awareness of this trend is projected to result in a notable rise in $H_2$ consumption [3]. Still, the current production setting presents a major obstacle. Mostly via means of coal gasification and steam reformation of natural gas, most hydrogen generation is now used as industrial feedstock and comes from fossil fuels. The biggest challenge in achieving real clean energy targets is the reliance on fossil fuels for hydrogen generation, which emphasizes the great need of scalable and reasonably priced renewable hydrogen alternatives [4].

Generated by the electrolysis of water using renewable energy sources, green hydrogen has grown to be a vital part of the global effort toward a change to sustainable energy systems. Particularly important in addressing climate change and improving

 | https://doi.org/10.1515/9783112181072-010

energy security is this clean energy carrier since it can significantly lower greenhouse gas emissions and lessen reliance on fossil fuels. Green hydrogen is growing more and more important in several areas, including transportation, manufacturing, and power generation as nations aim for climate neutrality [5–6]. Although green hydrogen has great potential, its manufacturing raises environmental issues that need careful control. Main concerns include prospective changes in land use that can compromise biodiversity and food security, major water resource needs, and the consequences on local ecosystems resulting from large renewable energy installations [7–8]. Moreover, even if the manufacturing of green hydrogen aims to lower lifetime emissions, the carbon footprint might vary greatly depending on the energy sources used, and emissions could still develop during the hydrogen transportation and conversion operations [9–10]. The arguments about green hydrogen mostly concern its environmental impact and financial viability. The high costs of production, influenced by electricity prices and technology limitations, pose significant barriers to its widespread adoption [11–12].

The International Energy Agency "Renewable Energy Report 2021" projects notable global increase in renewable energy capacity. According to Figure 10.1, from 2021 to 2050 the 10 biggest renewable energy producers in the world will account for 80% of the installed capacity worldwide. The worldwide energy revolution will depend much on these countries. China, leading in the increase of renewable energy, is predicted to account for roughly 43% of this capacity; followed by the United States at 12% and India at 7%. These figures show the strategic investment in renewable energy development made by several of the big economies.

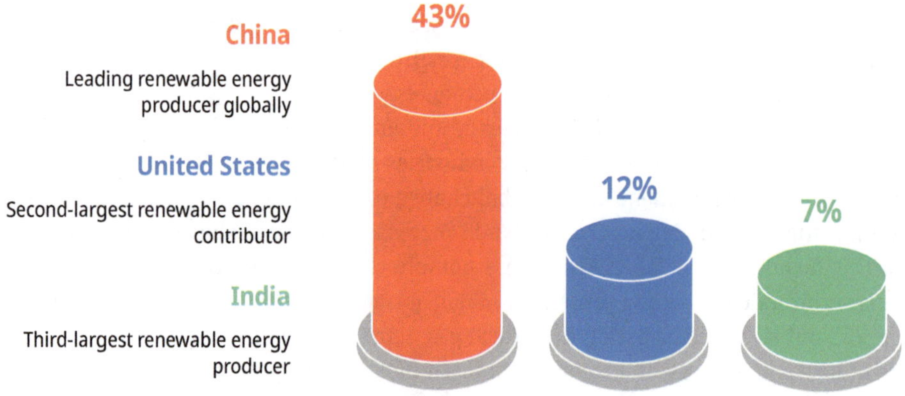


**Figure 10.1:** Projected renewable energy capacity in major countries by 2050.

Furthermore, the environmental impacts of sourcing materials for hydrogen production and the management of associated waste raise questions about the sustainability of green hydrogen as a long-term solution [13–14]. In light of these complexities, ongoing research, technological advancements, and supportive government policies are essential to maximize the benefits of green hydrogen while addressing its environmental and economic challenges. The future outlook for green hydrogen is promising, with expectations of decreasing production costs and increased efficiency, paving the way for its integration into a more sustainable energy landscape [15–16].

Drawing worldwide interest, hydrogen production technology is essential for a hydrogen energy economy and the global energy revolution [17]. Main hydrogen generating technologies are shown in Figure 10.2, developing techniques, water electrolysis, biological hydrogen synthesis, and fossil fuel reforming. Natural gas steam reforming is the recommended hydrogen generating technique since of its economic efficiency. This method generates a lot of carbon dioxide, which runs against efforts at greenhouse gas reduction. Biodegradable organic compounds can be broken down by algae and bacteria, therefore promoting environmental friendly hydrogen generation. Although ecologically friendly, this method suffers with efficiency. Renewable energy allows water electrolysis to generate hydrogen free of emissions. Falling renewable energy costs and electrolysis technology help to make this process more practically and financially feasible. For a sustainable energy change and general use as an energy carrier, hydrogen generating technologies have to get better. Technical challenges must be addressed by research on fresh approaches to boost hydrogen production efficiency, reduce prices, and hasten commercialization and sustainable development of hydrogen technology. Solving world energy and environmental challenges depends on combining renewable energy with water electrolysis for hydrogen generation [18]. This method greatly lowers greenhouse gas emissions and uses of fossil fuels.

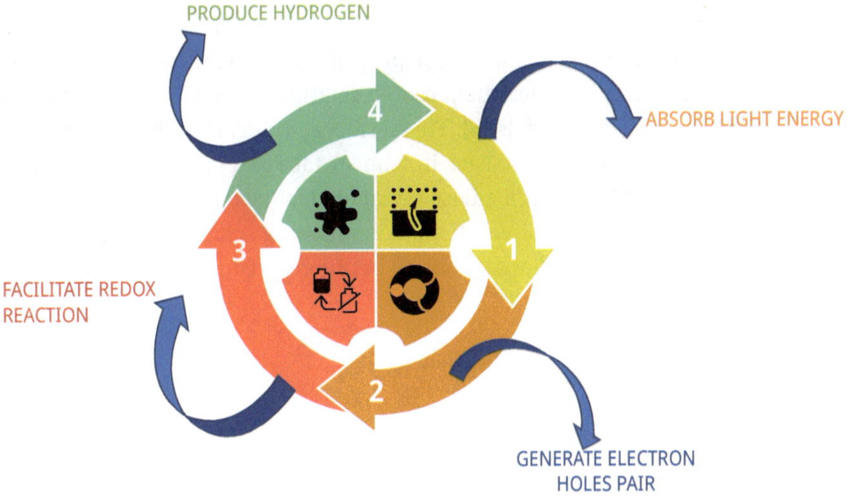


**Figure 10.2:** Cycle of photocatalytic hydrogen production.

Direct solar energy conversion into chemical energy is made possible by photoelectrochemical (PEC) hydrogen generation technology, therefore offering a convincing method for sustainable energy. Sunlight helps the breakdown of water into its basic components: hydrogen and oxygen [19], in a PEC cell. The design of the photoanode and careful material choosing determine the efficiency of this technique. This fundamental component must efficiently grab sunlight, generate electron-hole pairs, and then help water break down on its surface, as well as at the counter electrode (refer Figure 10.2). Strong light absorption across the solar spectrum, effective charge separation and transport, and great chemical stability in aqueous settings should all be ideally shown by photoanode materials. Notable examples of materials looked at for this use are cadmium sulfide and titanium dioxide. While addressing important issues including material cost and long-term stability to enable more general deployment, the present studies in this field are mostly focused on improving photoelectric conversion efficiency.

## 10.2 Environmental Impacts

The production and use of green hydrogen, while promising for reducing greenhouse gas emissions, carry a range of environmental impacts that must be carefully managed.

### 10.2.1 Ecosystem and Biodiversity Effects

Particularly solar and wind farms, the building of large renewable energy projects for the generation of green hydrogen can significantly affect the nearby ecosystems and biodiversity. Often requiring the destruction of vegetation, solar farms disturb habitats and migration routes for animals [5]. Furthermore, influencing the survival and reproduction of native species are changes in land surface and microclimate brought about by solar panels. Developers are advised to apply biodiversity-friendly strategies include the preservation of natural vegetation surrounding solar arrays and the design of installations that enable wildlife mobility in order to reduce these effects [6].

### 10.2.2 Water Resource Considerations

Usually, the synthesis of green hydrogen involves the water electrolysis process, which calls for significant water supplies. Especially in arid regions, the extraction of freshwater or seawater for this procedure can put great pressure on local water bodies [7]. Furthermore, the discharge of brine from desalination plants connected to the generation of green hydrogen could contaminate marine habitats, therefore endangering aquatic life, as well as the populations depending on these ecosystems [7].

### 10.2.3 Land Use Changes

The large-scale production of renewable energy for hydrogen synthesis has the capacity to change land use, therefore influencing food security and biodiversity. This usually means the modification of agricultural or natural ecosystems, therefore raising vulnerability to environmental hazards such landslides and flooding [7]. Eliminating mangroves and wetlands, among other natural buffer zones, greatly increases hazards to community safety and fuels environmental degradation.

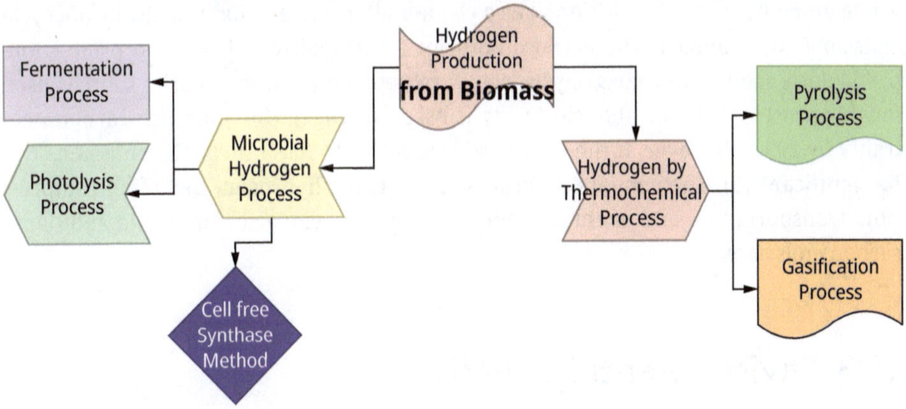


**Figure 10.3:** Hydrogen production from biomass.

As a clean energy carrier, biomass hydrogen generating technologies make hydrogen from several biomass resources. Reducing greenhouse gas emissions and dependency on limited fossil fuels depends critically on resources such crop wastes, forestry waste, urban organic waste, and other kinds of renewable organic matter. This technology provides environmental benefits and improves energy diversification by efficiently recycling waste biomass. Diversification is crucial for improving energy security, protecting the environment, and fostering economic growth. Biomass hydrogen production is primarily classified into two categories: thermochemical conversion and microbial processes. The categories comprise seven distinct methods, as detailed in [20] and illustrated in Figure 10.3.

### 10.2.4 Worker Safety and Environmental Risks

The hydrogen production process may also pose risks to workers. Exposure to intense electromagnetic fields within electrolyzer facilities, as well as potential toxins from conversion and storage units, raises safety concerns. The use of rare earth elements in catalysts can lead to labor risks if these materials are not sourced responsibly [7]. Additionally, the production of ammonia and methanol, often associated with hydrogen production, can result in harmful waste that threatens water sources if not managed appropriately.

### 10.2.5 Life Cycle Emissions and Carbon Footprint

While green hydrogen is positioned as a cleaner alternative to fossil fuels, its life cycle emissions are complex. The carbon footprint associated with hydrogen production, particularly from water electrolysis, can vary depending on the energy sources used for this process. If renewable electricity is used, hydrogen can approach carbon neutrality by 2050. However, if the energy source relies on fossil fuels, the emissions can be significant [8]. The conversion processes for using hydrogen, including compression, transportation, and eventual energy conversion, can also contribute additional carbon emissions.

## 10.3 Environmental Benefits

Green hydrogen production presents several significant environmental benefits (refer Figure 10.4), contributing to climate change mitigation and promoting a transition toward sustainable energy systems.

### 10.3.1 Renewable Energy Source

One of the primary advantages of green hydrogen is that it is produced using renewable energy sources, such as solar and wind power. Unlike fossil fuels, which are finite and contribute to greenhouse gas emissions when combusted, green hydrogen harnesses abundant and sustainable energy, thereby significantly reducing the carbon footprint associated with energy production [6, 9]. The shift toward utilizing green hydrogen is vital for combating climate change and promoting a greener future.

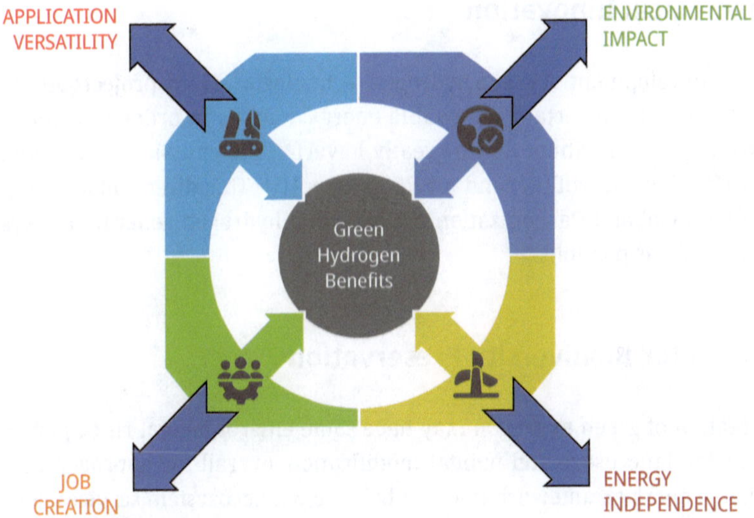


**Figure 10.4:** Benefits of green hydrogen..

## 10.3.2 Decreased Reliance on Fossil Fuels

Investing in green hydrogen allows countries to lessen their dependence on fossil fuels. This transition not only enhances energy security by minimizing vulnerability to international oil and gas price fluctuations but also has the potential to improve a nation's trade balance by reducing the need for imported fossil fuels [10]. Industries that are difficult to decarbonize, such aviation, heavy-duty transportation, and particular industrial operations, where alternative low-carbon solutions remain under development, depend mainly on this shift [11].

## 10.3.3 Support for Local Economies

Encouragement of local green hydrogen generation from renewable energy helps countries to develop their economies in line. Using local resources increases energy independence and helps to lower the economic risks associated with shifting world fuel prices, hence improving trade balances and lowering of foreign currency outflow [12].

### 10.3.4 Technological Innovation

Notable also is the development of green hydrogen technologies. Many projects aim to link electrolyzers directly to certain renewable energy sources in order to achieve low-emission hydrogen generation, hence greatly lowering the emissions connected with grid electricity that can still depend on fossil fuels [11]. This direct relationship ensures, in line with global decarbonization projects, that hydrogen generation stays as ecologically friendly as possible.

### 10.3.5 Potential for Biodiversity Preservation

Although the creation of green hydrogen may have some environmental risks, particularly with regard to land usage and habitat modification, overall the approach promotes a sustainable energy framework that can balance with ecosystem services. Perfect design and implementation of green hydrogen projects can help to restore environmental damage and preserve natural ecosystems, so improving biodiversity and satisfying energy needs [6–7].

### 10.3.6 Economic Implications

Green hydrogen (GH2) production has significant financial consequences that affect local economies, as well as the dynamics of global trade. By means of foreign investment, countries adopting GH2 policies stand to benefit from potentially exceeding local decarbonization needs, therefore enhancing foreign exchange earnings, tax revenues, and local economic activity [16]. Still, the creation of special economic zones with tax benefits for investors could lower host country tax collections.

### 10.3.7 Domestic Economic Development

By improving the manufacturing sector's contribution to GDP, so fostering skills development, and so creating sustainable employment possibilities, the switch to GH2 production could help to boost domestic economic growth [16]. The possibility for value addition of GH2 emphasizes the need of developing downstream businesses especially in the manufacturing of green steel or fertilizer. Strong industrial frameworks help countries to achieve these benefits; meanwhile, those without such frameworks may find it difficult to change their industrial systems, a process that can take many years [17].

### 10.3.8 International Trade and Market Integration

Engaging in international GH2 trade not only enhances a country's trade balance but also increases its autonomy and political significance within the global energy framework [12]. This trade can facilitate knowledge transfer, further promoting socioeconomic development in exporting countries. However, many potential exporting nations may heavily rely on imports of industrial equipment, which can diminish net export revenues despite the overall gains in technological capabilities and domestic innovation [12].

### 10.3.9 Cost Challenges and Technological Advancements

Despite the promising economic outlook, GH2 production faces significant cost challenges. The cost of hydrogen produced via electrolysis currently ranges from $4 to $7 per kilogram, heavily influenced by electricity prices and the efficiency of electrolyzers [13–14]. The U.S. Department of Energy aims to reduce this cost to $2 per kilogram by 2025 through technological advancements and economies of scale [13]. As production costs remain a critical barrier, strategic investments in technology and infrastructure are essential for establishing a seamless green hydrogen ecosystem [15].

### 10.3.10 Government Support and Investment

Advancement of the green hydrogen industry depends on government participation. Assuming risks and maybe producing profits through stock shares in early movers allows governments to drive private investment and gain a multiplier impact [16]. Aligning public investments with more general public goals depends on conditions including profit reinvestment and share buy-back limits [16]. This cooperative approach can help to create conditions favorable for energy transition and steady economic development.

## 10.4 Policy and Regulation

### 10.4.1 Regulatory Framework for Hydrogen

Advancement of the exploration and commercialization of hydrogen facilities in the United States depends on the building of a regulatory framework. Since it guarantees cost recovery for their expenditures, regulatory certainty helps utilities to invest with assurance in the manufacturing, storage, and transportation of hydrogen [17]. Historically, this has been essential for the growth of electric, natural gas, and water utilities, inspiring discussions about the necessity of a similar framework for hydrogen given the growing industry interest and investment. However, the creation of complementary federal and state regulations for hydrogen lags behind the investment initiatives, which poses a risk of diminishing returns for taxpayer-funded projects [17]. As private industries capitalize on funding opportunities, there is a growing recognition that a clear regulatory landscape is crucial for broader commercialization and operational consistency in the hydrogen market.

### 10.4.2 International Trade and Cooperation

Countries are increasingly adopting dual strategies regarding hydrogen, focusing on both domestic use and international trade. These strategies often define a nation's intent to export, import, or achieve self-sufficiency in hydrogen production, which influences their role in the global market [13]. International cooperation is pivotal for establishing trade partnerships, sharing knowledge, and developing hydrogen technology, which ultimately enhances local economies through increased investment and technological advancement [16]. Furthermore, countries that actively engage in international hydrogen markets can reap economic benefits, including foreign investment and enhanced political significance. However, they must balance tax incentives for investors against the potential loss of tax revenues for the host country, which can arise from special economic zones set up for hydrogen production [18].

### 10.4.3 Supporting Policies for Infrastructure and Electrolyzer Deployment

To effectively scale up hydrogen production, particularly green hydrogen, comprehensive policies are needed to support the development of hydrogen infrastructure and the deployment of electrolyzers. Long-term planning is crucial; defined strategies will provide investors with insights into future market pathways, thereby attracting necessary capital [18]. Current government incentives for electrolyzers and hydrogen infrastructure are limited, indicating a need for more robust policies that promote the expansion of electrolyzer capacity and reduce associated costs [18]. This includes setting specific capacity targets within national hydrogen policies, which can signal commitment to the energy transition and stimulate market growth.

### 10.4.4 Sustainability Considerations

As hydrogen production expands, ensuring the sustainability of the processes is vital. The energy used for producing green hydrogen must be derived from renewable sources, and efforts should be made to keep costs competitive against traditional, carbon-intensive hydrogen production methods [19]. Aligning national research agendas and infrastructure development with sustainable practices will be important in maintaining the viability of hydrogen production and transportation.

## 10.5 Challenges and Limitations

Though it has significant promise, the generation of green hydrogen has several challenges that limit its wider acceptability and integration into the scene of global energy.

### 10.5.1 Cost Barriers

Gray hydrogen [15, 16] is expected to have production costs three to seven times lower than those of green hydrogen at present. The higher costs are mostly related to the present phase of technical development and the economies of scale still unmet. The financial obstacle seriously compromises the economic viability of green hydrogen projects and impedes the change toward better energy sources [11].

### 10.5.2 Efficiency Challenges

Apart from cost, efficiency remains a top issue. The general inefficiency of green hydrogen systems is largely influenced by energy losses occurring during storage, conversion, and production [24]. This inefficiency seriously hinders the realization of the environmental benefits green hydrogen promises and affects its financial feasibility. Thus, to increase the efficiency of electrolysis technologies and lower energy losses [25], constant research and creativity are necessary.

### 10.5.3 Infrastructure Requirements

Another great difficulty is the building of a strong infrastructure for hydrogen transportation and storage. New construction of extensive hydrogen networks calls both strategic planning and significant investment [23]. The complex network of production, storage, and distribution infrastructure calls for careful management to build a coherent green hydrogen ecosystem, therefore complicating the more general application of these technologies.

### 10.5.4 Regulatory and Permitting Delays

Long-standing permission procedures and regulatory uncertainty might hinder the deployment of green hydrogen projects even more. These delays could result in frozen funding during construction, therefore raising project risks and perhaps leading to financial losses [17]. The simplification of licensing procedures and the development of open regulatory systems will help determine the fast introduction of green hydrogen projects and the procurement of necessary investments [26].

### 10.5.5 Environmental Considerations

Although green hydrogen reduces greenhouse gas emissions, its production has negative effects on the surroundings. Especially in areas with limited water, the water needs for electrolysis are a major problem [27]. Moreover, the land use connected to renewable energy sources such as solar farms for the generation of green hydrogen may cause problems with local ecosystems and agricultural territory [28]. Thus, the sustainability of green hydrogen initiatives depends critically on careful planning and evaluation of environmental implications.

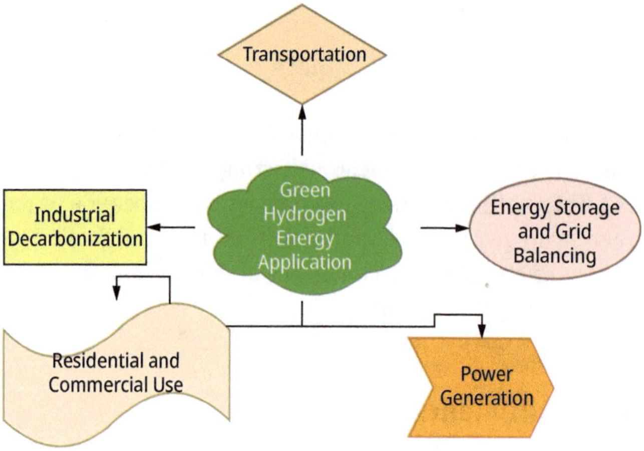


**Figure 10.5:** Prospect of the application of green hydrogen energy.

As shown in Figure 10.5, green hydrogen is a sustainable and clean energy source covering several spheres including thermal energy consumption, transportation, industrial production, and energy storage. Its flexibility makes it a desirable replacement for fossil fuels since it offers a path to decarbonization in difficult places. Anticipated to be essential in energy transformation and climate change mitigation, as well as in building a more sustainable and resilient energy system is green hydrogen. Realizing its full potential calls for coordinated efforts in technological innovation to improve efficiency and scalability, major cost reductions to guarantee competitiveness with conventional energy sources, strong policy support through incentives and regulations, and the strengthening of the industrial chain to enable the broad adoption of green hydrogen across many sectors. Attaching ambitious climate targets and accelerating its inclusion into the worldwide energy framework depend on this all-encompassing approach [35].

## 10.6 Future Outlook

The future of green hydrogen production is promising, with significant potential to contribute to global climate goals and energy transition. As nations strive for climate neutrality by 2050, low-carbon hydrogen is projected to be an essential element in achieving these targets [29]. The International Energy Agency's scenarios suggest that greenhouse gas emissions per kilogram of hydrogen could decrease significantly, indicating a positive trend toward cleaner hydrogen production methods.

### 10.6.1 Technological Advancements

Ongoing technological advancements are expected to enhance the efficiency and reduce the costs associated with green hydrogen production. The development of various electrolysis technologies such as alkaline, proton exchange membrane, and solid oxide electrolysis demonstrates the potential for improved performance and scalability [30, 31]. As manufacturers prepare for a green hydrogen expansion, they are reportedly increasing the electrolyzer pipeline, which could significantly bolster production capacity and reduce costs [32].

### 10.6.2 Economic Viability

Currently, the cost of green hydrogen production ranges between \$3 and \$6 per kilogram, influenced by electricity prices and capital expenditures [33]. Although this remains more expensive than gray hydrogen, the cost gap is expected to narrow as renewable energy sources become more prevalent and cheaper. Predictions suggest that the cost gap between green hydrogen and grey hydrogen could narrow to about three times in 2024, improving its competitiveness in the energy market [34].

### 10.6.3 Policy Support and Market Development

Government initiatives and international commitments are crucial for fostering a robust market for green hydrogen. Policies aimed at promoting renewable energy and low-carbon technologies are likely to create a conducive environment for investment and innovation in the hydrogen sector [34]. Notably, strategic frameworks like the Medium and Long-Term Plan for the Development of Hydrogen Energy Industry (2021–2035) from China illustrate how national strategies can guide the growth of this industry.

## 10.7 Conclusion

This chapter emphasizes the great relevance of green hydrogen as a transforming clean energy source, essential for lowering reliance on fossil fuels and enabling the worldwide energy transition toward sustainability. Emphasizing the fast developments and increasing policy support in this vital sector, this paper investigates the present scene of green hydrogen development both locally and internationally. Water electrolysis is the most exciting approach for future hydrogen creation as a thorough study of several green hydrogen generating technologies underlined the special advantages and disadvantages of each. Although green hydrogen has great promise, its general acceptance will be hampered in various ways. Significant momentum is shown by continuous research, technical developments, and industry and government cooperative initiatives. This chapter shows that green hydrogen has great potential to change energy systems and offers a sustainable, flexible, and clean option to meet growing worldwide energy needs and satisfy long-term environmental targets. A sustainable worldwide energy future depends on investments in green hydrogen technologies, as well as creativity in them.

# References


[1] Zainal, B.S.; Ker, P.J.; Mohamed, H.; Ong, H.C.; Fattah, I.M.R.; Rahman, S.M.A.; Nghiem, L.D.; Mahlia, T.M.I. Recent advancement and assessment of green hydrogen production technologies. Renew. Sustain. Energy Rev. 2024, 189, 113941.

[2] IEA, Net Zero by 2050, International Energy Agency, 2021.

[3] I. Staffell, D. Scamman, A. Velazquez Abad, P. Balcombe, P. E. Dodds, P. Ekins, N. Shah and K. R. Ward, Energy Environ. Sci., 2019, 12, 463–491.

[4] IEA, Global Hydrogen Review 2021, International Energy Agency, Paris, 2021.

[5] The Environmental Impact of Large-Scale Solar Farms: What You Need to know: https://www.solarfeeds.com/mag/the-environmental-impact-of-large-scale-solar-farms-what-you-need-to-know/

[6] Key aspects for managing the environmental and social risks of green hydrogen: https://blogs.iadb.org/sostenibilidad/en/key-aspects-for-managing-the-environmental-and-social-risks-of-green-hydrogen/

[7] Estimating The Carbon Footprint Of Hydrogen Production Forbes https://www.forbes.com/sites/rrapier/2020/06/06/estimating-the-carbon-footprint-of-hydrogen-production/

[8] What Are the Benefits of Green Hydrogen on the Environment https://hyenr.com/blogs/news/what-are-the-benefits-of-green-hydrogen-on-theenvironment

[9] New study examines green hydrogen fuel production in Washington: https://ecology.wa.gov/blog/january-2025/new-study-examines-green-hydrogen-fuel-production-in-washington

[10] Economic benefits of investing in green hydrogen: https://renovables.blog/en/green-hydrogen/Economic-benefits-of-investing-in-green-hydrogen/

[11] True Environmental Impact of Green Hydrogen: Emission Intensity: https://www.hydrogennewsletter.com/true-environmental-impact-of-green-hydrogen-emission-intensity-and-water-usage-rewarding-the-lowest-impactor/

[12] Policies | UNIDO Green Hydrogen https://hydrogen.unido.org/policies

[13] Hydrogen Electrolysis: The Future of Cost-Effective Green Hydrogen: https://energycentral.com/c/pip/hydrogen-electrolysis-future-cost-effective-green-hydrogen-production

[14] Green Hydrogen vs Traditional Energy Sources inPower Generation https://prismecs.com/blog/green-hydrogen-vs-traditional-energy-sources-in-power-generation

[15] Green hydrogen: Breakthroughs in clean energy innovation: https://www.policycircle.org/environment/green-hydrogen-sustainability/

[16] Financial instruments for a green hydrogen transition: https://iap.unido.org/articles/financial-instruments-green-hydrogen-transition

[17] Ge, L.; Zhang, B.; Huang, W.; Li, Y.; Hou, L.; Xiao, J.; Mao, Z.; Li, X. A review of hydrogen generation, storage, and applications in power system. J. Energy Storage 2024, 75, 109307.

[18] Ham, K.; Bae, S.; Lee, J. Classification and technical target of water electrolysis hydrogen production. J. Energy Chem. 2024, 95, 554–576.

[19] Chen, S.; Liu, T.; Zheng, Z.; Ishaq, M.; Liang, G.; Fan, P.; Tang, J. Recent progress and perspectives on Sb2Se3-based photocathodes for solar hydrogen production via photoelectrochemical water splitting. J. Energy Chem. 2022, 67, 508–523.

[20] Osman, A.I.; Deka, T.J.; Baruah, D.C.; Rooney, D.W. Critical challenges in biohydrogen production processes from the organic feedstocks. Biomass Convers. Biorefin. 2020, 13, 8383–8401.

[21] Considerations for a Hydrogen Regulatory Framework in the U.S.: https://guidehouse.com/insights/energy/2024/hydrogen-framework-in-the-united-states

[22] Green hydrogen supply: A guide to policy making REGlobal – https://reglobal.org/green-hydrogen-supply-a-guide-to-policy-making/

[23] Green Hydrogen Explained – IDTechEx's Primer on Water Electrolysis: https://renewable-carbon.eu/news/green-hydrogen-explained-idtechexs-primer-on-water-electrolysis/

[24] The benefits and challenges of green hydrogen: Findings from a new report: https://envirotecmagazine.com/2024/12/10/the-benefits-and-challenges-of-green-hydrogen-findings-from-a-new-report/

[25] Franco, A., & Giovannini, C. Recent and Future Advances in Water Electrolysis for Green Hydrogen Generation: Critical Analysis and Perspectives. *Sustainability*, 2023, *15*(24), 16917.

[26] Financial Considerations for Green Hydrogen Development: https://reglobal.org/financial-considerations-for-green-hydrogen-development/

[27] The energy water nexus: balancing water needs for hydrogen production: https://www.arup.com/insights/the-energy-water-nexus-balancing-water-needs-for-hydrogen-production/

[28] Wei, Shijie and Sacchi, Romain and Tukker, Arnold and Suh, Sangwon and Steubing, Bernhard, Future environmental impacts of global hydrogen production, Energy Environ. Sci., 2024, 17,6, 2157–2172, doi 10.1039/D3EE03875K,

[29] Electrolysis- reen hydrogen production. Process and areas of application: https://seshydrogen.com/en/electrolysis-green-hydrogen-production-process-and-areas-of-application/

[30] Green Hydrogen, Meaning, Benefits, Production, Applications: https://vajiramandravi.com/quest-upsc-notes/green-hydrogen/

[31] Green Hydrogen: How It Works and Applications –renewablesadvice.com https://renewablesadvice.com/energy/greenhydrogen/

[32] The Economics of Green Hydrogen: Costs, Benefits, and the Path to: https://www.n51.io/news/the-economics-of-green-hydrogen-costs-benefits-and-the-path-to-commercial-viability

[33] Green hydrogen Wikipediahttps://en.wikipedia.org/wiki/Green_hydrogen

[34] Tuluhong, A., Chang, Q., Xie, L., Xu, Z., & Song, T. Current Status of Green Hydrogen Production Technology: A Review. *Sustainability*, 2024, *16*(20), 9070. https://doi.org/10.3390/su16209070

[35] Morales-Espana, G.; Hernández-Serna, R.; Tejada-Arango, D.A.; Weeda, M. Impact of large-scale hydrogen electrification and retrofitting of natural gas infrastructure on the European power system. Int. J. Electr. Power Energy Syst. 2024, 155, 109686.

Shraddha Awasthi, Mohammad Izhar Alam, Ambneesh Mishra,
Amit Kumar Rathoure, Dan Bahadur Pal

# 11 Energy Conservation Strategies for Sustainable Hydrogen Systems

**Abstract:** Hydrogen is becoming a major component of the energy mix of the future due to the growing challenges of climate change and the need to decarbonize global energy systems. Its special qualities, such as its high-energy density and clean burning capabilities, create a lot of opportunities for its use as an energy carrier in a variety of economic sectors, including transportation, industry, home heating, and electricity generation. To provide more cost-effective, environmentally friendly, and sustainable technologies for the conservation strategies for hydrogen systems, this chapter focuses on the recently developed smart energy systems that combine renewable energy (RE) sources and hydrogen energy systems. These technologies include improvements in electrolysis efficiency, waste heat recovery, hydrogen storage optimization, and integration with RE sources. In particular, the most important way to raise industrial productivity, sustainable technological development, and the standard of living for people in a society is through the use of affordable, environmentally friendly green energy. To optimize hydrogen's potential as a sustainable energy source, this chapter also highlights the necessity of thorough strategic planning and transdisciplinary efforts. Overcoming these infrastructure and technological obstacles, backed by strong regulatory frameworks, will be essential to improve hydrogen's economic viability and market integration. This all-encompassing strategy will guarantee that hydrogen energy can make a substantial contribution to a low-carbon and sustainable future.

**Keywords:** Climate change, energy density, smart energy systems, hydrogen storage systems, sustainable technologies

## 11.1 Introduction

Hydrogen, being the lightest and most abundant element of the universe, has become one of the main carriers of energy in the global quest to establish sustainable energy systems. It can be produced with a plethora of primary energy sources, fossil fuels being one of the popular bases; biomass can be used, or it can be produced by renewable electricity, displaying the utmost in versatility [30]. Hydrogen production can be used for industrial purposes, such as the manufacture of ammonia or in petroleum refining, its use in fuel cells along with grid-based energy genera-

 | https://doi.org/10.1515/9783112181072-011

tion and backup is still evolving [65]. In the background of governments and industries seeking low-carbon alternatives, hydrogen allows for an option to decarbonize those sectors which are otherwise difficult to electrify, such as heavy transportation and some industrial operations. Depending on how it was produced and the impact it has on the environment, hydrogen can be categorized. Gray hydrogen is produced when fossil fuels such as natural gas are used, but emissions are not captured; blue hydrogen uses carbon capture and storage (CCS) to reduce carbon emissions, while green hydrogen is obtained by water electrolysis using electricity from RE sources such as wind and solar [9]. Green hydrogen is heralded as the most important hydrogen in the pursuit of net-zero carbon, although the current challenges in respect to its cost and efficiency obstruct its widespread adoption. Given the challenges to hydrogen storage and transport were explained in some detail, let us see: Under normal ambient conditions, hydrogen has a low volumetric energy density. Hence for use, the gas has to be compressed, liquefied, or chemically converted [18]. Energy is consumed at every step, from production to storage, all the way to the end-use, creating an urgent need to improve efficiencies within the system so as not to lose much of it to wastage. There are different storage systems under ongoing development: compressed gas cylinders, cryogenic hydrogen liquid tanks, and advanced ones such as metal hydrides or liquid organic hydrogen carriers (LOHCs) [21]. Hydrogen does not restrict the energy systems to storage or direct fuel but enables coupling in many ways. For example, excess RE can be converted into hydrogen via power-to-gas (P2G) technologies and be used to balance grids and contribute to energy security [24]. Furthermore, hydrogen can be integrated into synthetic fuel production, opening new opportunities for a circular carbon economy. Nevertheless, hydrogen systems are highly energy-intensive. The conversion processes do not matter much if they occur in production, storage, transportation, or use, since there are huge energy losses occurring [1]. Hence, hydrogen technologies should focus on energy-saving approaches upon deployment to be able to sustain themselves better.

### 11.1.1 Importance of Energy Conservation

Energy conservation within hydrogen systems is critical for multiple reasons: improving efficiencies, reducing environmental footprints, lowering operation costs, and affording better competitiveness to hydrogen technologies compared to traditional energy systems based on fossil fuels. Without effective energy conservation measures, the energy consumption involved in hydrogen production and use could, in effect, negate any environmental advantages sought out for [79]. In the process of splitting water through electrolysis, for instance, a large quantity of electricity is required. If this electricity is not generated from renewable sources, this activity will be contributing to the emissions of greenhouse gases when it should act as a mitigating force. The efficiency of electrolyzers is about 60–70%, meaning the majority of input energy is released to the environment as heat [11]; therefore, through waste heat recovery, system integration, and improvements in electrolyzer technology, these inefficiencies ought to be reduced. Energy conservation also influences the economics of hydrogen systems. At present, green hydrogen is significantly more expensive than gray or blue hydrogen due to a multitude of factors, one prominent factor being the huge energy requirements [31]. Hence, increased energy efficiency will pave the way for cost reductions, thus giving hydrogen the possibility to be adopted on a greater scale. For instance, linking hydrogen manufacturing with a system for RE where efficiency appears paramount can bring down both production and operation costs, thereby fostering competitiveness in emerging markets.

Pressure-raising and storing, however, add another dimension to hydrogen's energy profile. At 700 bars, compression consumes about 10–15% of the energy embedded in the hydrogen gas [70], which is typically needed for fuel cell electric vehicles (FCEVs). Liquefaction, however, is more awesome in terms of energy, demanding almost 30% of the energy available in hydrogen [8]. Hence, innovations in storage, including materials science solutions such as solid-state hydrogen storage, will hugely contribute to energy conservation on a larger scale. End-use applications also get to gain energy conservation measures. For example, in hydrogen fuel cells, maximizing the efficiency of electrochemical conversion processes ensures energy conservation, increased lifespan of the systems, and a reduction in life cycle emissions [16]. The integration of intelligent control systems, optimization of operational conditions, and application of advanced catalysts are all proven ways to increase fuel cell efficiency.

In addition, in the context of the global sustainability agenda, enhancing the energy efficiency of hydrogen systems is in line with a series of policy initiatives that promote reductions in energy intensity and better utilization of RE. The United Nations' (UN) Sustainable Development Goal 7 (SDG7), Affordable and Clean Energy, highlights that doubling the global rate of improvement in energy efficiency is central in ensuring access to affordable, reliable, sustainable energy for all [71]. From a systems viewpoint, an energy balance for hydrogen processes relates to the well-known energy return on investment (EROI), which is defined as the amount of usable energy obtained from a fuel resource relative to the energy used to get it. A high energy loss would result in a low EROI, which renders the hydrogen system less appealing than alternatives such as direct electrification using renewables [76].

In summary, the sustainability of hydrogen systems, while related to the source of energy, is determined predominantly by the efficacy with which this energy can be used at each step of the hydrogen life cycle. The future hydrogen infrastructure needs to have energy efficiency built into its design and operational concepts, in order to make it possible for hydrogen to deliver on its potential as key element of clean energy transitions around the world.

# 11.2 Fundamentals of Hydrogen Production and Utilization

Hydrogen is a versatile and clean energy carrier that holds immense promise for supporting global energy transitions. However, its widespread adoption and sustainability hinge upon a thorough understanding of its production methods, storage and transport technologies, and diverse utilization pathways. This section provides a foundational overview of these components, which form the backbone of any energy conservation strategy applied to hydrogen systems.

## 11.2.1 Major Hydrogen Production Methods

Hydrogen is not a primary energy source but an energy carrier that must be produced using energy-intensive processes. These production methods are often classified by color codes representing their environmental impact: Gray, blue, and green hydrogen being the most prominent.

### 11.2.1.1 Steam Methane Reforming (SMR)

The most dominant hydrogen production method is steam methane reforming (SMR). It accounts for about 75% of global hydrogen production [30]. This process involves methane (natural gas) reacting with high-temperature steam (700–1,000 °C). It produces hydrogen and carbon monoxide.

Then, a water-gas shift reaction happens. This reaction produces more hydrogen and carbon dioxide:

$$CH_4 + H_2O \rightarrow CO + 3H_2$$

$$CO + H_2O \rightarrow CO_2 + H_2$$

The efficiency of SMR ranges from 65% to 75%. However, it emits large amounts of $CO_2$. This can be changed to blue hydrogen with CCS. Without CCS, it is called gray hydrogen and has high greenhouse gas emissions.

#### 11.2.1.2 Electrolysis of Water

Electrolysis uses electricity to split water into hydrogen and oxygen in an electrochemical cell:

$$2H_2O(l) \rightarrow 2H_2(g) + O_2(g)$$

When powered by renewable electricity, this process yields green hydrogen. It's considered the most sustainable form. Electrolysis technologies include alkaline electrolysis, proton exchange membrane (PEM) [73], and solid oxide electrolysis cells (SOECs). PEM is favored for its high current density and responsiveness, though it remains expensive due to the need for noble metal catalysts [11].

While current electrolyzers have efficiencies between 60% and 70%, system improvements and cheaper renewable electricity are gradually making electrolysis more competitive and scalable [31].

#### 11.2.1.3 Biomass Gasification

Biomass gasification involves converting organic materials into a hydrogen-rich syngas through partial oxidation at high temperatures (800–1,000 °C). The resulting syngas is a mixture of hydrogen, carbon monoxide, and carbon dioxide. A subsequent water-gas shift reaction increases hydrogen yield.

Biomass gasification offers a renewable hydrogen source with lower carbon intensity, particularly when using waste biomass or sustainable feedstocks [19]. Integration with CCS technologies could result in negative carbon emissions, positioning it as a future-ready technology.

However, challenges related to tar formation, feedstock heterogeneity, and purification of hydrogen limit its current large-scale deployment.

#### 11.2.1.4 Emerging Methods

Other promising technologies under investigation include photoelectrochemical (PEC) water splitting, thermochemical cycles using concentrated solar energy, and microbial electrolysis. While these technologies offer high theoretical efficiencies and minimal environmental impact, they remain at experimental or pilot stages [70].

### 11.2.2 Hydrogen Storage and Transportation

Unlike fossil fuels, hydrogen has a low volumetric energy density (about 3 $kWh/m^3$ at atmospheric pressure). This poses substantial challenges for its storage and transport.

These stages often consume significant amounts of energy, thereby reducing the net efficiency of hydrogen systems.

#### 11.2.2.1 Physical Storage

**Compressed Hydrogen Gas**: Storing hydrogen as a high-pressure gas (typically at 350 or 700 bar) is the most common method, particularly in transportation and portable applications. While technically mature, this method requires high-strength composite tanks and consumes 10–15% of the energy content of hydrogen for compression [8].

**Liquid Hydrogen**: Hydrogen can be stored as a cryogenic liquid at − 253 C. This increases its energy density. But it takes about 30–40% of the hydrogen's energy to cool it down. Also, there are big losses during storage and transport [70].

#### 11.2.2.2 Material-Based Storage

Advanced materials can absorb and release hydrogen. These include:

- **Metal Hydrides**: Absorb hydrogen in a solid-state matrix. They offer compact storage but add weight and need thermal management.
- **Carbon-Based Materials** (e.g., carbon nanotubes): Show promise due to high surface area but remain under development.
- **LOHCs**: Compounds like methylcyclohexane or dibenzyl toluene chemically bind hydrogen. They require dehydrogenation at the point of use, consuming additional energy [21].

**Hydrogen Transportation:** Hydrogen is transported via pipelines, trucks, rail, or maritime routes. Each method has its efficiency:

- **Pipelines**: Ideal for high-volume, continuous supply. Retrofitting existing natural gas infrastructure is a cost-effective option in some regions [18].
- **Trucks and Rail Tankers**: Typically carry compressed or liquefied hydrogen. More flexible but energy-intensive, especially for long distances.
- **Hydrogen Carriers (Ammonia and Methanol)**: Hydrogen can be chemically converted to ammonia or methanol for easier transportation and reconverted at the destination. This adds conversion losses but simplifies logistics [30].

### 11.2.3 Hydrogen Applications in Energy Systems

Hydrogen's flexibility allows it to be used across a range of sectors. It makes energy decarbonization easier.

**Industrial Use:** Hydrogen has long been used in:

- **Ammonia production** via the Haber–Bosch process
- **Petroleum refining**, for hydrocracking and desulfurization
- **Methanol production** and **metal treatment**

Replacing gray hydrogen with green hydrogen in these industries offers immediate decarbonization opportunities [30].

**Transportation Sector: FCEVs** convert hydrogen into electricity through electrochemical reactions, emitting only water vapor. While FCEVs face competition from battery-electric vehicles, they offer advantages in heavy-duty, long-range, and fast-refueling applications such as trucks, buses, and trains [24].

**Power and Grid Services:** Hydrogen can act as an energy buffer through P2G or power-to-hydrogen (P2H) pathways. It converts surplus renewable electricity into hydrogen during low demand periods. This hydrogen can later be reconverted to electricity using fuel cells or gas turbines, enabling seasonal energy storage and grid balancing [79].

**Combined Heat and Power (CHP)** systems using hydrogen fuel cells also offer high-efficiency energy solutions for residential or commercial buildings.

**Emerging Applications:** Hydrogen is increasingly being considered for:

- High-temperature industrial heat (e.g., cement and glass)
- Synthetic fuels for aviation or marine use
- Decentralized off-grid systems in remote areas
- Blending with natural gas in existing pipelines for low-carbon heating

## 11.3 Energy Losses in Hydrogen Systems

Increasingly reputed as a panacea for sustainable energy paths, hydrogen systems are energy-intensive due to their fuel inefficiencies throughout the overall life cycle. Designing efficient systems means that one needs to understand energy losses at every step (production, storage, distribution, and end-use) and to quantify them. We now discuss the hydrogen technologies' energy loss mechanisms and efficiency considerations.

### 11.3.1 Points of Energy Loss

**Production:** Hydrogen generation, being the first step, is the biggest energy sink of the process. The processes are very efficient in some technologies but quite the opposite in others. For instance, one of the most frequently used methods, SMR, produces gray hydrogen at only about 65–75% efficiency, reporting losses due to heat rejection, incomplete reactions, and energy-intensive separation of gases [17]. Electrolysis for green hydrogen is on the order of 60–70% efficiency for both alkaline and PEM systems, with losses mainly due to electrical resistance, overpotential at electrodes, and heat generation [11]. SOECs, which hold the potential benefit of high efficiency – up to 90% under optimum conditions – are still, however, hindered by high operating temperatures and material degradation, and these result in higher energy losses in operation [79]. Biomass gasification, another bio-based method for loss of energy majorly due to tar formation, incomplete conversion, and energy consumption to produce a clean syngas include feedstock pretreatment and syngas cleaning [19].

**Storage:** Due to hydrogen's low volumetric energy density, storage methods tend to be energy-intensive. Compressing hydrogen at 350 or 700 bar (standard for fuel cell vehicles) requires up to 10–15% of the hydrogen (due to work done by compressors) for compression [8]. Liquefaction for high-density storage is more energy-hungry, needing 30–40% of the hydrogen's energy to bring it down to − 253 °C; more energy is lost due to boil-off upon storage and transportation [70]. Storage by materials like metal hydrides or LOHCs witness energy losses during hydrogen absorption and release. For instance, for LOHCs, there is an energy need for dehydrogenation, which can take up to 10–20% of the stored hydrogen's energy [21].

**Distribution:** Transport of hydrogen delivers extra energy losses, pipelines being the favored means for determining them. Pipeline transport is quite efficient for large volumes but does require compression energy to keep the flow going, especially for long-distance transportation [18]. Transport of compressed or liquidities hydrogen by truck or rail is less efficient owing to energy requirements in compression or liquefaction and fuel use during transport. Chemical carriers such as ammonia or methanol are easier to transport but have energy penalties on either side of synthesis and conversion, with round-trip efficiencies often below 50% [30]. Maritime carriage of liquid hydrogen or ammonia is, therefore, subject also to boil off losses and the energy consumed in mitigations through re-liquefaction.

**End-Use:** Energy loss mechanisms depend on the technology in question for end-user applications. The electrochemical inefficiencies characterize the polarization losses subjected to hydrogen fuel cells used in cars and stationary power systems, while parasitic loads are generated by auxiliary requirements from cooling and air supply setups, limiting its in-use electrical efficiency between 50% and 60%. Thermodynamic limits and incomplete combustion, on the other hand, can push the efficiency down to less than 40%, mostly between 30% and 40%, during the combustion of industrial burners or hydrogen turbines [24]. On the other hand, natural gas heating pipes could be vulnerable to hydrogen mixtures preventing adequate mixing and reducing combustion rates.

## 11.3.2 Efficiency Challenges in Hydrogen Technologies

Being a multifaceted challenge, efficiency is affected by technologies, materials, and systems in hydrogen systems. In production, the main efficiency-improving challenge is energy conversion. Regarding electrolysis, system efficiency is reduced by ohmic losses and overpotentials at electrodes, as well as gas crossover, and these issues must be resolved by research into catalysts, membranes, and cell design [11]. For SMR, heat losses must be minimized while also improving carbon capture for blue hydrogen purposes, as current CCS technologies are only capable of capturing about 85–90% of emissions with an additional energy penalty to do so [30]. For biomass gasification, a major problem is that feedstock varies a lot and that purification of the syngas is so energy-intensive that it reduces efficiency altogether [19].

Storage technologies contend with finding a balance between energy efficiency and practicality. Compression and liquefaction are mature but rather energy-intensive methods; their energy penalties must be cut down through more efficient compressors and cryogenic systems [70]. Material-based stores, including metal hydrides, suffer from slow kinetics, heavy weight, and complicated energy application for thermal management during hydrogen release [21]. Though promising for safe transport, LOHC catalysts need to find ways to increase their activity to reduce the energy needed in dehydrogenation. Distribution systems must work to surmount challenges related to infrastructural compatibility and energy losses. To retrofit pipelines for hydrogen transport, issues relating to material embrittlement and leakage must be dealt with, which could cause energy losses and significant safety concerns [18]. Energy losses in the long-distance transport of liquid hydrogen or ammonia, due to highly energy-intensive processes of re-liquefaction and conversion, call for innovations in insulation and carrier molecules [30].

In end-use applications, fuel cell efficiency is hindered by catalyst performance and membrane durability, as well as by system integration issues. Platinum-based catalysts are used widely and effectively but are expensive and have the problem of catalyst poisoning, which decreases their efficiency over time [16]. Combustion-based systems also suffer from low efficiency caused by hydrogen's low flame speed and the creation of specialized burners for the purpose. On the other hand, systemic integration of hydrogen technologies within existing energy infrastructures, such as power grids or industrial plants, face challenges against energy conversion and system balancing inefficiencies [79]. Resolution of these efficiency concerns must take place in tandem with enhancements in the EROI of hydrogen systems. High energy losses cause low EROI values, thereby making hydrogen less competitive relative to direct electrification or other RE solutions [76]. Energy-conserving techniques will be discussed in later sections.

## 11.4 Principles of Energy Conservation

Energy conservation is fundamental in sustainable hydrogen systems, in particular, to ensure the optimal exploitation of investment costs in production, storage, distribution, and utilization of energy. Energy conservation and the increased environmental and economic attractiveness of hydrogen as a clean energy carrier are promoted through minimizing energy losses and optimizing system operation. This chapter introduces the basic principles of energy efficiency and saving, and the role that energy saving can play in realizing sustainable hydrogen systems through energy management.

### 11.4.1 The Energy Efficiency and Conservation Concept

Energy efficiency, energy conservation, related yet different. Efficiency and conservation are conceptually inseparable and form the cornerstone of sustainable energy systems. Energy efficiency means that the same amount of energy input can deliver the same output or service, and is generally achieved by technological innovation or design optimization [41]. In terms of hydrogen production, for instance, if an electrolyzer that has 60% electrolysis efficiency is improved to 80%, more hydrogen is formed and thus the energy loss as heat or chemical energy is decreased [78]. Enhancement of efficiency typically occurs in terms of materials (high performance catalysts, etc.), process flow sheet concepts (waste heat recovery systems, etc.).

On the other hand, energy saving refers to reduction of total energy consumption through changes in operational procedures, system layout or user behavior, while maintaining the intended purpose [64]. In a $H_2$ system, conservation may consist of measures such as demand side management to match the production of $H_2$ with periods of surplus RE, to prevent storage or energy-consuming conversion processes [51]. The industry must consider life cycle aspects of conservation such as by curtailing energy in secondary operations, e.g., in gas purification or compression. Energy efficiency and conservation together attempt to increase the ratio of useful energy or EROI, that measures the energy yield from a system relative to the energy necessary for the system to be built, maintained, and operated [26]. In the case of hydrogen, it is important to have a high EROI in order to be competitive with other energy carriers like direct electrification, as well as fossil fuels. It should be the same for hydrogen systems where low EROI numbers (harvested on electrolysis inefficacies, compression loss or the inefficiency on the fuel cell) can make these hydrogen systems barely sustainable, and mostly not for the green hydrogen using renewable energies (that may be scarce or expensive [23]. The science of energy efficiency and conservation has its basis in the first and second laws of thermodynamics. The first law (energy balance) states that energy can neither be created nor be destroyed, only converted among forms and that it predicts that the energy inputs to a hydrogen system must either be useful output, losses, or waste [64]. The second law emphasizes that energy conversions are bound to be inefficient because of the entropy that energy cannot be created or destroyed, and therefore every process has some amount of waste, such as the waste of heat during the process of electrolysis or fuel cells [63]. Energy system management aims to mitigate these losses by maximizing energy streams, recuperation, waste, and by minimizing or avoiding unnecessary energy consumption.

In the case of hydrogen, applying these ideas requires a "cradle-to-grave" perspective, which covers the entire life cycle: from how and where the gas is produced to how and where it is consumed. The development of more efficient PEM water electrolyzers has so far been facing the challenge of lowering overpotentials and ohmic losses by utilizing advanced electrode materials [78]. On the way to sustainable power generation, energy systems might benefit from conservation of hydrogen by deploying the hydrogen production close to renewable sources [20], where no or limited loss is noticed in transmission and/or exploit P2G technologies to store surplus RE as hydrogen, thus limiting curtailment [23].

## 11.4.2 Role of Energy Management in Sustainable Systems

Energy management the discipline of managing energy by identifying opportunities and ensuring that energy demand and its use are met in the most cost-effective manner and in the future [2]. Energy management in hydrogen systems energy management is paramount for sustainability of hydrogen system, uncovering inefficient segments, prioritizing saving energy and applying RE sources in optimal way. Successful energy management requires technology and operations, as well as data-informed decisions and cutting-edge controls.

**Monitoring and Auditing of the Energy Used:** The initial stage in the management of energy is an energy audit so as to determine the flows of energy and pinpoint the losses throughout the life cycle of hydrogen. For instance, in an electrolysis plant, an energy audit might indicate 20–30% of input energy is vented as heat and would prompt waste heat retrofit applications [78]. Monitoring systems with sensors and real-time data analytics make it possible to track energy performance in real time, and while monitoring efforts can easily map deficient energy performance, such as less efficient operation of compressors and/or boil-off losses at LH2 storage [38]. These findings can guide focused interventions to improve system effectiveness.

**Optimization and Control of Process Systems:** Energy control systems unitize advanced control techniques like model predictive control (MPC) and programmable logic controllers to optimize operating conditions at online basis. In hydrogen synthesis, MPC can optimize the operation of the HS in response to different RE inputs (e.g., current density and temperature of the electrolyzer) [51]. Similarly, with fuel cell systems, an integrated control algorithm adjusts air/hydrogen flow rates to reduce the parasitic losses due to auxiliary equipment such as pumps, fans, and compressors available in the fuel cells [75]. These control methods enable energy use only when and where it is required, consistent with conservation approaches.

**Renewable Energy Integration:** RE sources like sun, wind, or water are essential building blocks for energy management in sustainable hydrogen systems. Renewable power sources tend to be intermittent, needing effective scheduling of hydrogen production when such resources are in abundance. Forecasting models and demand response techniques may be utilized by energy management systems for scheduling production at maximum renewable generation, thereby lowering dependence on grid electricity or energy-intensive storage [23]. For example, P2G systems have the capability to convert excess wind energy into hydrogen that can be stored for later use for grid balancing or industrial processes, saving energy that might otherwise be curtailed [39].

**Life Cycle Energy Management:** For hydrogen systems that thrive towards sustainability, energy management should be considered all along the hydrogen life cycle, from the raw material extraction, to the system decommissioning. This encompasses the selection low-energy materials of electrolyzer and fuel cell, and optimization of supply chain for transportation energy reduction and system design for recyclability to minimize embodied energy [52]. The Cumulative Energy Demand (CED) is also a key factor considered in life cycle assessment (LCA) for energy and life cycle analysis of hydrogen technologies; it represents the total energy required for the production, operation, and disposal of a hydrogen energy system. Energy management, by reducing CED, guarantees that hydrogen systems provide net environmental advantages over fossil fuel options [2].

**Human and Organizational Factors:** Energy management is not a solely technical endeavor; it entails many human and organizational factors. Training operators to engage in energy-conscious practices, such as shutting down equipment not in use, or running maintenance to optimize energy utilization, will help minimize energy wastage [68]. Organizational policies that set energy performance targets or provide incentives toward energy efficiency improvements help further the cause of conservation. Industrial facilities using hydrogen, for example, may be equipped to implement energy management standards such as ISO 50001 to further energy conservation in a systematic fashion and continual improvement [32]. Energy management, therefore, supports hydrogen systems' sustainability by reducing energy losses, cutting operational costs, and improving environmental performance. As hydrogen technologies grow, solid energy management will be key to solving efficiency challenges and ensuring hydrogen lives to its promise as a prime mover in the global energy transition.

## 11.5 Energy Conservation Strategies

Energy conservation is important for improving the sustainability of hydrogen systems as it lessens the energy lost in their production, storage, distribution, and utilization processes. In this way, they can greatly raise the efficiency, give it a less harmful profile concerning the environment, and make it an economically worthy contender for a clean energy carrier. This section discusses five energy conservation measures: process optimization for hydrogen production, waste heat recovery systems, advanced materials for energy efficiency, smart storage technologies, and integration with RE sources. Each of the conservation methods attacks specific inefficiencies in the hydrogen life cycle, thus contributing to the overall sustainability of the hydrogen systems.

### 11.5.1 Process Optimization in Hydrogen Production

Making hydrogen through process optimization requires altering operational parameters, equipment design and control systems to lessen energy consumption while improving – or at least maintaining – efficient production levels. According to Nikolaidis and Poullikkas [47], energy losses occur in all hydrogen production methods like electrolysis, SMR, and biomass gasification due to energy conversion and auxiliary process inefficiencies. Strategies of optimization try to counter losses occurring due to undesired actions.

In case of electrolysis, mainly green hydrogen related, minimization of electrical resistance and overpotentials is optimized. Example: The efficiency of alkaline electrolyzers may increase by 8–12% by optimizing the operating temperature and pressure of the electrochemical cell in the electrolyzer, which can reduce the energy loss [47]. Advanced control systems would have control algorithms adaptive enough to modify the current density in real time according to the variable energy input ensuring maximum efficiency [59]. Designing stack configurations, as well as gas management systems in PEM electrolyzers can reduce parasitic losses. In SMR the cooling of reactor is optimized to avoid heat loss through means of heat transfer enhancement. Thus, up to 15–25% of energy input is lost [62]. Insulation improvements of reformers; preheating installations for feedstocks can help to lessen energy needs. According to Sikarwar et al. [61], the design of reactors, such as fluidized bed reactors will help with the optimization of biomass gasification and improve the yield of syngas and will also help in reducing the energy-intensive tar removal. Building production facilities near energy sources also saves energy. For example, the distance from the electrolyzers to hydropower plants has a transmission loss of 4 Å, ~8% over long distances [47]. The hydrogen production process has been optimized to be more energy efficient and lower costs.

### 11.5.2 Waste Heat Recovery Systems

Systems that use hydrogen result in waste heat. Waste heat recovery systems are used to utilize that heat. The process of creating waste heat during electrolysis and running a fuel cell consumes account for about 25–35% of the input energy [13]. We can achieve overall energy savings and improve the sustainability of hydrogen systems by heat recovery. Heat exchangers that recover waste heat to preheat water or for space heating can be utilized during electrolysis to recover the heat from electrochemical reactions. The auxiliary energy requirements can drop by 10–15% [54]. The efficiency of a system can be high, about 3–7%, if the low-grade heat from electrolyzers is converted to useful electricity through thermoelectric generators [59]. According to Simpson and Lutz, in high-temperature cases of SMR, waste heat can be used to drive a steam turbine or preheat methane thus reducing energy input [62].

Waste heat recovery is especially advantageous when used for CHP applications within fuel cell systems. "CHP can have total efficiencies of 80–90% by recovering heat from the fuel cell stacks, while the efficiency for just electricity generation is 50–60% range" [13]. In the case of liquid hydrogen, waste heat from industrial processes could be used to supply the heat that would otherwise be required in order to avoid boil-off, thus minimizing losses between 2% and 5% per day [61]. Waste heat recovery systems usage guarantees thermal energy is used optimally, which provides hydrogen systems with greater energy efficiency.

### 11.5.3 Advanced Materials for Energy Efficiency

Materials like catalysts, membranes, and storage media where energy barriers and performance are the top priority to enhance energy efficiency of hydrogen systems. The above electrolysis, fuel cells, and storage are inefficient due to material limitations with respect to energy (high activation energy) and flow (low conductivity) [42]. In electrolysis, using nickel or iron-based transition metal oxides as catalysts reduces energy consumption for water splitting due to overpotential reduction. These smart catalysts are 5–10% more efficient than conventional catalysts. High conductivity membranes like perfluorosulfonic acid-based membranes enhance ion transport in PEM electrolyzers, reducing electrical losses [59]. As per Chi and Yu [13], for fuel cells, platinum-alloy nanoparticles enhance the reaction kinetics and durability [13].

According to Marban and Valdes (2007), using advanced materials like lightweight metal hydrides or porous carbon structures allows hydrogen to be stored at lower pressures, thus needing 10–20% less storage energy [42]. Research shows that LOHCs with optimized catalysts can achieve round-trip efficiencies of 85–90% [61]. Advanced materials can save a lot of energy in hydrogen systems. This makes them competitive and sustainable.

### 11.5.4 Smart Storage Technologies

Smart storage materials make use of monitoring and automation technologies, as well as advanced materials, to reduce energy losses within hydrogen storage, which is a challenge due to hydrogen's low volumetric energy density. Compression, liquefaction, and storage on materials are processes that require a lot of energy and incur losses between 10% and 40% of hydrogen's energy content [47]. Smart technologies are optimized to make these processes more efficient.

For compressed hydrogen, smart systems employ sensors and machine learning algorithms to track pressure and temperature in order to improve the operation of compressors, saving 8–12% of energy consumption in the process [59]. This kind of compressor will avoid over-compressing because they vary speed according to the dynamic needs of the system. Smart insulation systems with integrated sensors that enable a cooled only when needed reduce the boil-off losses that can reach 1–2% per day in liquid hydrogen storage [62], while metal hydrides as a material-based storage option exploit smart thermal management systems using phase-change materials in order to maintain the temperature throughout the hydrogen absorption and release cycle, thus decreasing energy losses by 10–15% [42]. In the case of LOHC systems, the reactors are automated and improve the catalyst performance in real time, which allows minimization of the energy related to dehydrogenation and efficiencies of 80–85% can be achieved [61]. "Moplusthey eliminate the energy and cost waste associated with other storage technologies, and make hydrogen storage practical [50]."

## 11.5.5 Integration with Renewable Energy Sources

Joining hydrogen systems with renewable energies such as solar, wind, and hydropower is one of the main approaches that saves energy and promotes sustainability. This method also minimizes dependence on fossil fuel electricity generation while making use of excess renewables and avoiding curtailment [54]. The process of P2H entails the production of hydrogen via electrolysis from surplus electricity generated by renewable sources, storing energy for use later either in fuel cells or in industrial processes [47]. Energy management systems implementing forecasting have production scheduled when renewables are expected to peak, and lead to a 10–15% increase in system efficiency over a system connected to the grid [59]. By avoiding the necessity to store energy, they can reach efficiencies between 60% and 70%, for example when electrolyzers are combined with solar photovoltaic technologies [13].

Having hybrid renewable systems that integrate different kinds of sources would guarantee an uninterrupted energy supply for hydrogen generation. Energy balancing algorithms maximize input from solar and wind and minimize lost energy due to intermittency [62]. Also, positioning manufacturing plants near the renewable ones saves energy, as transmission losses, which do occur even in non-remote areas, can be considerable in that distant place [61]. This hydrogen produced from renewables is used in fuel cells in off grid applications thereby eliminating the distribution losses. It is a way of maximizing energy conservation and supporting the decarbonization of the hydrogen systems.

## 11.6 System Design and Optimization Approaches

However, the smart design and development of hydrogen technologies are key to lowering energy requirements, increasing energy efficiency, and promoting sustainability in the hydrogen "cycle" from production to end-use. The best place to embed energy efficiency into system design is during all stages of system design, while systems optimization utilizes the latest techniques to enhance system efficiency. The three main strategies addressed in this section are minimum energy design, process integration and intensification (PII), and modeling and simulation tools. The focus is on overall energy efficiency and minimizing energy losses, reducing costs in operation, and aiding scalability of sustainable hydrogen systems.

### 11.6.1 Design for Minimum Energy Consumption

Ideally, hydrogen systems would be designed to minimize energy use from the outset by incorporating energy efficiency considerations into the structure of the production, storage, distribution, and end-use infrastructure. It focuses on low-energy technologies, efficient choices of components, and system designs that minimize energy loss [3]. The design of systems for hydrogen production involves choosing energy efficient technologies, based on the desired application. For instance, PEM electrolyzers have efficiencies between 65% and 70% and are more suited for dynamic operation alongside of variable RE since they have a high dynamic response [12]. Energy losses can be reduced by the use of efficient power electronics and thermal management systems within the design of the system. For steam methane reforming, small reformers can be designed with recuperators for energy savings of 10–15% over conventional designs [55]. Storage systems "strike a balance between energy efficiency and practicalities". The compression energy could be decreased by 5–10% if hydrogen is stored at 350 bars instead of 700 bars, which would still be sufficient for a number of applications including industrial applications [33]. When it comes to liquid hydrogen, modern cryogenic tanks with multilayer insulation reduce boil-off losses and therefore the energy required to reliquefy it [36]. Material-based storage systems, like those based on metal hydrides, have the advantage of being light-weight and operate under low pressure, which reduces the energy requirements for hydrogen adsorption/desorption [36].

Pipelines are then designed to minimize pressure drops and leak risks in distribution systems. Hydrogen can be transported through existing natural gas pipelines if proper materials are used to avoid embrittlement, and is energetically favorable to be transported in this way over long distances [43]. Fuel cell systems are configured of modular stacks and employ efficient balance of plant equipment, for example, air compressors, to reduce parasitic losses and result in end-use electrical efficiencies of 55–60% [12]. Designing for low overall energy consumption also includes life cycle decisions like using materials that are recyclable and employing modular systems to allow for upgrades and reduce embodied energy [10]. Hydrogen systems can be made much more sustainable and economically viable by designing them to be energy efficient and can result in substantially fewer energy requirements for realization [3].

### 11.6.2 Process Integration and Intensification

PII is defined as the merging of multiple processes, or an improvement of processes within a system in order to reduce energy and/or enhance process performance. Ideally, within hydrogen systems, PII approaches would allow for more efficient energy flow, reduced auxiliary energy requirements, and a more compact system resulting in a more efficient overall system [15]. The aim of process integration is to link complementary processes in order to recover and reuse energy. For instance, if waste heat recovery systems are integrated with electrolysis, process heat from the water splitting can be harnessed to preheat feedwater or to run auxiliary processes, resulting in an increase in system efficiency of 8–12% [12]. This approach uses a water-gas shift reactor combined with a heat exchanger that recovers waste heat at high temperature to generate additional steam, thereby minimizing the energy input by 10–20% [55]. P2G projects combine the production of hydrogen with RE grids, so that the electricity beyond demand is converted to hydrogen gas to stabilize supply and demand and eliminate energy curtailment from the grid [43].

In process intensification, efficiency is achieved by a reengineering of processes so they can occur under more favorable conditions or with smaller, more efficient sized units. In this regard, microchannel reactors used for SMR minimize heat and mass transfer limitations, are more efficient in terms of reaction and 15–20% less energy is lost in the reaction compared to conventional reactors [15]. For example, high-pressure electrolyzers can generate hydrogen directly at high pressures, such as 30 bar, eliminating the need for compression later on, which saves 5–10% of energy costs [33]. Highly intensified designs like the planar solid oxide fuel cell allow for higher power density and lower material consumption and can be as efficient as 65% [36]. PII also considers hybrid hydrogen and other energy carriers' systems. As an example, coupling hydrogen production with ammonia synthesis creates a closed-loop energy storage system where hydrogen is converted to ammonia, a much easier to store and transport energy carrier, thus minimizing energy losses during transport [3]. The use of PII by hydrogen systems leads to improved energy efficiency, smaller systems, and lower costs of operation, thus supporting the scale and sustainability of hydrogen systems.

### 11.6.3 Modeling and Simulation Tools

Hydrogen systems require modeling and simulations to best design and operate them; engineers need to know how systems will perform, where efficiencies can be found, and how to conserve energy before they attempt to do so. These tools, based in computer simulations of energy flows, process dynamics and interactions within systems, generate insights that are useful for design and improvements in operations [35]. Aspen Plus or HYSYS are examples of process simulators able to describe thermodynamic and kinetic behavior for hydrogen production processes such as SMR or electrolysis. Simulations, for instance, can help selecting the parameters of the electrolyzer operation such as temperature and pressure so that losses in energy are reduced and, hence, performance increased by 5–10% [12]. Computational fluid dynamics modeling of gas flow and heat transfer in reactors and fuel cells highlights potential for improving designs to lower energy losses through changes in how the electrodes are constructed to reduce overpotentials, for example [36]. System level modeling programs like HOMER or Energy PLAN examine the coupling of hydrogen systems with RE grids, optimizing productions schedules to when renewables are available. They can also lower the energy that has to be curtailed by 10–20% by enabling hydrogen generation at times of high renewable generation [43]. In addition, machine learning based models are used to predict how the system performs under different conditions, allowing to improve parameters such as the operation of the compressor or the load of fuel cells in real time, achieving reductions in energy consumption of 5–15% [35].

Hydrogen-related systems embrace technologies and activities with widely ranging degrees of energy requirements and environmental impacts. LCA is to analyze these impacts from cradle to grave so that low-energy materials and processes are preferred in the decision selection. To cite one such situation, the LCA would show that, in some applications, the least cumulative energy demand for storage is metal hydrides and not liquid hydrogen [33]. Digital twins optimize operating systems on a continuous basis using real-time data and simulations with the objective of mitigating inefficiencies such as excessive boil-off losses in liquid hydrogen storage [15]. On the other hand, engineering modeling and simulation tools are used to design and operate hydrogen systems with minimum energy consumption and maximum efficiency and sustainability. These tools allow rapid experimentation concerning new designs and operational strategies, thereby hastening the journey toward next-generation hydrogen technologies [55].

## 11.7 Emerging Technologies and Innovations

As the new technologies and concepts for hydrogen systems are being developed, they bring with them awaited transformations in energy performance enhancement, cost reduction, and sustainability improvement. These breakthroughs address the key issues along the hydrogen chain – from production to use – toward a low-carbon energy ecosystem. This covers AI and machine learning for energy optimization; newer hydrogen production processes like photocatalytic water splitting [5], or advanced energy storage like metal hydrides and LOHCs. These technologies bear great promise in reducing the energy losses and accelerating in a big way the uptake of sustainable hydrogen systems.

### 11.7.1 Artificial Intelligence and Machine Learning for Energy Optimization

By means of data-driven decision-making and predictive analytics, AI and ML are transforming the design, control, and optimization of hydrogen systems. These technologies minimize energy consumption throughout the hydrogen life cycle by spotting inefficiencies, predicting system behavior, and automating control operations [14]. For hydrogen production, AI and ML algorithms govern electrolyzer operation, thus predicting optimal operating conditions, such as current density or temperature, from real-time data of RE inputs. As an example, ML models allow PEM electrolyzers to reduce electrical losses dynamically by adjusting operating parameters, thus improving their efficiency by 5–10% [45]. AI-based process control systems regulate conditions in SMR reactors for optimum conduction, achieving a 10–15% reduction in energy consumption, thereby optimizing heat integration and catalyst performance [77]. Compression and liquefaction processes are optimized through the AI for hydrogen storage and distribution; the AI forecasts demand patterns and adjusts compressor operation to avoid energy wastage. ML models predict the optimal pressure settings and can reduce compression energy by 8% to 12% in high-pressure hydrogen storage systems [14]. In contrast, transport algorithms based on AI optimize hydrogen transport via pipeline or truck in distribution networks, thereby reducing energy losses due to long-distance delivery by between 5% and 10% [58].

In the end-use domain, AI increases fuel cell efficiency by predicting degradation patterns and optimizing operating parameters such as hydrogen flow rates and stack temperature. Due to the ML control system, fuel cell efficiency is increased by between 3% and 7%, with the system deepening its aging and lessening energy losses [45]. Digital twins powered by AI reproduce hydrogen systems virtually to monitor and optimize the system in real time to minimize energy consumption during production, storage, and usage [77]. By ensuring AI and ML simultaneously act on hydrogen systems, such technologies make it possible to enable them to be highly energy efficient, adaptive, and automated; hence, these abilities make them more competitive with fossil energy systems and facilitate large-scale deployment [58].

### 11.7.2 Novel Hydrogen Production Techniques (e.g., Photocatalytic Water Splitting)

Hydrogen production methods, such as photocatalytic water splitting [66] and others are not only cleaner in terms of $CO_2$ emissions but also show promise to be more energy efficient than traditional SMR or electrolysis. Providing hydrogen from sources such as sunlight, these technologies can also be energy-efficient compared to traditional processes that require high-cost energies [4]. Photocatalytic water splitting uses semi conducting materials (e.g., titanium dioxide ($TiO_2$) or cadmium sulfide (CdS)) as sun light absorption catalysts to catalyze the dissociation of water into hydrogen and oxygen gases. In sharp contrast to electrolysis, which requires a huge amount of input electrical energy not only for splitting each water molecule [80] but also due to inefficiencies arising from low reaction stoichiometry, leading to higher overpotential requirements even at lower current densities—theoretically limited to ~1–2 V on average in real-life operation—photocatalysis relies on near-abundant solar energy (>20 times higher if harvested effectively with well-customized bandgaps, as studied in PV cells that we have evidently focused on in this review), thereby enabling a reduction in power consumption to an estimated range below ~100 $W/cm^2$ without planned strategies, depending upon groups. However, innovative nanostructured photocatalysts such as metal organic framework and perovskite-based materials have increased QEs to up to 10–15% improving the prospects for large-scale $H_2$ production challenges [34, 80].

Related to this method is the PEC water splitting, which merges photocatalysis with electrochemical systems for increased efficacy. PEC systems employing tandem photoelectrodes can operate at efficiencies of 15–20%, higher than those achievable in traditional photocatalysis, yet by requiring less energy as compared to the standard electrolysis [29]. These can be installed along with solar concentrators to enhance capture of energy, further minimizing the energy footprint. Some new approaches are thermochemical water splitting that uses high-temperature heat from concentrated solar power to drive chemical cycles with efficiencies of 30–40% [22]. Microbial electrolysis cells, powered by organic waste, are potentially the most energy efficient biohydrogen production process for decentralized application due to their low carbon plants relative to conventional hydrogen generation processes with moderate high energetic inputs [34]. The emerging methods reduce the reliance on fossil fuels and grid electricity, as expected from energy conservation, congruent with sustainable hydrogen production [27].

### 11.7.3 Advanced Energy Storage Methods (e.g., Metal Hydrides and LOHCs)

Metal hydrides and LOHCs hold great promise as advanced energy storage techniques, which are essential to bridge the high-energy intensity of compressing or liquefying gaseous $H_2$. These novel methods can bring enhanced energy efficiency, safety, and scalability to hydrogen storage for extensive applications [53]. An example of an adsorption method is metal hydrides, in which hydrogen atoms are stored as a hydride compound within the interstitial atomic sites of a solid-state metal lattice. Unlike in the case of storage as compressed hydrogen, which takes 10–15% of its energy contents for compression [57], metal hydrides function near ambient conditions with an efficiency drop ~8% to ~12%. Hydrogen is an attractive and powerful energy-carrier, thanks to advanced hydride materials, e.g., magnesium-based alloys or complex hydrides which exhibit high storage capacities with kinetics suitable for low-cost processing along excellent round-trip efficiencies at 85–90% [53]. This is accelerated by utilization of thermal management systems, which are coupled with waste heat recovery to recover the generated wasted heat during hydrogen absorption and release [34].

LOHCs store hydrogen as liquid organic compounds, using them for a fuel is that they can chemically bond the molecular hydrogen to dibenzyl toluene and be affected in an equivalent way that regular fuels are. However, design of catalyst materials has provided an increase in round-trip efficiency by 10–15%, today reaching levels between 80% and 85% [46], suggesting that lower renewable integration costs might be achieved with LOHCs. Furthermore, while liquid hydrogen has boil-off losses of 1–3% per day [57]. Novel storage approaches, such as carbonaceous nanomaterials (graphene/carbon nanotubes), provide a highly available specific surface area for hydrogen adsorption, which decreases compression energy by 5–10% in comparison to possible traditional alternatives [29]. The energy efficiency of hydrogen systems can be improved through the use of these advanced storage technologies, allowing for supporting transportation, industrial processes, as well as grid storage while also helping to align those with sustainability goals [53].

## 11.8 Case Studies and Real-World Applications

The case studies and examples of energy saving measures in hydrogen systems show how they work in real life. They highlight the benefits and challenges of using these measures. By looking at successful projects and how systems perform before and after optimization, we see how these strategies improve efficiency, cut costs, and help the environment.

### 11.8.1 Successful Implementation of Energy Conservation Measures

Many real-world projects have shown that energy conservation measures in hydrogen systems can make them more sustainable and cost-effective. Here are three examples of how strategies like process optimization, waste heat recovery, and RE integration have been applied.

#### 11.8.1.1 $H_2$ Future Project, Austria

The $H_2$ Future project started in 2017 at the Voestalpine steel plant in Linz, Austria. It combines green hydrogen production with RE. A 6-MW PEM electrolyzer, powered by hydropower and wind energy, makes hydrogen for steelmaking. This reduces the need for fossil fuel-based hydrogen. Energy-saving steps include using predictive control algorithms to optimize electrolyzer operation [40]. This adjusts current density to match RE availability, boosting efficiency to 68–72% from 60% to 65% before. Waste heat from the electrolyzer is also recovered to preheat water, cutting energy input by 8–10%. The project cuts carbon emissions by 280,000 tons a year, showing the environmental gains of efficient hydrogen production.

#### 11.8.1.2 $H_2$ Balance Project, Denmark

The $H_2$ Balance project in Hobro, Denmark, turned surplus wind energy into hydrogen for industrial and mobility use. Since 2018, it has used a 1.2 MW alkaline electrolyzer optimized for dynamic operation, reaching 65–70% efficiency with advanced process control systems. It also recovers waste heat for district heating, saving 5–7% of energy. By aligning hydrogen production with wind energy peaks, the project avoids wasting 2–3 GWh of RE a year. This shows how effective energy conservation can be through renewable integration. The hydrogen supports fuel cell vehicles and industrial processes, helping Denmark meet its decarbonization goals.

#### 11.8.1.3 Fukushima Hydrogen Energy Research Field (FH2R), Japan

The FH2R project, started in 2020 in Namie, Japan, is a large green hydrogen production facility. It has a 10-MW PEM electrolyzer powered by a 20-MW solar photovoltaic plant. Energy-saving steps include high-pressure electrolysis and smart storage systems using metal hydrides to reduce energy losses. AI-driven energy management systems optimize production schedules based on solar forecasts, boosting system efficiency by 7–9%. The project supplies hydrogen for fuel cell vehicles and stationary power, cutting fossil fuel use by 1,500 tons of $CO_2$ equivalent a year. FH2R shows how advanced technologies can make hydrogen systems more efficient.

These case studies demonstrate how energy conservation can significantly improve the energy efficiency and sustainability of hydrogen systems. They offer examples to be emulated in other regions of the world [7, 37].

## 11.8.2 Comparative Analysis of Optimized and Non-optimized Systems

Pre- and post-optimization comparison for hydrogen systems reveals the advantages of energy-saving measures. It takes into account their impact on efficiency, energy consumption, costs, and environmental footprint. The following two examples are instances where optimization has had an actual impact on real-world hydrogen systems.

**Case Study: Orkney Islands Hydrogen Project, Scotland**: The Surf'n'Turf initiative in the Orkney Islands got under way in 2017. It generates green hydrogen using tidal energy and wind energy. This hydrogen is employed for local power and transport. Originally, the project featured a 0.5 MW alkaline electrolyzer with an efficiency of 58–60%. However, it confronted with drawbacks of low temperature conditions and high losses in electrical energy [6]. Energy audits revealed issues with power conversion and heat dissipation. The project underwent a major overhaul in 2019. It upgraded to high-efficiency power converter and added in waste heat recovery for heating. It also employed ML powered control systems to align production with peaks of RE. For his modifications, the electrolyzer became up to 66–70% efficient. This intensification reduced energy consumption by 12–15% [25]. The hydrogen production cost was reduced by 10%, from €8/kg of hydrogen to €7.2/kg of hydrogen. In addition, there was a 200-ton reduction in annual $CO_2$ emissions due to less grid electricity use. This demonstrates the importance of targeted optimization in the improvement of performance and reduction in cost [7].

**Case Study: Utsira Island Hydrogen System, Norway**: The Utsira Island project has been operating since 2004, and was updated in 2018. It is using wind power to produce hydrogen and power fuel cells. The original configuration was equipped with a 0.6 MW wind turbine and an alkaline electrolyzer with efficiency of 55–60%. It stored hydrogen at 200 bar, consuming 10–12% of the energy of the hydrogen energy for compression [56]. Improvement came in 2018, with a high-pressure electrolyzer at 15 bar, and metal hydride storage. These modifications reduce the compression energy by about 8% and the hydrogen release losses by 5–7% [44]. In addition, a waste heat recovery system was installed to conserve 6% of the energy input. Following these improvements, the efficiency of the system increased from 50% to 62%. The power required for a kilogram of hydrogen to drop from 55 kwh to 48 kwh. The costs of the system decreased by 15% while $CO_2$ emissions per year were reduced by 150 tons, and the efficacy of energy-saving measures was demonstrated [37].

## 11.9 Policy, Regulations, and Economic Considerations

Whether sustainable hydrogen systems succeed will rely on good policies, strong rules and a robust economy. These are the things that save energy, grow hydrogen tech, and put it on the world's throat. That's the big-picture view we'll present in this installment, with a focus on government support for saving energy, the economic benefits of efficient hydrogen systems (Table 11.1), and the challenges and potential advantages for deploying them. Leaders and groups can create a more hospitable environment for energy-saving hydrogen systems by focusing on these problems. The relevant policy frameworks and their economic impacts on energy-efficient hydrogen systems are summarized in Table 11.1 [67–72].

**Table 11.1:** Policy frameworks and economic impacts on energy-efficient hydrogen systems [67–72].

| Region | Policy framework | Energy conservation measures | Economic benefits | Challenges | Incentives |
|---|---|---|---|---|---|
| EU | EU Hydrogen Strategy [81] | Waste heat recovery and integration of renewables | 10–15% cost reduction, creation of 2 million jobs by 2050 | High initial costs and regulatory hurdles | Innovation Fund, Horizon Europe (covering up to 50% of project costs) |
| Japan | Basic Hydrogen Strategy (2017, 2023) | AI-powered optimization and advanced electrolyzers | 7–9% boost in efficiency and cost reduction of ¥4/kg | Technology readiness and shortage of skilled workers | NEDO R&D grants, various subsidies |
| USA | Bipartisan Infrastructure Law (2021) | High-pressure electrolysis and metal hydride storage | 8–12% energy savings and $8B in total investment | Competitive markets and labor and skills gaps | Tax credits, funding for regional hydrogen hubs |

### 11.9.1 Government Policies Supporting Energy Conservation

Government actions are key in making hydrogen systems more energy-efficient. They offer money, set rules, and encourage new ideas. Around the world, rules are being made to help make hydrogen production, storage, and use more efficient, fitting with big climate and energy goals.

In the EU, the [81] also includes energy saving as an integral component of green hydrogen. It is targeting 40 GW of electrolyzer capacity by 2030. The plan in question has funding such as the Innovation Fund for waste heat and renewables projects, saving energy by 10–15% [69]. The EU has energy audit requirements for hydrogen facilities as well and it is calling for improved control and smart storage [72]. Basic Hydrogen Strategy (2017, updated 2023) of Japan helps make savings in energy through subsidies for effective electrolyzer and fuel cells.

It supports initiatives such as the Fukushima Hydrogen Energy Research Field, which uses AI to improve efficiency by 7–9% [60]. Japan also has tight energy use criteria for hydrogen facilities, demanding a minimum of 65% efficiency [48].

In the United States, the Bipartisan Infrastructure Law (2021) provides $8 billion for hydrogen hubs centered on energy affordability. These include high pressure electrolysis and metal hydride storage, reducing energy waste by 8–12% [49]. The DOE's Hydrogen Program Plan provides tax incentives for waste heat and renewable based projects to achieve greater efficiency [67]. States like California also have guidelines that pay hydrogen producers to conserve energy, encouraging even more conservation. These policies illustrate global attempts to conserve energy in hydrogen systems. They deploy money, rules, and incentives to drive the growth of efficient tech. But universalizing and stabilizing a framework for the long term is essential for sustained progress [69].

### 11.9.2 Economic Benefits of Energy-Efficient Hydrogen Systems

Energy-efficient hydrogen systems bring big economic wins by cutting costs, making things more competitive, and opening new markets. These gains are big in making, storing, and using hydrogen, where saving energy directly lowers costs. In making hydrogen, saving energy makes green hydrogen cheaper. It's now 2–3 times pricier than gray hydrogen because of high electricity costs [48]. For example, better PEM electrolyzers can go from 60% to 70% efficient, saving 12–15% on energy and cutting costs to €4.3/kg from €5/kg [72]. Systems that use waste heat also save 5–10% on costs in factories [49]. In storage and distribution, new tech like LOHCs and metal hydrides cut energy loss. This reduces storage costs by 10–20% compared to old methods [60]. Smart systems with real-time monitoring also help, cutting costs by 8–12% in big facilities [67]. Efficient distribution systems, like updated pipelines, lower transport costs by 5–10% [69]. These improvements make hydrogen use more cost-effective. Using hydrogen in fuel cells for cars and power plants also saves money. For example, AI-optimized fuel cells use 60–65% less hydrogen. This cuts fuel costs for cars by €0.05–0.07/km [48]. Energy-efficient hydrogen systems also create jobs and grow markets. The hydrogen economy could add 2 million jobs by 2050 [72]. This boosts the economy and makes hydrogen more appealing to investors.

## 11.9.3 Barriers and Incentives

Despite benefits, many barriers slow down the use of energy-efficient hydrogen systems. We need special incentives to overcome these hurdles. Knowing these barriers and incentives helps us grow sustainable hydrogen tech.

### 11.9.3.1 Barriers

- **High Capital Costs**: New tech costs 20–30% more upfront, making it hard to start [67].
- **Technological Maturity**: New tech like photocatalytic water splitting is not yet ready for market [60].
- **Regulatory Gaps**: Different rules for hydrogen infrastructure and energy efficiency confuse investors [69].
- **Market Competition**: Hydrogen faces tough competition from natural gas and batteries, which are cheaper and more established [48].
- **Skills Shortage**: Not enough trained people to work on these systems slows down their use [49].

### 11.9.3.2 Incentives

- **Financial Support**: Grants, subsidies, and low-interest loans help with energy-efficient hydrogen projects. For example, the EU's Horizon Europe program offers up to 50% funding for projects that use RE and waste heat recovery [72].
- **Tax Credits and Rebates**: Tax incentives make high-efficiency technologies cheaper. The U.S. Investment Tax Credit can cut the cost of electrolyzers and fuel cells by 20–30% [67].
- **Carbon Pricing**: Carbon taxes and emissions trading schemes make fossil fuels more expensive. This encourages the use of energy-efficient hydrogen [48].
- **Research and Development (R&D) Funding**: Government-backed R&D programs support innovation. Japan's New Energy and Industrial Technology Development Organization (NEDO) helps speed up the use of new technologies [60].
- **Training Programs**: Training programs help address skills shortages. The EU's Skills Agenda trains technicians in hydrogen system optimization and maintenance [49].

## 11.10 Conclusion

Producing clean hydrogen is essential for the transition to a clean energy system. Saving energy is a key to making hydrogen systems efficient, affordable and effective for the environment. This chapter has presented a large number of the options and technologies that make hydrogen systems environmentally friendly. It discusses the ways to minimize energy loss in the process of making, storing, and using hydrogen. For instance, waste heat and advanced materials can reduce energy loss of 10–20% [28]. Projects in Austria and Denmark demonstrate that such steps can make hydrogen systems 65–72% efficient and reduce $CO_2$ emissions by several thousand tons annually [51]. But the combination of hydrogen with RE and smart grids is a huge step up. This helps co-ordinate hydrogen use with clean energy and reduces waste by 15–20% [65]. The fact that the EU and Japan have introduced hydrogen strategies, including funding support and standards, has also lifted the global picture. But we need to focus on making sure these efforts are consistent and sustainable [74]. High costs, unfamiliar technology, rules that still need to be made – those are a few of them. We still need to have continuous availability of cheaper and accessible hydrogen [51]. And we will be also needing new mechanisms of payment to pay for these systems. Hydrogen systems will be a big deal in the future. If we work on saving energy and put resources into new concepts, hydrogen can play a major role in a low-carbon world. The conclusions of this chapter serve as a timely reminder of the centrality of energy efficiency and the necessity of working collaboratively more about that later if hydrogen is to make it [65].

## References

[1] Aakko-Saksa, P., Cook, C., Kiviaho, J., & Repo, T. 2018. Liquid organic hydrogen carriers for transportation and storing of renewable energy: Review and discussion. *Journal of Power Sources*, 396, 803–823. https://doi.org/10.1016/j.jpowsour.2018.06.011.

[2] Abdelaziz, E. A., Saidur, R., & Mekhilef, S. 2011. A review on energy saving strategies in industrial sector. *Renewable and Sustainable Energy Reviews*, 15(1), 150–168. https://doi.org/10.1016/j.rser.2010.09.003.

[3] Abe, J. O., Popoola, A. P. I., Ajenifuja, E., & Popoola, O. M. 2019. Hydrogen energy, economy and storage: Review and recommendation. *International Journal of Hydrogen Energy*, 44(29), 15072–15086. https://doi.org/10.1016/j.ijhydene.2019.04.068.

[4] Acosta-Herazo, R., Mueses, M. A., & Machuca-Martinez, F. 2020. Fundamentals and applications of photo-catalytic water splitting for hydrogen production: A review. *Environmental Science and Pollution Research*, 27(26), 32409–32427. https://doi.org/10.1007/s11356-020-09232-8.

[5] Ahmad, H., Kamarudin, S. K., Minggu, L. J., & Kassim, M. 2015. Hydrogen from photo-catalytic water splitting process: A review. *Renewable and Sustainable Energy Reviews*, 43, 599–610. https://doi.org/10.1016/j.rser.2014.10.101.

[6] Alves, I., Ferreira, J., & Salvador, C. 2019. Power-to-X projects in Europe: An overview of current developments. *Energy Procedia*, 158, 1657–1662. https://doi.org/10.1016/j.egypro.2019.01.389.

[7] Bailera, M., Lisbona, P., Peña, B., & Romeo, L. M. 2021. Power-to-gas projects in Europe: Technical and economic assessment. *International Journal of Hydrogen Energy*, 46(5), 4137–4152. https://doi.org/10.1016/j.ijhydene.2020.10.197.

[8] Bartels, J. R., Pate, M. B., & Olson, N. K. 2010. An economic survey of hydrogen production from conventional and alternative energy sources. *International Journal of Hydrogen Energy*, 35(16), 8371–8384. https://doi.org/10.1016/j.ijhydene.2010.04.035.

[9] Bhandari, R., Trudewind, C. A., & Zapp, P. 2014. Life cycle assessment of hydrogen production via electrolysis–a review. *Journal of Cleaner Production*, *85*, 151–163. https://doi.org/10.1016/j.jclepro.2013.07.048.

[10] Boudellal, M. 2018. Power-to-gas: Renewable hydrogen economy for the energy transition. *Renewable Energy*, 128, 466–478. https://doi.org/10.1016/j.renene.2018.05.089.

[11] Buttler, A., & Spliethoff, H. 2018. Current status of water electrolysis for energy storage, grid balancing and sector coupling via power-to-gas and power-to-liquids: A review. *Renewable and Sustainable Energy Reviews*, *82*, 2440–2454. https://doi.org/10.1016/j.rser.2017.09.003.

[12] Carmo, M., Fritz, D. L., Mergel, J., & Stolten, D. 2013. A comprehensive review on PEM water electrolysis. *International Journal of Hydrogen Energy*, 38(12), 4901–4934. https://doi.org/10.1016/j.ijhydene.2013.01.151.

[13] Chi, J., & Yu, H. 2018. Water electrolysis based on renewable energy for hydrogen production. *Chinese Journal of Catalysis*, 39(3), 390–394. https://doi.org/10.1016/S1872-2067(17)62949-8.

[14] Daoutidis, P., Zachar, M., & Jogwar, S. S. 2018. Sustainability and process control: A survey and perspective. *Journal of Process Control*, 68, 189–206. https://doi.org/10.1016/j.jprocont.2018.05.003.

[15] Demirhan, C. D., Tso, W. W., Powell, J. B., & Pistikopoulos, E. N. 2020. Sustainable ammonia production through process synthesis and optimization. *AIChE Journal*, 66(6), e16941. https://doi.org/10.1002/aic.16941.

[16] Dincer, I., & Acar, C. 2015. A review on clean energy solutions for better sustainability. *International Journal of Energy Research*, 39(5), 585–606. https://doi.org/10.1002/er.3329.

[17] Dincer, I., & Acar, C. 2015. Review and evaluation of hydrogen production methods for better sustainability. *International Journal of Hydrogen Energy*, *40*(34), 11094–11111. https://doi.org/10.1016/j.ijhydene.2014.12.035.

[18] Dodds, P. E., & Demoullin, S. 2013. Conversion of the UK gas system to transport hydrogen. *International Journal of Hydrogen Energy*, *38*(18), 7189–7200. https://doi.org/10.1016/j.ijhydene.2013.03.070.

[19] Dufour, A., Girods, P., Masson, E., & Zoulalian, A. 2012. Syngas production from wood gasification using downdraft reactors. *Energy*, *36*(3), 1782–1792. https://doi.org/10.1016/j.energy.2010.12.053.

[20] Dufour, J., Serrano, D. P., Gálvez, J. L., Moreno, J., & González, A. 2012. Life cycle assessment of processes for hydrogen production: Environmental feasibility and energy efficiency. *International Journal of Hydrogen Energy*, 37(2), 1153–1163. https://doi.org/10.1016/j.ijhydene.2011.09.147.

[21] Eberle, U., Felderhoff, M., & Schüth, F. 2012. Chemical and physical solutions for hydrogen storage. *Angewandte Chemie International Edition*, *48*(36), 6608–6630. https://doi.org/10.1002/anie.200806293.

[22] Fletcher, E. A. 2001. Solarthermal processing: A review. *Journal of Solar Energy Engineering*, 123(2), 63–74. https://doi.org/10.1115/1.1351814.

[23] Gielen, D., Taibi, E., & Miranda, R. 2019. Hydrogen: A renewable energy perspective. *International Renewable Energy Agency (IRENA)*. https://www.irena.org/publications/2019/Sep/Hydrogen-A-renewable-energy-perspective.

[24] Glenk, G., & Reichelstein, S. 2019. Economics of converting renewable power to hydrogen. *Nature Energy*, *4*(3), 216–222. https://doi.org/10.1038/s41560-019-0326-1.

[25] Grube, T., Robinius, M., Stolten, D., & Reuß, M. 2020. A techno-economic perspective on green hydrogen production: Opportunities and challenges. *Chemie Ingenieur Technik*, 92(9), 1172–1182. https://doi.org/10.1002/cite.202000068.

[26] Hall, C. A. S., Lambert, J. G., & Balogh, S. B. 2014. EROI of different fuels and the implications for society. *Energy Policy*, 64, 141–152. https://doi.org/10.1016/j.enpol.2013.05.044.

[27] Hisatomi, T., Kubota, J., & Domen, K. 2014. Recent advances in semiconductors for photocatalytic and photoelectrochemical water splitting. *Chemical Society Reviews*, 43(22), 7520–7535. https://doi.org/10.1039/C3CS60378D.

[28] Hosseini, S. E., & Wahid, M. A. 2016. Hydrogen production from renewable and sustainable energy resources: Promising green energy carrier for clean development. *Renewable and Sustainable Energy Reviews*, 57, 850–866. https://doi.org/10.1016/j.rser.2015.12.112.

[29] Hu, S., Xiang, C., Haussener, S., Berger, A. D., & Lewis, N. S. 2019. An analysis of the optimal band gaps of light absorbers in integrated tandem photoelectrochemical water-splitting systems. *Energy & Environmental Science*, 12(11), 3267–3278. https://doi.org/10.1039/C9EE01854J.

[30] IEA. 2019. The Future of Hydrogen: Seizing Today's Opportunities. *International Energy Agency*. https://www.iea.org/reports/the-future-of-hydrogen.

[31] IRENA. 2020. Green Hydrogen: A Guide to Policy Making. *International Renewable Energy Agency*. https://www.irena.org/publications/2020/Nov/Green-hydrogen.

[32] ISO 2018. *ISO 50001:2018 – Energy management systems – Requirements with guidance for use.* International Organization for Standardization. https://www.iso.org/standard/69426.html.

[33] Jensen, J. O., Vestbø, A. P., Li, Q., & Bjerrum, N. J. 2007. The energy efficiency of onboard hydrogen storage. *Journal of Alloys and Compounds*, 446–447, 723–728. https://doi.org/10.1016/j.jallcom.2007.04.051.

[34] Kadier, A., Simayi, Y., Abdeshahian, P., Azman, N. F., Chandrasekhar, K., & Kalil, M. S. 2016. A comprehensive review of microbial electrolysis cells (MEC) reactor designs and configurations for sustainable hydrogen gas production. *Biofuel Research Journal*, 3(1), 372–381. https://doi.org/10.18331/BRJ2016.3.1.5

[35] Keipi, T., Tolvanen, H., Konttinen, J., & Karppinen, M. 2018. Economic analysis of hydrogen production by methane thermal decomposition: Comparison to competing technologies. *Energy Conversion and Management*, 159, 264–273. https://doi.org/10.1016/j.enconman.2017.12.063.

[36] Klell, M. 2010. Storage of hydrogen in the pure form. In *Handbook of Hydrogen Storage* (pp. 1–38). Wiley-VCH. https://doi.org/10.1002/9783527629817.ch1.

[37] Koj, J. C., Wulf, C., Zapp, P., & Schreiber, A. 2019. Deploying power-to-gas in Europe: Economic potential and technological barriers. *Frontiers in Energy Research*, 7, 132. https://doi.org/10.3389/fenrg.2019.00132.

[38] Kojima, Y., Miyaoka, H., & Ichikawa, T. 2017. Hydrogen storage materials for energy-efficient systems. *Materials Science and Engineering: B*, 225, 114–121. https://doi.org/10.1016/j.mseb.2017.07.006.

[39] Lehner, M., Tichler, R., Steinmüller, H., & Koppe, M. 2014. *Power-to-gas: Technology and business models*. Springer. https://doi.org/10.1007/978-3-319-03995-4.

[40] Li, B., Roche, R., & Miraoui, A. 2017. Microgrid sizing with combined evolutionary algorithm and MILP unit commitment. *Applied Energy*, 188, 547–562. https://doi.org/10.1016/j.apenergy.2016.12.008.

[41] Lovins, A. B. 2005. Energy efficiency, taxonomic overview. *Encyclopedia of Energy*, 2, 383–401. https://doi.org/10.1016/B0-12-176480-X/00038-8.

[42] Marbán, G., & Valdés-Solís, T. 2007. Towards the hydrogen economy? *International Journal of Hydrogen Energy*, 32(12), 1625–1637. https://doi.org/10.1016/j.ijhydene.2006.12.017.

[43] Melaina, M. W., Antonia, O., & Penev, M. 2013. Blending hydrogen into natural gas pipeline networks: A review of key issues. *National Renewable Energy Laboratory (NREL)*, Technical Report NREL/TP-5600-51995. https://doi.org/10.2172/1068610.

[44] Mitsushima, S., Nagasawa, K., & Ota, K. 2021. Green hydrogen production in Japan: Current status and future perspectives. *Electrochemistry*, 89(2), 123–129. https://doi.org/10.5796/electrochemistry.20-00143.

[45] Nguyen, H. Q., Aris, A. M., & Shabani, B. 2019. PEM fuel cell heat recovery for preheating inlet air in standalone solar-hydrogen systems for telecommunication applications: An exergy analysis. *International Journal of Hydrogen Energy*, 44(29), 14983–14993. https://doi.org/10.1016/j.ijhydene.2019.04.087.

[46] Niermann, M., Beckendorff, A., Kaltschmitt, M., & Bonhoff, K. 2019. Liquid organic hydrogen carriers (LOHC) – Techno-economic analysis of LOHC-based hydrogen supply chains. *International Journal of Hydrogen Energy*, 44(29), 15094–15107. https://doi.org/10.1016/j.ijhydene.2019.04.088.

[47] Nikolaidis, P., & Poullikkas, A. 2017. A comparative overview of hydrogen production processes. *Renewable and Sustainable Energy Reviews*, 67, 597–611. https://doi.org/10.1016/j.rser.2016.09.044.

[48] Noussan, M., Raimondi, P. P., Scita, R., & Hafner, M. 2021. The role of green and blue hydrogen in the energy transition – A technological and geopolitical perspective. *Sustainability*, 13(1), 298. https://doi.org/10.3390/su13010298.

[49] Ochu, E., Shah, N., & Fulton, L. 2021. Policy and market frameworks for advancing hydrogen in the United States. *Environmental Research Letters*, 16(7), 073002. https://doi.org/10.1088/1748-9326/ac0f6b.

[50] Orimo, S., Nakamori, Y., Eliseo, J. R., Züttel, A., & Jensen, C. M. 2007. Complex hydrides for hydrogen storage. *Chemical Reviews*, 107(10), 4111–4132. https://doi.org/10.1021/cr0501846.

[51] Parra, D., Valverde, L., Pino, F. J., & Patel, M. K. 2019. A review on the role, cost and value of hydrogen energy systems for deep decarbonisation. *Renewable and Sustainable Energy Reviews*, 101, 279–294. https://doi.org/10.1016/j.rser.2018.11.010.

[52] Pehnt, M. 2006. Life cycle assessment of fuel cell systems. *International Journal of Life Cycle Assessment*, 11(1), 55–64. https://doi.org/10.1065/lca2005.09.224.

[53] Preuster, P., Papp, C., & Wasserscheid, P. 2017. Liquid organic hydrogen carriers (LOHCs): Toward a hydrogen-free hydrogen economy. *Accounts of Chemical Research*, 50(1), 74–85. https://doi.org/10.1021/acs.accounts.6b00474.

[54] Rashid, M. M., Al Mesfer, M. K., Naseem, H., & Danish, M. 2015. Hydrogen production by water electrolysis: A review of alkaline water electrolysis, PEM water electrolysis and high temperature water electrolysis. *International Journal of Engineering and Advanced Technology*, 4(3), 80–93. https://doi.org/10.13140/RG.2.1.2388.8488.

[55] Rostrup-Nielsen, J. R. 2005. Advances in catalysts for hydrogen production by steam reforming of hydrocarbons. *Catalysis Today*, 111(1–2), 4–11. https://doi.org/10.1016/j.cattod.2005.10.003.

[56] Rozmi, A., Osman, N., & Borhan, M. N. 2020. Hydrogen energy for sustainable development: A review of production, storage, and applications. *Journal of Cleaner Production*, 268, 122192. https://doi.org/10.1016/j.jclepro.2020.122192.

[57] Sakintuna, B., Lamari-Darkrim, F., & Hirscher, M. 2007. Metal hydride materials for solid hydrogen storage: A review. *International Journal of Hydrogen Energy*, 32(9), 1121–1140. https://doi.org/10.1016/j.ijhydene.2006.11.022.

[58] Salvi, B. L., & Subramanian, K. A. 2015. Sustainable development of road transportation sector using hydrogen energy system. *Renewable and Sustainable Energy Reviews*, 51, 1132–1155. https://doi.org/10.1016/j.rser.2015.07.030.

[59] Santos, D. M. F., Sequeira, C. A. C., & Figueiredo, J. L. 2013. Hydrogen production by alkaline water electrolysis. *Química Nova*, 36(8), 1176–1193. https://doi.org/10.1590/S0100-40422013000800017.

[60] Sharma, S., & Ghoshal, S. K. 2015. Hydrogen the future transportation fuel: From production to applications. *Renewable and Sustainable Energy Reviews*, 43, 1151–1158. https://doi.org/10.1016/j.rser.2014.11.093.

[61] Sikarwar, V. S., Zhao, M., Clough, P., Yao, J., Zhong, X., Memon, M. Z., Shah, N., Anthony, E. J., & Fennell, P. S. 2016. An overview of advances in biomass gasification. *Energy & Environmental Science*, 9(10), 2939–2977. https://doi.org/10.1039/C6EE00935B.

[62] Simpson, A. P., & Lutz, A. E. 2007. Exergy analysis of hydrogen production via steam methane reforming. *International Journal of Hydrogen Energy*, 32(15), 3237–3247. https://doi.org/10.1016/j.ijhydene.2006.12.025.

[63] Singhal, S. C. 2014. Solid oxide fuel cells for stationary, mobile, and military applications. *Solid State Ionics*, 257, 1–16. https://doi.org/10.1016/j.ssi.2013.12.026.

[64] Sørensen, B. 2017. *Renewable energy: Physics, engineering, environmental impacts, economics and planning* (5th ed.). Academic Press. https://doi.org/10.1016/B978-0-12-804567-1.00001-9.

[65] Staffell, I., Scamman, D., Velazquez Abad, A., Balcombe, P., Dodds, P. E., Ekins, P., Shah, N., & Ward, K. R. 2019. The role of hydrogen and fuel cells in the global energy system. *Energy & Environmental Science*, 12(2), 463–491. https://doi.org/10.1039/C8EE01157E.

[66] Takanabe, K. 2017. Photocatalytic water splitting: Quantitative approaches toward photocatalyst by design. *ACS Catalysis*, 7(12), 8006–8022. https://doi.org/10.1021/acscatal.7b02662.

[67] Talebian, H., Herrera, O. E., & Mérida, W. 2019. Policy and economic barriers to the deployment of clean energy technologies: A review. *Energy Policy*, 134, 110950. https://doi.org/10.1016/j.enpol.2019.110950.

[68] Thollander, P., & Ottosson, M. 2010. Energy management practices in Swedish energy-intensive industries. *Journal of Cleaner Production*, 18(12), 1125–1133. https://doi.org/10.1016/j.jclepro.2010.04.011.

[69] Trencher, G., Taeihagh, A., & Yarime, M. 2020. Overcoming barriers to developing and diffusing fuel cell vehicles: Governance strategies and experiences in Japan. *Energy Policy*, 142, 111533. https://doi.org/10.1016/j.enpol.2020.111533

[70] Turner, J. A. 2004. Sustainable hydrogen production. *Science, 305*(5686), 972–974. https://doi.org/10.1126/science.1103197.

[71] UN. 2015. Transforming our world: The 2030 Agenda for Sustainable Development. *United Nations*. https://sdgs.un.org/2030agenda.

[72] Velazquez Abad, A., & Dodds, P. E. 2020. Green hydrogen characterisation initiatives: Definitions, standards, guarantees of origin, and challenges. *Energy Policy*, 138, 111300. https://doi.org/10.1016/j.enpol.2020.111300.

[73] Vincent, I., & Bessarabov, D. 2018. Low-cost hydrogen production by anion exchange membrane electrolysis: A review. *Renewable and Sustainable Energy Reviews*, 81, 1690–1704. https://doi.org/10.1016/j.rser.2017.05.258.

[74] Wang, M., Wang, G., Sun, Z., Zhang, Y., & Xu, D. 2020. Review of renewable energy-based hydrogen production processes for sustainable energy innovation. *Global Energy Interconnection*, 3(5), 436–443. https://doi.org/10.1016/j.gloei.2020.11.005

[75] Wang, Y., Chen, K. S., Mishler, J., Cho, S. C., & Adroher, X. C. 2011. A review of polymer electrolyte membrane fuel cells: Technology, applications, and needs on fundamental research. *Applied Energy*, 88(4), 981–1007. https://doi.org/10.1016/j.apenergy.2010.09.030.

[76] Weissbach, D., Huke, A., Czerski, K., Gottlieb, S., & Hussein, A. 2013. Energy intensities, EROIs (energy returned on invested), and energy payback times of electricity generating power plants. *Energy, 52*, 210–221. https://doi.org/10.1016/j.energy.2013.01.029.

[77] Yoon, S., Kim, M., Kim, Y., & Lee, J. 2020. Machine learning for real-time optimization of chemical processes: A review. *Chemical Engineering Research and Design*, 160, 209–224. https://doi.org/10.1016/j.cherd.2020.05.019.

[78] Zeng, K., & Zhang, D. 2010. Recent progress in alkaline water electrolysis for hydrogen production and applications. *Progress in Energy and Combustion Science*, 36(3), 307–326. https://doi.org/10.1016/j.pecs.2009.11.002.

[79] Zhang, X., Bauer, C., Mutel, C. L., & Volkart, K. 2020. Life cycle assessment of power-to-gas: Approaches, system variations and their environmental implications. *Applied Energy*, 190, 326–338. https://doi.org/10.1016/j.apenergy.2016.12.098.
[80] Zhang, X., Bauer, C., Mutel, C. L., & Volkart, K. 2020. Life cycle assessment of power-to-gas: Approaches, system variations and their environmental implications. *Applied Energy*, *251*, 113–360. https://doi.org/10.1016/j.apenergy.2019.113360.
[81] Menner, Martin, & Götz Reichert. "EU hydrogen strategy." 2020.
[82] Iida, S., & Sakata, K., 2019. Hydrogen technologies and developments in Japan. *Clean Energy*, 3(2), 105–113.
[83] Zhang, M., & Batjargal, T., 2022. Review on new spending of United States bipartisan infrastructure bill. *Journal of Infrastructure, Policy and Development*, 6(2), 1507.

Alok Tiwari, Vishal Sandhwar, Shivendu Saxena, Diksha Saxena, Gajendra Kumar Gaurav, Gourang Damle

# 12 Challenges in Scaling Up Hydrogen and Fuel Cell Technologies

**Abstract:** The world is adopting hydrogen and fuel cell technologies because they are essential for establishing low-carbon energy systems. Combining these technologies across transportation, power generation, and industrial processing sectors proves their ability to cut greenhouse gas emissions substantially. The paper provides a complete analysis of the hydrogen energy system by evaluating the advantages and challenges of gray, blue, and green hydrogen production methods. The economic advantage of gray hydrogen production persists because of existing infrastructure, yet its substantial carbon emissions create an immediate need for emission-reducing alternatives. Blue hydrogen presents a middle-ground solution by implementing carbon capture and storage, yet its implementation faces obstacles regarding cost and storage feasibility. The cleanest hydrogen production method exists through renewable-powered electrolysis, but green hydrogen faces challenges due to expensive manufacturing and technical obstacles. The chapter moves beyond production to explore the storage and distribution logistics of hydrogen systems. The scalability of hydrogen technology remains restricted because of three main obstacles: low energy density, material degradation, and inadequate infrastructure. The analysis examines critical performance issues of fuel cells, which include catalyst degradation, membrane failure, and thermal management problems. The evaluation assesses economic factors through cost competitiveness analysis and investment risk assessment, together with policy and societal elements, which include public perception, regulatory readiness, and workforce development. New developments in nanomaterials and catalyst design, and international partnerships, are creating a more sustainable hydrogen economy of the future. The research demonstrates that major obstacles exist, but targeted investments combined with policy coordination and technological progress will propel hydrogen into sustainable energy systems.

**Keywords:** Green hydrogen, fuel cells, hydrogen storage, electrolysis, infrastructure, decarbonization, nanomaterials

## 12.1 Introduction

The global decarbonization strategies heavily rely on hydrogen and fuel cell technologies because of their importance. The worldwide climate change response efforts have elevated hydrogen to a primary energy transmission agent because it enables significant emissions reductions in transportation systems, industrial facilities, and

 | https://doi.org/10.1515/9783112181072-012

power plants. The energy system requires hydrogen integration to achieve deep decarbonization levels. The multiple uses of hydrogen as a fuel enable fuel cells to generate efficient electricity from stored energy, which produces cleaner power options than fossil fuels [1, 2]. The international community supports hydrogen as the primary decarbonization agent to meet climate targets, which establishes its essential role in sustainable energy development [3, 4].

The present situation regarding hydrogen production and utilization demonstrates progress alongside existing production obstacles. The majority production of hydrogen through steam methane reforming (SMR) of natural gas exceeds 90% of global output, yet this method relies on fossil fuels, which results in carbon emission generation [5]. The production of green hydrogen from renewable energy sources, including solar power and wind power, through electrolysis processes has received increasing attention. The hydrogen economy transition stems from the dual requirement of environmental protection alongside growing clean energy requirements [6, 7]. Hydrogen finds diverse uses in fuel cell electric vehicles alongside industrial processes because of its energy vector capabilities across multiple sectors, yet its adoption faces obstacles from current production and storage, and distribution limitations [8, 9].

The upcoming sections will discuss the technical barriers, economic barriers, and societal barriers to the adoption of hydrogen and fuel cell technologies. We will begin by reviewing different hydrogen production strategies, which include traditional approaches and renewable technologies alongside technological advancements to boost productivity while lowering expenses [5, 10]. The following section will focus on the vital aspects of hydrogen storage and distribution systems because adequate infrastructure development is essential for hydrogen adoption across the board. The chapter will examine fuel cell efficiency technologies, which are critical components for better performance and lower operational costs [11, 12]. The economic assessment of hydrogen technologies will evaluate their financial viability relative to existing energy systems and determine required investments and market dynamics [4, 13]. The analysis will examine societal barriers through public perception studies and regulatory framework evaluation to determine their effect on hydrogen solution deployment [14, 15]. The final section will examine the prospects of hydrogen technology while discussing its ability to reshape the worldwide energy system through sustainable energy transformation.

## 12.2 Hydrogen Production Methods

### 12.2.1 Gray Hydrogen

The term "gray hydrogen" describes hydrogen produced through SMR of natural gas because this technology remains the dominant method of hydrogen generation globally at 99% [16]. The process consists of methane's reaction with high-temperature steam, which produces hydrogen while simultaneously creating carbon monoxide and carbon dioxide. The process works efficiently while being cost-effective, but it produces substantial $CO_2$ emissions, which create substantial environmental problems. The industrial sector has adopted this carbon-intensive production system across fertilizer and petroleum refining operations because of established infrastructure and cost efficiency [17].

The economic benefits of gray hydrogen manufacturing led to severe environmental consequences. The SMR process creates between 8 and 12 kg of $CO_2$ atmospheric emissions during the production of 1 kg of hydrogen [18, 19]. The high emission levels of gray hydrogen production make it rank as one of the most environmentally damaging hydrogen generation methods. The use of fossil fuels through this method runs counter to global carbon neutrality goals while creating additional environmental problems [20]. The advancement of decarbonization efforts and the achievement of long-term climate objectives depend on minimizing gray hydrogen use [21].

### 12.2.2 Blue Hydrogen

The production of blue hydrogen functions as a transitional solution that bridges traditional fossil fuel-based methods with renewable hydrogen supply. The blue hydrogen production via SMR follows the same steps as gray hydrogen manufacturing but includes carbon capture and storage (CCS) to decrease emissions. The systems implement carbon dioxide capture from hydrogen production to either use the captured $CO_2$ in industrial applications or store it underground to prevent atmospheric $CO_2$ emissions [22]. Blue hydrogen shows advantages over traditional gray hydrogen, but it experiences various operational and economic challenges. The implementation of CCS requires suitable geological storage sites as well as efficient capture system technology. CCS implementation leads to higher production costs, which makes the technology less competitive [23, 24]. Theoretically, blue hydrogen generates the same level of emissions reductions as green hydrogen, but its implementation faces obstacles because of the required infrastructure and operational complexities. Safety issues and long-term $CO_2$ permanence need thorough evaluation as part of strategic decarbonization plans [25].

### 12.2.3 Green Hydrogen

The electrolysis of water with solar and wind renewable energy powers the production of green hydrogen. The process of splitting water molecules through this method produces hydrogen and oxygen without producing direct greenhouse gas emissions [26]. Future low-carbon energy systems depend on green hydrogen for environmentally friendly hydrogen production.

The adoption of green hydrogen at scale proves difficult because of major obstacles. The main hurdles to the wide adoption of green hydrogen include high expenses for both electrolyzer systems and renewable power systems needed for operation [27]. The current market prices of green hydrogen surpass those of gray and blue hydrogen, but experts predict technological advancements will decrease costs in the future [28].

The main obstacle to implementation stems from electrolysis systems that show low-efficiency rates by transforming 60–70% of input energy into usable hydrogen [29]. The development of improved electrolyzer systems and electrocatalytic materials, and better hydrogen system integration with renewable energy sources, remains a focus of research [30]. The future global energy transition toward sustainability will heavily depend on green hydrogen development through consistent technological progress and financial support [31, 32].

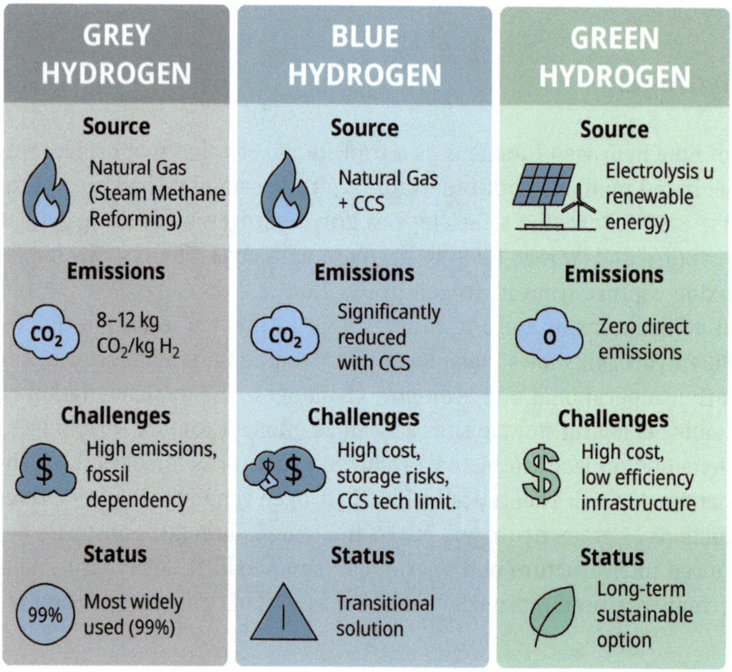


**Figure 12.1:** Comparative overview of hydrogen production pathways.

The comparative overview of hydrogen production pathways is shown in Figure 12.1. The infographic compares the three main hydrogen production methods: Gray, blue, and green hydrogen. Each route is compared based on its source of power, carbon footprint, technical difficulty, and position in current and future energy systems. The largest hydrogen source comes from gray hydrogen production by SMR of natural gas, yet this method produces significant $CO_2$ emissions. Blue hydrogen employs CCS technology as a solution to environmental problems, yet green hydrogen created by renewable energy-powered electrolysis represents a zero-emission solution. The system exists in a limited state due to both expensive manufacturing expenses and restrictions in infrastructure deployment. This illustration displays essential factors influencing the hydrogen economy transformation between sustainable and environmental factors and economic considerations.

## 12.3 Storage and Distribution Challenges

The development of hydrogen and fuel cell technologies faces major barriers because hydrogen storage and distribution systems remain limited. The following section describes active storage methods while discussing their technical constraints, along with hurdles that restrict hydrogen deployment through infrastructure systems.

### 12.3.1 Storage Technologies

The storage of hydrogen uses three main methods: compressed gas, cryogenic liquid, and solid-state storage, with their technical advantages and limitations.

Specially designed tanks store compressed hydrogen when compressed to specific pressures. This technique continues to be widely used because it can be easily implemented with existing infrastructure, and the technology is already mature [33, 34]. Its main weakness exists in its low volumetric energy storage capacity. High-pressure maintenance of hydrogen storage systems poses severe safety dangers, including tank ruptures and storage material embrittlement problems that result from hydrogen [35, 36].

The method of storing liquid hydrogen provides improved energy storage capability, although it requires expensive cryogenic cooling procedures. The storage system needs temperatures lower than −253 °C for hydrogen to maintain its state, thus requiring substantial thermal insulation components and substantial energy consumption [34, 37]. The storage technique presents multiple technical obstacles, including fast boiling and flammability dangers, and complicated handling protocols, which result in elevated costs [38].

Metal hydride technology shows promising development in solid-state storage systems. The absorption properties of magnesium hydride ($MgH_2$) materials provide beneficial hydrogen storage capabilities that exceed gas and liquid hydrogen storage capacity [39]. The materials experience unresolved thermodynamic stability problems together with slow desorption kinetics difficulties [34, 37]. Scientists work to create materials that store hydrogen safely and efficiently at room temperature conditions to boost their usage potential.

The current hydrogen storage systems face multiple major barriers because of their limited energy density, safety concerns, storage durability issues, and high infrastructure and maintenance costs. Scientists now focus on creating new porous materials and nanostructured frameworks that enhance hydrogen storage capabilities and safety aspects and decrease costs [40–42].

The main challenge for hydrogen technology adoption exists because of distribution network restrictions. The absence of developed hydrogen transport systems generates various logistical and economic challenges that prevent the deployment of hydrogen technologies. The two primary methods for hydrogen distribution include dedicated pipeline transport of high-pressure gas and specialized storage equipment needed for cryogenic liquid transportation [43, 44]. The integration of hydrogen with existing energy systems creates intricate technical requirements that act as barriers to straightforward implementation.

Multiple elements create obstacles in expanding hydrogen distribution networks. The main obstacle involves expensive pipeline materials alongside system modification requirements for existing infrastructure. The physical and chemical properties of hydrogen make most natural gas pipelines unsuitable for hydrogen transport since the material suffers from hydrogen embrittlement [42, 45]. The deployment of new hydrogen transport lines faces significant obstacles in urban areas where population density is high, which limits the expansion of refueling networks [43, 44].

Developing a dedicated hydrogen distribution network requires significant financial investment, according to economic studies. The project requires supportive energy policies and regulatory frameworks that stimulate investment and infrastructure [42, 44]. The hydrogen economy faces substantial distribution bottlenecks that restrict its ability to become a mainstream energy solution unless substantial investment is made in technology and infrastructure development.

**Table 12.1:** Comparison of hydrogen storage technologies.

| Parameter | Compressed hydrogen gas | Cryogenic liquid hydrogen | Metal hydrides | Liquid organic hydrogen carriers (LOHCs) | References |
|---|---|---|---|---|---|
| Physical state and storage conditions | Gaseous state; typically stored at 350–700 bar | Liquid state; maintained at −253 °C | Solid-state; typically requires heating | Liquid state; requires specific organic solvents | [46–49] |
| Energy density | Gravimetric: ~33.6 kWh/kg; volumetric: ~ 2.4 kWh/L | Gravimetric: 39.4 kWh/kg; volumetric: 2.36 kWh/L | Gravimetric: varies (~10–20 kWh/kg); volumetric: ~ 0.6 kWh/L | Gravimetric: 6–9 kWh/kg; volumetric: ~6–10 kWh/L | [46, 48, 50, 51] |

**Table 12.1** (continued)

| Parameter | Compressed hydrogen gas | Cryogenic liquid hydrogen | Metal hydrides | Liquid organic hydrogen carriers (LOHCs) | References |
|---|---|---|---|---|---|
| Storage efficiency and system capacity | Efficiency: ~90% under optimal conditions; capacity limited by gas compressibility | Efficiency: roughly 85%; capacity depends on liquefaction energy | Efficiency: ~80%; capacity highly variable depending on metal alloy used | Efficiency: ~70%; capacity influenced by extraction processes | [49, 52] |
| Safety concerns | High flammability; risk of gas leaks and explosions | Risks include boil-off, cryogenic burns, and pressure hazards | Thermal management required; risk of hydrogen embrittlement | Moderate flammability; safety concerns regarding chemical interaction | [47, 49, 53] |
| Infrastructure and material costs | High initial costs of compression equipment; ongoing operational costs | Significant liquid hydrogen infrastructure required; high liquefaction costs | Moderate costs; ongoing research to improve materials and reduce costs | Moderate to high costs; still in developmental stages for widespread use | [48, 49, 52] |

Table 12.1 comprehensively compares four key hydrogen storage technologies – Compressed hydrogen gas, cryogenic liquid hydrogen, metal hydrides, and liquid organic hydrogen carriers (LOHCs)-based on recent peer-reviewed literature. Compressed gas offers high gravimetric energy density and fast refueling but faces safety and cost challenges due to high-pressure requirements. Cryogenic liquid hydrogen achieves the highest energy density but suffers from boil-off losses and requires complex, costly infrastructure. Metal hydrides provide solid-state storage with moderate efficiency, though they are material-dependent and require thermal management. LOHCs, while safer and easier to handle, have relatively lower efficiencies and are still under development for large-scale deployment. Overall, each method presents trade-offs between energy density, safety, system complexity, and cost, making the optimal choice application-specific. Continued research is needed to improve materials, reduce costs, and enhance system-level integration for future hydrogen-based energy systems.

To establish hydrogen as an alternative to conventional fuel options, the successful adoption of hydrogen requires resolving its distribution and storage problems. Materials science innovations, together with strategic infrastructure development, will help solve current barriers to the effective use of hydrogen technologies throughout various sectors.

## 12.4 Fuel Cell Technologies: Performance, Efficiency, and Longevity

Multiple forms of hydrogen fuel cells exist, which match the distinct operational requirements of their applications. Transportation benefits from proton-exchange membrane fuel cells (PEMFCs) because these cells require minimal heat for operation, while they start quickly [54]. The higher operational temperature of solid oxide fuel cells (SOFCs) makes them suitable for stationary power generation and combined heat and power (CHP) systems [54]. The efficiency-driven alkaline fuel cells (AFCs) operate in aerospace systems, but phosphoric acid fuel cells (PAFCs) function best in providing a continuous, stationary power supply [55].

The different fuel cell technologies have their own particular performance and durability restrictions, which affect their use. PEMFCs develop unstable electrochemical activity through material variations along with operating conditions, which shortens their operational lifespan [56]. The anode of SOFCs faces two major degradation sources because nickel coarsening happens simultaneously with sulfur poisoning [57, 58]. All fuel cell types undergo severe deterioration from contaminants, including carbon monoxide (CO) and hydrogen sulfide ($H_2S$), which diminish catalytic activity and shorten component operational time [59–61].

The specific degradation of catalysts causes PEMFCs problems because it represents the main challenge. Platinum-based catalysts demonstrate significant performance deterioration from minimal CO exposure, which requires the development of CO-tolerant catalyst materials [62]. Electrochemical catalyst design methods, which include alloying techniques and surface modification approaches, show promise to solve these problems [63, 64]. The system faces another crucial weakness because of membrane integrity issues. The performance of PEMFCs declines when water management becomes insufficient since it results in membrane flooding or drying, which reduces ion conductivity and minimizes cell output [65]. SOFCs need precise thermal management since high operating temperatures boost material degradation, yet insufficient temperature's lower reaction output [58].

Materials engineering research has developed various effective solutions for existing problems. Researchers examine new electrocatalysts alongside hybrid membranes and high-temperature materials to achieve longer operation times and higher efficiency in fuel cells. The research demonstrates that nanocomposite membranes composed of organic-inorganic materials display superior gas separation characteristics, together with enhanced water management properties [65]. The improved sulfur tolerance of SOFCs catalysts enables better resistance against fuel stream contaminants, as observed by researchers [57, 58].

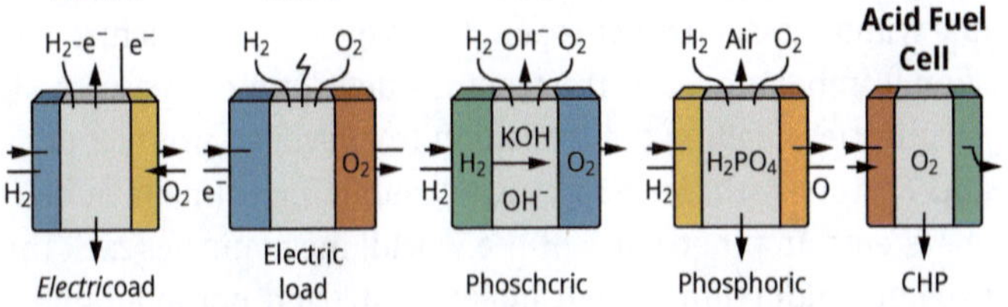


| | PEMFC | SOFC | AFC | PAFC | Phosphoric Acid Fuel Cell |
|---|---|---|---|---|---|
| Operating Temperature | 60–100 °C | 500–1,000 °C | 60–90 °C | 180–200 °C | 180–200 °C |
| Electrolyte | Polymer membrane | Ceramite | Kotanium-hydroxide(KOH) | Backylote | CHP |
| Efficiency | 40–60 % | 50–65 % | 50–70 % | 50–70 % | 40–50 % |
| Use Cases | Automotive Portable | Automotive portable | Space Backup power | CHP | CHP |

**Figure 12.2:** Comparison of five fuel cell types by temperature, electrolyte, efficiency, and applications.

Fuel cell commercial success depends on resolving three key challenges, which include efficiency problems, catalyst stability issues, and water and thermal management needs. The development of fuel cells needs sustained financial support for research, materials science, and system integration technologies that boost their operational reliability in various sectors.

The assessment based on Figure 12.2 of hydrogen fuel cell technologies shows that each fuel cell type, including PEMFCs, SOFCs, AFCs, and PAFCs, offers application-specific advantages along with unique limitations. PEMFCs operate at low temperatures and start quickly, which makes them suitable for transportation and portable power applications, yet their membranes and catalysts experience durability issues. SOFCs, though offering high efficiency and suitability for large-scale stationary applications, demand complex thermal management due to their high operating temperatures. AFCs, once prominent in space missions, suffer from $CO_2$ sensitivity, limiting broader deployment. PAFCs strike a balance for stationary CHP systems but contend with corrosive electrolytes and modest efficiency. This comparison underscores the importance of selecting fuel cell types based on targeted use-cases, while ongoing research must address their respective technical barriers to enhance performance and longevity.

## 12.5 Infrastructure and Supply Chain Constraints

Hydrogen and fuel cell system deployment requires a complete infrastructure network that covers every stage from production to storage and distribution, and final usage. The scalability of these technologies remains restricted by significant infrastructure deficits, which mainly affect emerging economies. Multiple developed nations have started building hydrogen production facilities together with refueling stations, but most of the world remains without essential hydrogen deployment infrastructure [66, 67].

This regional imbalance in infrastructure development poses a significant challenge. In the countries, leading the transition to hydrogen production, mainly located in Europe and East Asia, there is significant public investment in electrolysis plants and pipeline systems. In other parts of the world, the countries lack the proper facilities for hydrogen production, which makes hydrogen not available in the country and, therefore, less favorable as a reliable energy resource. These differences in the development of the hydrogen infrastructure hinder both global cooperation and the unification of markets. The experts point out the necessity of a coordinated approach to planning that combines environmental, economic, and social elements to create a hydrogen-based system that is inclusive and integrated.

As well as infrastructure, the hydrogen supply chain has other difficulties with the logistics and material aspects of the hydrogen supply chain. Hydrogen is a gas with low density and flammable properties, which makes it difficult and expensive to store and transport. In many areas, the construction of safe and economically sound hydrogen infrastructure is not fully developed, which leads to higher operational costs and delays in deployment of the system [67]. Furthermore, there are supply chain bottlenecks because there are shortages of essential components such as electrolyzers and the demand for catalysts and storage systems, which are rare [68, 69].

Real-life illustrations show the potential of hydrogen systems, but also the problems that come with expanding the hydrogen infrastructure. The Netherlands is an example of a country that opted for decentralized hydrogen integration into its energy plan at the national level [70]. These cases demonstrate that the development of infrastructure should be done according to the specifics of the region. On the other hand, regions without such strategic approaches to development are likely to experience fragmented development and miss out on possible synergies.

Therefore, the improvement of the hydrogen infrastructure and the supply chain is essential for the broad use of hydrogen and fuel cell technologies. To establish a strong and sustainable hydrogen economy, substantial funding will be needed, together with international cooperation and the development of standardized policies and technical requirements.

## 12.6 Economic Challenges

### 12.6.1 Cost Competitiveness

The price of hydrogen as an energy carrier depends on its production cost compared to traditional fossil fuels and other renewable energy sources. There are three ways to obtain hydrogen: Gray, blue, and green. Gray hydrogen is produced through the

SMR process without carbon capture, and it is the cheapest, with a price range of $1.50–2.50/kg. Blue hydrogen is gray hydrogen that has been treated with CCS, and its price is slightly higher than that of gray hydrogen, which is from $2.50–3.50/kg. The production of green hydrogen through water electrolysis with renewable energy sources results in the most expensive outcome, which costs between $4.00/kg and $7.00/kg. The main factors driving these costs are renewable power expenses, together with equipment expenses [71, 72]. The technology cost will decrease because of improved technology development and expanding economies of scale [73].

The production expenses of hydrogen stem from three main factors: The first factor includes the initial capital outlay, the second factor includes operational energy costs, and the third factor includes the size and complexity of the supporting facilities. The establishment of green hydrogen production facilities that integrate renewable energy systems with electrolyzers and network infrastructure demands substantial upfront investments [74]. Moreover, the uncertainty of electricity prices for electrolysis is a major challenge to the economic viability of green hydrogen compared to fossil fuel competitors [75].

Furthermore, there is the challenge of market uncertainty. The demand for hydrogen in the future is quite uncertain because of the changing energy landscape, the emergence of new technologies, and changes in market demand and consumer behavior. The volatility may act as a deterrent to long-term investment in hydrogen technologies and hinder their uptake unless there are supportive policies that are forward-looking and promote innovation as well as flexibility [73, 76].

## 12.6.2 Investment Barriers

The acquisition of investments from government entities to private investors remains essential yet challenging for promoting hydrogen technologies. The sector faces investor reluctance because of its perceived high risk and technological challenges. Strategic incentives, together with subsidies and specific policy measures implemented by governments, have proven effective in encouraging investors to take action. South Korea and Japan have established complete national hydrogen plans through which they provide direct state funding for research and infrastructure development to lead hydrogen development [77, 78].

Multiple factors continue to restrict private sector involvement despite such promising models. The main obstacles include the long periods needed to generate returns on investment, the small short-term profits, and the insufficient infrastructure that exists in various regions [74, 76]. The combination of transparent policies and sustained government backing acts as a major confidence booster for investors and drives capital flow into hydrogen value chain operations [73].

Leading nations' successful investment models provide essential learning from their approaches. Germany and the Netherlands use public-private partnerships to fund both pilot projects and commercial deployments through partnerships that balance the financial burden and risk associated with innovation [75, 78]. The programs demonstrate how established incentive structures combined with dependable regulatory systems, along with pricing guarantees, create powerful investment environments [76].

# 12.7 Societal and Policy-Related Challenges

## 12.7.1 Public Perception

Society-wide expansion of this technology depends heavily on how public entities perceive hydrogen. Previous incidents receive continuous media attention, which creates ongoing safety issues. The public does not support hydrogen technologies because widespread misconceptions about hydrogen flammability and volatility block community support and affect political decisions [79]. The technology keeps advancing, but the lack of public understanding and incorrect information continues to block its social acceptance [80].

Public opinion patterns create delays in political backing, which results in regulatory slowdowns. Organizations need to create active public outreach programs that use clear communication methods to resolve these concerns. People develop a better understanding of hydrogen technologies through open discussion forums and educational campaigns, which build their trust in these systems [3].

## 12.7.2 Regulatory Frameworks

The current hydrogen regulatory system shows major disorder in how various rules connect. Different regions lack unified safety transportation and storage procedures because their regulatory frameworks do not align. A fragmented regulatory system prevents large-scale deployment and creates investment uncertainties for developers and investors [82].

Complete regulatory reforms need to be implemented as a matter of urgency. Standardizing safety regulations with modernized procedural protocols that adapt to new technologies and stakeholder involvement throughout policy development form essential steps for the future. The integration of hydrogen across multiple industries requires researchers to demand better emergency response systems and investment-oriented regulations [81, 82].

### 12.7.3 Workforce Readiness

The deployment of hydrogen systems requires workers who hold specialized skills to succeed. Experts with specialized knowledge for implementing hydrogen technology are in short supply throughout the current labor market. The deficiency of qualified personnel in this area creates an obstacle to industrial growth [83].

The development of educational programs requires academic institutions to join forces with industry partners. Theoretical instruction should be combined with practical technical training that includes hydrogen safety and fuel cell maintenance, and system integration courses. The growth of these programs will become essential because they will develop the employees needed for the quickly expanding hydrogen economy [82].

## 12.8 Future Outlook

The development of hydrogen technology continues to grow swiftly because of strategic policy cooperation among nations and technological breakthroughs in nanomaterials. Research on nanostructured titanium dioxide nanograins has improved electrolysis and photocatalysis efficiency through higher yields at reduced production costs.

Nanotechnology continues to improve hydrogen storage solutions. Researchers examine both high-capacity materials and dual-atom catalysts for their application in seawater electrolysis and other advanced production methods. These developments help solve major challenges regarding energy density, together with safety and material expenses. The European Union, together with other government entities, demonstrates increasing support for their strategies. The EU's hydrogen roadmap demonstrates strong decarbonization objectives through policy mechanisms that promote both infrastructure growth and technological development. The frameworks play a crucial role in drawing long-term investments, which drive the fuel transition process. The IEA and IPHE join international organizations to serve as leading forces in creating unified global hydrogen initiatives. The partnerships create a foundation for a worldwide hydrogen market through their support of collaboration, knowledge sharing, and funding initiatives. The convergence of technological advancements with political initiatives and institutional frameworks creates a solid foundation for hydrogen to become a primary component in the clean energy transition. The initiatives toward a net-zero future become more concrete as they continue to develop.

## 12.9 Conclusion

The development of hydrogen and fuel cell technologies stands as a vital advancement toward creating sustainable, low-carbon energy systems. Several significant barriers need to be overcome before hydrogen technology can reach its maximum potential. The growth of hydrogen production through gray, blue, and green pathways has progressed significantly, yet challenges from infrastructure constraints, market uncertainty, and high costs persist, especially in transportation sectors. The advancement of the hydrogen economy depends on simultaneous efforts to improve infrastructure while receiving focused investments, supportive policies, and strong international collaboration. Stakeholders can speed up hydrogen solution deployment through technology, regulatory, and workforce development improvements to create a sustainable energy system with resilience.

## References

[1] Hassan, Q., Azzawi, I.D., Sameen, A.Z. and Salman, H.M., 2023. Hydrogen fuel cell vehicles: opportunities and challenges. Sustainability, *15*(15), pp.11501.

[2] Rath, R., Kumar, P., Mohanty, S. and Nayak, S.K., 2019. Recent advances, unsolved deficiencies, and future perspectives of hydrogen fuel cells in transportation and portable sectors. International Journal of Energy Research, 43(15), pp.8931–8955.

[3] Li, F., Liu, D., Sun, K., Yang, S., Peng, F., Zhang, K., Guo, G. and Si, Y., 2024. Towards a future hydrogen supply chain: A review of technologies and challenges. Sustainability, 16(5), pp. 1890

[4] Griffiths, S., Sovacool, B.K., Kim, J., Bazilian, M. and Uratani, J.M., 2021. Industrial decarbonization via hydrogen: A critical and systematic review of developments, socio-technical systems and policy options. Energy Research & Social Science, 80, pp.102208.

[5] Dash, S.K., Chakraborty, S. and Elangovan, D., 2023. A brief review of hydrogen production methods and their challenges. Energies, 16(3), pp.1141.

[6] Dutta, S., 2021. Review on solar hydrogen: Its prospects and limitations. Energy & Fuels, 35(15), pp.11613–11639.

[7] Maihan, R. and Uysal, S. (2023). A brief overview of renewable hydrogen production: prospects and challenges. IJEAST, 7(9), pp.21–28.

[8] Abadi, V.M.M., Beheshti, A., Mirhabibi, M., Askari, M.B. and Bahnamiri, S.B., 2018. Proposing a Novel Economical and Reversible Method for Hydrogen Production Using Solar Oven. Journal of Energy and Natural Resources, 6(6), pp. 69–72.

[9] Tayarani, H. and Ramji, A., 2022. Life cycle assessment of hydrogen transportation pathways via pipelines and truck trailers: Implications as a low carbon fuel. Sustainability, 14(19), pp.12510.

[10] Singla, M.K., Gupta, J., Nijhawan, P., Oberoi, A.S., Alsharif, M.H. and Jahid, A., 2023. Role of a unitized regenerative fuel cell in remote area power supply: a review. Energies, 16(15), pp.5761.

[11] Staffell, I., Scamman, D., Abad, A.V., Balcombe, P., Dodds, P.E., Ekins, P., Shah, N. and Ward, K.R., 2019. The role of hydrogen and fuel cells in the global energy system. Energy & Environmental Science, 12(2), pp.463–491.

[12] Huang, E., 2022, November. Design of Hydrogen Fuel Cell: Methods to Higher Efficiency. In Proceedings of the 2nd International Conference on Green Environmental Materials and Food Engineering (GEMFE 2022), Tianjin, China, pp. 19–20.

[13] Zhao, F., Mu, Z., Hao, H., Liu, Z., He, X., Victor Przesmitzki, S. and Ahmad Amer, A., 2020. Hydrogen fuel cell vehicle development in China: An industry chain perspective. Energy Technology, 8(11), pp.2000179.

[14] Sharma, S., Agarwal, S. and Jain, A., 2021. Significance of hydrogen as economic and environmentally friendly fuel. Energies, 14(21), pp.7389.

[15] Alzoubi, A., 2021. Renewable Green hydrogen energy impact on sustainability performance. International Journal of Computations, Information and Manufacturing (IJCIM), 1(1), 90–110.

[16] Hu, T., Song, Y., Zhang, X., Lin, S., Liu, P., Zheng, C. and Gao, X., 2025. A mini review for hydrogen production routes toward carbon neutrality. Propulsion and Energy, 1(1), pp.1–20.

[17] Rafi, F.A., Rana, M.M., Agili, B. and Al Shahrany, A.S., 2025. From Grey to Green: Hydrogen Storage for a Net-Zero Earth. European Journal of Energy Research, 5(2), pp.1–15.

[18] Agarwala, N., 2024. Is Hydrogen a decarbonising fuel for maritime shipping?. Maritime Technology and Research, 6(4), 271244.

[19] Omid, M.A., Şahin, M.E. and Cora, Ö.N., 2024. Challenges and future perspectives on production, storage technologies, and transportation of hydrogen: a review. Energy Technology, 12(4), pp.2300997.

[20] Beschkov, V. and Ganev, E., 2023. Perspectives on the Development of Technologies for Hydrogen as a Carrier of Sustainable Energy. Energies, 16(17), pp.6108.

[21] Elshafei, A.M. and Mansour, R., 2023. Green hydrogen as a potential solution for reducing carbon emissions: a review. Journal of Energy Research and Reviews, 13(2), pp.1–10.

[22] Melnyk, O.M., Shumylo, O.M., Onishchenko, O.A., Mykhailova, I.V., Obniavko, T.S. and Korobko, T.O., 2023. Concept and Prospects for the Use of Hydrogen Fuel in Maritime Transport. Collect. Sci. Work. Ukr. State Univ. Railw. Transp, 203, pp.96–105.

[23] Zhang, S. (2023). Diverse sustainable methods for future jet engine. Applied and Computational Engineering, 11(1), pp. 143–148.

[24] Singh, H., Li, C., Cheng, P., Wang, X. and Liu, Q., 2022. A critical review of technologies, costs, and projects for production of carbon-neutral liquid e-fuels from hydrogen and captured CO 2. Energy Advances, 1(9), pp.580–605.

[25] Wei, T. and Glomsrød, S., 2023. Cost reduction in low-carbon hydrogen: effective but insufficient to mitigate carbon emissions. Discover Energy, 3(1), pp.2.

[26] Xu, Z.J. and Wang, X., 2020. Electrocatalysis: A Core Technique for a Sustainable Future. Chemistry (Weinheiman der Bergstrasse, Germany), 26(18), pp.3897–3897.

[27] Chavez-Angel, E., Castro-Alvarez, A., Sapunar, N., Henríquez, F., Saavedra, J., Rodríguez, S., Cornejo, I. and Maxwell, L., 2023. Exploring the potential of green hydrogen production and application in the antofagasta region of Chile. Energies, 16(11), pp.4509.

[28] Liu, X., Huang, Y., Shi, X., Bai, W., Huang, S., Li, P., Xu, M. and Li, Y., 2025. Offshore Wind Power-Seawater Electrolysis-Salt Cavern Hydrogen Storage Coupling System: Potential and Challenges. Energies, 18(1), pp.169.

[29] Subedi, A. and Thapa, B.S., 2022, June. Parametric modeling of re-electrification by green hydrogen as an alternative to backup power. In IOP Conference Series: Earth and Environmental Science, 1037(1), pp. 012057.

[30] Moore, C. (2025). Modeling climate impacts of hydrogen transition pathways.

[31] Awadh, K. and Al-Qaheem, Y., 2023. Hydrogen generation using low carbon technologies. KimyaProblemleri, 21(4), pp.307–322.

[32] Nagarajan, D., Dong, C.D., Chen, C.Y., Lee, D.J. and Chang, J.S., 2021. Biohydrogen production from microalgae – Major bottlenecks and future research perspectives. Biotechnology journal, 16(5), pp.2000124.

[33] Dong, S., Li, C., Wang, J., Liu, H., Ding, Z., Gao, Z., Yang, W., Lv, W., Wei, L., Wu, Y. and Li, H., 2022. The "burst effect" of hydrogen desorption in MgH 2 dehydrogenation. Journal of Materials Chemistry A, 10(42), pp.22363–22372.

[34] Xu, Y., Yang, X., Li, Y., Zhao, Y., Shu, X., Zhang, G., Yang, T., Liu, Y., Wu, P. and Ding, Z., 2024. Rare-Earth Metal-Based Materials for Hydrogen Storage: Progress, Challenges, and Future Perspectives. Nanomaterials, 14(20), pp.1671.

[35] Yadav, V., Sivakumar, G., Gupta, V. and Balaraman, E., 2021. Recent advances in liquid organic hydrogen carriers: an alcohol-based hydrogen economy. ACS Catalysis, 11(24), pp.14712–14726

[36] Chaturvedi, K.R., Bajpai, S., Trivedi, J. and Sharma, T., 2022. Air foams for mobility control and subsurface storage of hydrogen in porous media: an experimental study. Energy & Fuels, 36(9), pp.5036–5046.

[37] Sigen, A. and Li, X., 2024. Strategic Analysis of Past and Modern Hydrogen Storage Methods and Prospects for Its Future. Innovation in Science and Technology, 3(2), pp.14–20.

[38] Huang, Y., Cheng, Y. and Zhang, J., 2021. A review of high density solid hydrogen storage materials by pyrolysis for promising mobile applications. Industrial & Engineering Chemistry Research, 60(7), pp.2737–2771.

[39] Ren, L., Li, Y., Zhang, N., Li, Z., Lin, X., Zhu, W., Lu, C., Ding, W. and Zou, J., 2023. Nanostructuring of Mg-based hydrogen storage materials: recent advances for promoting key applications. Nano-Micro Letters, 15(1), pp.93.

[40] Le, T.H., Kim, M.P., Park, C.H. and Tran, Q.N., 2024. Recent developments in materials for physical hydrogen storage: a review. Materials, 17(3), pp.666.

[41] Zhang, C., Low, J. and Xiong, Y., 2024. Integration of Green Hydrogen Production and Storage via Electrocatalysis. Precision Chemistry, 2(6), pp.229–238.

[42] Li, H., Wang, J., Cui, M., Huang, R., Qi, W. and Su, R., 2024. Dynamic dual-atom synergistic catalysis boosted by liquid metal for direct seawater electroreduction. Journal of Materials Chemistry A, 12(22), pp.13466–13473.

[43] Khzouz, M. and Gkanas, E.I., 2020. Hydrogen technologies for mobility and stationary applications: hydrogen production, storage and infrastructure development. In Renewable energy-resources, challenges and applications. IntechOpen., pp. 255–278.

[44] Paliyal, P.S., Menon, S.K., Mondal, S. and Thapa, V., 2025. Location Analysis and Environmental Validation for Installation of Hybrid Solar-Wind Energy Generation System in Hilly Areas of Uttarakhand: Study Toward Forecasting. Forecasting Methods for Renewable Power Generation, pp.21–47.

[45] Heinemann, N., Alcalde, J., Miocic, J.M., Hangx, S.J., Kallmeyer, J., Ostertag-Henning, C., Hassanpouryouzband, A., Thaysen, E.M., Strobel, G.J., Schmidt-Hattenberger, C. and Edlmann, K., 2021. Enabling large-scale hydrogen storage in porous media–the scientific challenges. Energy & Environmental Science, 14(2), pp.853–864.

[46] Ghotia, S., Banerjee, S., Kumar, A., Dwivedi, N., Srivastava, A.K. and Kumar, P., 2025. Palladium-decorated unconventional graphene oxide for unprecedented hydrogen storage. Journal of Materials Chemistry A, 13, pp. 5621–5631.

[47] Gong, M., Jin, C. and Na, Y., 2023. Minimizing area-specific resistance of electrochemical hydrogen compressor under various operating conditions using unsteady 3D single-channel model. Membranes, 13(6), p.555.

[48] Ott, B., Delafontaine, L., Welchert, N.A., Dee, S. and Reza, A., 2023. Ensuring natural gas infrastructure is suitable for hydrogen service. Process Safety Progress, 42(2), pp.213–224.

[49] Yadav, V., Sivakumar, G., Gupta, V. and Balaraman, E., 2021. Recent advances in liquid organic hydrogen carriers: an alcohol-based hydrogen economy. ACS Catalysis, 11(24), pp.14712–14726

[50] Kasmani, R.M., Norazhar, N. and Kamaroddin, M.F.A., 2024. Numerical analysis of vapor dispersion from compressed hydrogen (H2) storage vessels. In E3S Web of Conferences (Vol. 516, p. 06002). EDP Sciences.

[51] Tang, J., Li, J., Pishva, P., Xie, R. and Peng, Z., 2023. Aqueous, rechargeable liquid organic hydrogen carrier battery for high-capacity, safe energy storage. ACS Energy Letters, 8(9), pp.3727–3732.

[52] Orlova, S., Mezeckis, N. and Vasudev, V.P.K., 2023. Compression of hydrogen gas for energy storage: a review. Latvian Journal of Physics and Technical Sciences, 60(2), pp.4–16.

[53] Seo, H.M. and Park, B.H., 2025. Application of EOS based on machine learning method on CFD study of rapid hydrogen refueling process. Korean Journal of Chemical Engineering, 42(8), pp.1–17.

[54] Ni, M., 2018. Why a more uniform fuel/oxygen distribution is critical for fuel cell stack performance improvement. International Journal of Energy Research, 42(14), pp.4259–4262.

[55] Strickland, K., Pavlicek, R., Miner, E., Jia, Q., Zoller, I., Ghoshal, S., Liang, W. and Mukerjee, S., 2018. Anion resistant oxygen reduction electrocatalyst in phosphoric acid fuel cell. Acs Catalysis, 8(5), pp.3833–3843.

[56] Chen, H., Shan, W., Liao, H., Zhang, T. and Pei, P., 2020. Study on the Influence of Segmented Fuel Cell on Internal Parameter Distribution.

[57] Steiger, P., Burnat, D., Madi, H., Mai, A., Holzer, L., Van Herle, J., Krocher, O., Heel, A. and Ferri, D., 2019. Sulfur poisoning recovery on a solid oxide fuel cell anode material through reversible segregation of nickel. Chemistry of Materials, 31(3), pp.748–758.

[58] Naeini, M., Lai, H., Cotton, J.S. and Adams, T.A., 2021. A mathematical model for prediction of long-term degradation effects in solid oxide fuel cells. Industrial & Engineering Chemistry Research, 60(3), pp.1326–1340.

[59] Valdés-López, V.F., Mason, T., Shearing, P.R. and Brett, D.J., 2020. Carbon monoxide poisoning and mitigation strategies for polymer electrolyte membrane fuel cells–A review. Progress in Energy and Combustion Science, 79, pp.100842.

[60] Ke, S., Sun, C., Cui, B., Qin, Y. and Dou, M., 2023. Operatable and efficient mitigation strategies for H2S poisoning in proton exchange membrane fuel cells: releasing Pt reactive sites for hydrogen oxidation. ACS Applied Energy Materials, 6(6), pp.3337–3346.

[61] Park, S., Oh, H. and Kim, H., 2024. Poisoning mechanism of the CO and H2S mixture on the anode of polymer electrolyte membrane fuel cell. ChemElectroChem, 11(9), pp.e202300828.

[62] Hegde, S., Wörner, R., and Shabani, B. (2025). Automotive PEM fuel cell catalyst layer degradation mechanisms and characterisation techniques, Part I: Carbon corrosion and binder degradation. International Journal of Hydrogen Energy, 118, 268–299.

[63] Wang, T., Li, L.Y., Chen, L.N., Sheng, T., Chen, L., Wang, Y.C., Zhang, P., Hong, Y.H., Ye, J., Lin, W.F. and Zhang, Q., 2022. High CO-tolerant Ru-based catalysts by constructing an oxide blocking layer. Journal of the American Chemical Society, 144(21), pp.9292–9301.

[64] Chen, Z., Shu, C., Gan, Z., Cao, J., Qiu, P., Sun, X., Deng, C., Wu, Y. and Tang, W., 2025. Research Progress and Perspectives on Anti-Poisoning Hydrogen Oxidation Reaction Electrocatalysts for Hydrogen Fuel Cells. Small, pp.2411049.

[65] Jalali, S., Seyedghayem, M.M., Saboor, F.H., Venkatramani, A. and Asgari, M., 2023. Proton Exchange Membrane Fuel Cells: Focused on Organic-Inorganic Nanocomposite Membranes. PeriodicaPolytechnica Chemical Engineering, 67(3), pp.472–489.

[66] Locke, K., 2024. The urgency of hydrogen: environmental issues and the need for change. Future Sustainability, 2(2), pp.46–58.

[67] Trencher, G. and Edianto, A., 2021. Drivers and barriers to the adoption of fuel cell passenger vehicles and buses in Germany. Energies, 14(4), pp.833.

[68] Alms, K., Ahrens, B., Graf, M. and Nehler, M., 2023. Linking geological and infrastructural requirements for large-scale underground hydrogen storage in Germany. Frontiers in Energy Research, 11, p.1172003.

[69] Vasbinder, P.W., de Vries, A.W. and Westerman, W., 2021. Hydrogen Infrastructure Project Risks in The Netherlands. Energies, 14(19), pp.6009.

[70] Ala, G., Colak, I., Di Filippo, G., Miceli, R., Romano, P., Silva, C., Valtchev, S. and Viola, F., 2021. Electric mobility in Portugal: current situation and forecasts for fuel cell vehicles. Energies, 14(23), pp.7945.

[71] Acharya, A., 2023. A Review of Renewable Electricity Cost and Capacity Factor Impact on Green Hydrogen Levelized Cost in Off-grid Configuration. Journal of Sustainable Development, 16(3), pp.106–118.

[72] Wu, S., Lin, J., Li, J. and Song, Y., 2022. Electricity allocation strategy for on-site hydrogen refueling stations in the forward and spot markets. IET Renewable Power Generation, 16(12), pp.2532–2546.

[73] Odenweller, A., Ueckerdt, F., Nemet, G.F., Jensterle, M. and Luderer, G., 2022. Probabilistic feasibility space of scaling up green hydrogen supply. Nature Energy, 7(9), pp.854–865.

[74] Lee, D. and Kim, K., 2021. Research and development investment and collaboration framework for the hydrogen economy in South Korea. Sustainability, 13(19), pp.10686.

[75] Kar, S.K., Sinha, A.S.K., Bansal, R., Shabani, B. and Harichandan, S., 2023. Overview of hydrogen economy in Australia. Wiley Interdisciplinary Reviews: Energy and Environment, 12(1), pp.e457.

[76] Yuan, M., Liu, C., Wang, B., Shang, W. and Zhang, H., 2023. Accelerating the net zero transition in Asia and the Pacific: Low-carbon hydrogen for industrial decarbonization (No. 1409). ADBI Working Paper.

[77] Cheon, K. and Kim, J., 2020. Hydrogen economy in major countries: policies of promotion and lessons learnt from them. Journal of the Korean Society of Mineral and Energy Resources Engineers, 57(6), pp.629–639.

[78] Stangarone, T., 2021. South Korean efforts to transition to a hydrogen economy. Clean technologies and environmental policy, 23, pp.509–516

[79] Sadat-Razavi, P., Karakislak, I. and Hildebrand, J., 2025. German media discourses and public perceptions on hydrogen imports: an energy justice perspective. Energy Technology, 13(2), pp.2301000.

[80] Stalker, L., Roberts, J.J., Mabon, L. and Hartley, P.G., 2022. Communicating leakage risk in the hydrogen economy: lessons already learned from geoenergy industries. Frontiers in Energy Research, 10, p.869264.

[81] Fetisov, V., Davardoost, H. and Mogylevets, V., 2023. Technological aspects of methane–hydrogen mixture transportation through operating gas pipelines considering industrial and fire safety. Fire, 6(10), pp.409.

[82] Wang, L., Lyu, X., Zhang, S., Zhang, J., Li, X., Chen, J., Song, Q., Lin, J. and Ma, T., 2024. The Simulation and Analysis of Leakage, Diffusion Behavior, and Risk Mitigation Measures in a Hydrogen-Refueling Station. Energy Technology, 12(8), pp.2400620.

[83] Vallejos-Romero, A., Cordoves-Sánchez, M., Cisternas, C., Sáez-Ardura, F., Rodríguez, I., Aledo, A., Boso, Á.,Prades, J. and Álvarez, B., 2022. Green hydrogen and social sciences: issues, problems, and future challenges. Sustainability, 15(1), pp.303.

# Index



 | https://doi.org/10.1515/9783112181072-013

# Green Tribology and Tribochemistry

**Volume 1**
Edited by: Yashvir Singh, Nishant Kumar Singh und Abhishek Sharma
Biolubricants – Feedstocks, Catalysts, and Nanotechnology, 2024
ISBN 978-3-11-114512-9, e-ISBN (PDF) 978-3-11-114513-6, e-ISBN (EPUB) 978-3-11-114555-6

**Volume 2**
Edited by: Yashvir Singh, Gagan Bansal
Biotribology – Friction, Wear and Lubrication in Biological Systems
ISBN 978-3-11-156294-0, e-ISBN (PDF) 978-3-11-156317-6, e-ISBN (EPUB) 978-3-11-156294-0